AF577081

Foundations of Comparative Genomics

Foundations of Comparative Genomics

Arcady R. Mushegian

AMSTERDAM • BOSTON • HEIDELBERG • LONDON
NEW YORK • OXFORD • PARIS • SAN DIEGO
SAN FRANCISCO • SINGAPORE • SYDNEY • TOKYO
Academic press is an imprint of Elsevier

ELSEVIER
ACADEMIC
PRESS

Academic Press is an imprint of Elsevier
30 Corporate Drive, Suite 400, Burlington, MA 01803, USA

Library of Congress Cataloging-in-Publication Data
Mushegian, Arcady.
Foundations of comparative genomics / Arcady Mushegian.
p. ; cm.
Includes bibliographical references and index.
ISBN-13: 978-0-12-088794-1 (hard cover : alk. paper)
ISBN-10: 0-12-088794-C (hard cover : alk. paper) 1. Genomics. 2. Gene mapping. 3. Physiology, Comparative.
I. Title.
[DNLM: 1. Genomics–methods. 2. Genetic Techniques. 3. Sequence Analysis–methods. QU 58.5 M987f 2007]
QH447.M8778 2007
572.8′6–dc22 2007003418

British Library Cataloguing-in-Publication Data
A catalogue record for this book is available from the British Library.

ISBN 13: 978-0-12-088794-1
ISBN 10: 0-12-088794-0

For information on all Academic Press publications
visit our Web site at www.books.elsevier.com

Transferred to Digital Printing 2009

Contents

To my parents

Preface

Many times I would find myself wondering how people choose what to do in their professional lives. After giving years of work to experimental virology, and some more to botany, microbial genetics, and teaching biological sciences, I settled at a research career in computational biology. And all this time, I was interested as much in biology as in the different ways in which people think about their research. How do we decide which problem to study? Why do some scientific questions sound interesting and important, and others do not? And why do different people have different opinions on what is interesting and important?

I noticed that much of what turns out to be interesting and important—in science and elsewhere—happens when two seemingly unrelated things suddenly reveal some sort of similarity. The pleasure of such discovery, of course, is only comparable to the joy of finding a difference between two things that were previously thought to be the same. Thus, I realized that I am interested in similarities and differences, and in patterns and motifs. And if this is what you are after, then computational biology is a good line of work.

As a "local bioinformatics specialist" at my institute, I spend a lot of time talking to the "noncomputational" biologists. My colleagues often tell me that they are more interested in ways to think about science than in actual applications and protocols. Remarks such as "I have read about database search statistics, and I think I understand how this algorithm works—but tell me how you decide which of these weak sequence similarities are more important than the others!" are common. So, it seems that the myriads of bioinformatics texts that are published these days need a reader's companion, which talks about prejudices, preferences, and priorities.

This is my attempt on such a companion. It is not intended as a comprehensive source on genome comparisons or other issues of computational biology. I wrote mostly about things that are of interest to me: For example, most of this book is concerned with the protein world, and there is almost no discussion of nucleotide sequence analysis. There is also very little mathematics, statistics, or computer science in the book, even though the practice of bioinformatics requires dealing with models, equations, and algorithms. Rather, this book is about scientific ideas that I believe to be the most important in computational biology and in its most accomplished branch, comparative genomics. I am also trying to show that the era of completely sequenced genomes is a truly novel age of biology, and that comparative genomics is the science for this age.

This book would not be possible without collaboration, friendship, and, throughout the years, many conversations with Eugene Koonin and Alexey Kondrashov. Luna Han, my editor, helped me to define where this book should be going and gently persuaded me to stay on track, and the members of the Bioinformatics Center at the Stowers Institute for Medical Research held the fort all the while.

Most important, my family put up with everything. I thank my wife Irina Sorokina—all the good things in my life for so many years are because of you, my love—and my children Alexandra, Nikolai, and Natalia.

1

The Beginning of Computational Genomics

Historians of science may disagree about when computational evolutionary genomics started in earnest. Some may associate the starting point with the work of geneticists Alfred Sturtevant and Theodosius Dobrzhansky or statistician Robert Fisher. Others may say that genomics is incomplete without the molecular-level analysis and mark the beginning of the era with the following citation from Francis Crick (1958):

> *Biologists should realize that before long we shall have a subject which might be called "protein taxonomy"—the study of amino acids sequences of proteins of an organism and the comparison of them between species. It can be argued that these sequences are the most delicate expression possible of the phenotype of an organism and that vast amounts of evolutionary information may be hidden away from them.*

However, I believe that most people would agree that several papers published from 1962 to 1965 by Linus Pauling and Emile Zuckerkandl were extremely important. One article in particular, "Molecules as Documents of Evolutionary History" (Zuckerkandl and Pauling, 1965), set the scene for most of the future work that is described in this book. The circumstances of its publication are also of some interest: Although written in 1963, it first appeared in 1964 as a Russian translation in a monograph dedicated to Alexei Nikolaevich Oparin, a true pioneer of experimental study of abiotic protein synthesis (Oparin, 1953) who, sadly, also endorsed and helped enforce Lysenkoist pseudo-science during his service at the Soviet Academy of Sciences from the 1940s to 1960s (Lewontin and Levins, 1976; Jukes, 1997).

The research first announced in that unlikely place (the original English language version of Zuckerkandl and Pauling's paper followed in 1965) sounds prophetic. If we outline the main ideas of that work, the density of novel ideas in that 10-page article is staggering:

1. The authors use the root "semantics" 72 times when speaking of genes and gene products. They called DNA, RNA, and proteins "semantides," or sense-carrying units. Unlike some of the modern uses of this word, which essentially equates semantics with postmodern relativism (e.g., "let us discuss the substance and not argue about semantics"), Pauling and Zuckerkandl took semantics seriously. So should we: By definition (and as understood by their readers in the early 1960s), *semantics* is the study of the meaning of sense-carrying units in a language or in other code. The meaning of words—and of genes—is exactly what we want to know.

2. There are dissimilarities between even closely related sense-carrying molecules. These dissimilarities are produced by genetic processes, such as nucleotide substitutions,

insertions, deletions, and rearrangements of large DNA fragments. Sense, or meaning, of genes and their products may be extracted by comparing related molecules, detecting the differences between them, and computing something about these differences.

3. Biopolymers contain information about evolution. It is threefold: (1) the time of existence of the ancestral molecule,(2) what the sequence was, and(3) the line of descent from the ancestor to each of the contemporary molecules.

4. Some sense-carrying units carry less sense than others. For example, simple biopolymers, build by repetition of a few blocks (nucleotides or amino acids), may not be a good source of information about complex evolutionary processes.

5. Changes in biopolymers may be of different types. Some of the changes are beneficial and favored by selection, whereas others have no phenotype and are "cryptic polymorphisms." One reason why some genetic changes have no phenotype is the degeneracy of genetic code: The same amino acid can be coded by different combinations of nucleotides. Another reason is degeneracy of protein sequence with regard to the three-dimensional structure and, ultimately, to the protein function: The same structure and function can be achieved by different combinations of amino acids. Analysis of these different solutions to the same problem may result in a better understanding of the relationships between genotype and phenotype.

6. Gene mutations and duplications of whole genes may put some genes into a "dormant" state. It is plausible that dormant genes may be reactivated after they accumulate changes, and this reactivation may be an important source of evolutionary novelty.

7. Sequences outside the protein-coding regions may have a regulatory function and may evolve differently than in the coding regions. Other noncoding regions may have no function, and mutations in these regions will be free of selection.

8. Chemical compounds may be synthesized by more than one biochemical pathway. Thus, functional convergence at the molecular level is expected, both at the level of the pathways and at the level of individual biochemical reactions.

Thus, the authors cast evolutionary molecular biology as information science and thought that particular attention should be given to distinguishing signals from noise in the sense-carrying units. Biologists, chemists, engineers, mathematicians, and computer scientists who work on in genome analysis today are in fact implementing the research program that, unbeknownst to some of them, was started by Zuckerkandl and Pauling.

This book is no exception. Nearly every chapter addresses an issue that can be traced back to an idea set forth in Zuckerkandl and Pauling's seminal paper. Chapters 2 and 3 discuss practical approaches to sequence comparison (point 2 as outlined previously). Evolutionary inferences from these comparisons (point 3) and the relationship between signal and noise in sequence comparison (point 4) are discussed in nearly every chapter. The issues of functional convergence (point 8) are of central importance in Chapters 6, 7, and 9. Cryptic polymorphism (point 5) is discussed in Chapters 9 and 10 in connection with sequence–structure–function degeneracy. Finally, "what the ancestors were" (point 3) is the central theme of Chapters 11–13. Even Chapter 14, which deals with genome-wide numerical data, draws inspiration from approaches to comparative sequence analysis foreseen by Pauling and Zuckerkandl.

The techniques of biological sequence comparison were not discussed at any length in "Molecules as Documents of Evolutionary History," but the central goal of *finding pairs of similar sequence fragments* was stated very clearly.

Sequence similarity lies at the heart of all biology, not just comparative genomics. The following statement has even been called "the first fact of biological sequence analysis" by Dan Gusfield (1997) at the University of California at Davis:

> *In biomolecular sequences high sequence similarity usually implies significant functional or structural similarity.*

This "first fact" may qualify as one of the most fundamental facts of our understanding of life. Most biologists, however, would not hesitate to add the following:

In biomolecular sequences, high sequence similarity also usually implies evolutionary relationship.

The two statements, though similar in form, are actually distinct, and in a quite fundamental way. The structure of a biological molecule, such as a protein, is something that can be physically defined. If we have a pure sample of this protein, a quiet place for growing crystals, and a synchrotron beamline, we can determine a structure of a protein molecule, at least in principle. Technical details aside, the same equipment would generally do the job for all proteins. Indeed, as I write this, the challenges of high-throughput protein structure determination are being met by the structural genomics projects (Chandonia and Brenner, 2006). Function, however, is not a physical characteristic but, rather, a description of some process, so function can be defined only in a biological context. At the bare minimum, function of a protein involves interactions with other molecules, which have to be identified and included in the description of function. Often, in order to define the biological function of a sequence, we need to monitor the interactions of many components in a cellular extract, in the whole cell, in a living organism, or in an ecosystem of which this organism is a part. As the protein function is performed, its structure may change. Thus, when we casually say "structure and function," in fact we are talking about many different things already. And the fact that sequence similarity can be used to make inferences about all those different properties of a sense-carrying unit—from physical properties of the molecule to its relationships with its environment—is not at all trivial. The "second fact" is also nontrivial: Unlike more or less directly observable structural and functional properties, the common ancestor of two molecules cannot be directly observed (with the exception of rare cases in which the ancestral DNA or protein have survived in ancient proteins or in biopsies), and yet we do not hesitate to infer such an ancestor from the sequence similarity.

Thus, on the basis of sequence similarity, we make conclusions about (1) similar structure, (2) similar function, and (3) common ancestry. These inferences are at the heart of computational biology; most biologists make them every day, and almost every theme in this book is based on such inferences. But how do we make them in practice?

At first glance, the statements about structure and function seem to follow from sequence similarity quite naturally. And without doubt, these statements are amenable to direct experimental corroboration. But in fact, structural and functional inference is inseparable from evolutionary inference. Indeed, when comparing sequences of two biopolymers, our path from sequence similarity to the conclusion about structural or functional similarity is never direct. Instead, we always infer common ancestry of these sequences first, and only from there can we proceed to making structural and functional inferences. This logic is not obvious when the similarity is very high, but if the two sequences are more distantly related to each other (as is the case with most sequence comparisons today), this chain of thought becomes explicit. Indeed, we measure similarity between sequences and immediately use statistics to compare the observed similarity with what would be expected by chance (discussed in Chapter 2). If the similarity is too high to occur by chance, this is usually sufficient for making predictions about protein function (discussed in Chapters 5–8) and structure (see Chapter 9). But the only reason why such reasoning works is because the only way for nonrandom sequence similarity to occur is by descent from a common ancestor of the two sequences. This is the homology inference (see Chapter 3). Thus, the inference of evolutionary relationship, which seems to be the least observable of all, turns out to be a prerequisite of proposing other, directly observable, relationships, such as similarity of structure and function.

Consider the alignment of three sequences, A′, A″, and A‴ (here and elsewhere in this book, I use capital letters in regular font to indicate genes and italicized capitals to indicate species in which these genes are found). Suppose that three sequences come from three different species, one from each, and only the function of A′ has been studied. Suppose that A′ and A″ are almost identical, and the third sequence, A‴, is less similar but still quite close to A′ and A″. Do we use the same information to infer common ancestry and common function of all these sequences? It seems that we do not really need every amino acid residue that is conserved between A′ and A″ to determine that they share a common ancestor; for example, we may not care about the sites conserved exclusively between A′ and A″ because we do not need these residues in order to recognize similarity between A′ and more distantly related A‴, as well as between A″ and A‴. On the other hand, when we are making the inference, "closely related A′ and A″ are more likely to have the same function, but a more distant A‴ may have different function," we, in effect, are using the information about the sites conserved exclusively between A′ and A″ but not between each of them and A‴. Thus, evolutionary, structural, and functional information is intertwined in sequence in subtle ways.

The reverse of the "first fact of sequence analysis" is not true: Functionally similar proteins do not have to have similar sequences, and proteins with similar structures also may have dissimilar sequences (this is discussed in much more detail in Chapters 6 and 9). Neither is the reverse of the "second fact" true: There may be an evolutionary connection between two sequences, but, if these sequences have diverged too far, the sequence similarity between them may not be discernible from the random-level similarity (this is discussed in more detail in Chapter 2). Note that in the case of the "reverse-second" fact, we are dealing with a relationship that still exists, even if the sequence similarity has already blended with the noise. The "reverse-first" fact, however, is more dramatic. Functionally similar proteins may have had lost sequence similarity, but, on the other hand, they may have never shared sequence similarity but converged to the same function from completely different, evolutionarily unrelated sequences. This principle applies to structures as well: Similarity of structures in the absence of sequence similarity may represent either extreme divergence of initially similar sequences or convergence of sequences that were not similar in the first place (discussed in Chapters 6, 9, and 10). Distinguishing between divergence and convergence at the molecular level is one of the most important problems of computational biology.

All these considerations are different facets of the most important postulate of Pauling and Zuckerkandl: Biopolymers contain information about their evolution, structure, and function, and these three types of signals may interact in different ways, sometimes enhancing and in other cases interfering with each other. In a sense, whole biology for the past few decades has been dominated by the quest for ways to extract and analyze signals contained in molecular sequences. Genomics is a continuation of these efforts for our times, when complete genetic makeups of many species are known. At the same time, genomics offers even more. Many times in this book, I will return to the argument that with complete genome sequences, we can answer many questions that we could not answer, or even could not think of asking, before. This is the new era in biology—the era of complete genomes.

Sequences of genes, genomes, and proteins are not the only kinds of data that are of interest to genomics. New technologies allow us to collect information about the occurrence and spatial organization of genes and regulatory sequences; the concentration of different molecules in cells, organs, and biological samples (measurement of mRNA levels, collected with the help of gene expression arrays, is the most famous, but by no means unique, example of this class of data); cellular morphology and physiological responses; and so on. This information often takes the form of rows and columns of numbers. It may seem that Zuckerkandl and Pauling did not have much to say about these data, which were not in the form of sense-carrying units anyway. But in Chapter 14, I argue that the analysis of these genomewide measurements also owes a lot to our experience in sequence comparison.

2

Finding Sequence Similarities

As discussed in Chapter 1, Pauling and Zuckerkandl in their seminal work outlined the research program of studying the complicated ways in which structural, functional, and evolutionary information is convoluted within a molecular sequence. It was clear to them that the comparison of sequences is a clue to uncovering all these types of information. Paraphrasing the famous quote from Theodosius Dobzhansky (1973), almost nothing in computational biology makes any sense except in light of sequence comparison.

Before the deciphering of genetic code and the advent of DNA cloning, the most common order of business in protein science was to isolate a protein, study its biological properties, and only then, motivated by its biological importance, attempt to sequence this protein using rather inefficient methods of direct peptide sequencing. The accumulation of novel protein sequences in those times was slow and deliberate. Even when methods of DNA cloning and sequencing came about in the late 1970s, they were applied mostly to one protein at a time, also guided by biological interest in the gene or its product or, in many cases, by the ease with which a gene could be isolated. Thus, proteins and mRNAs that were abundant or homogeneous, such as cytochrome C homologs, immunoglobulins, or virus capsid proteins, were studied at the sequence level much earlier than other families of proteins. And the biological, biochemical, and other properties of proteins usually were quite well studied by the time the sequence was determined.

But what about evolutionary relationships—how can we infer the common ancestry of the "sense-carrying units" without knowing their sequences? In fact, we can do it just fine in many cases. For example, the favorite subjects of comparative evolutionary biochemistry for most of the 20th century were globins, the main protein constituents of vertebrate red blood cells. Years of work in the lab have shown similarity of many physicochemical and biological properties of globins. At the same time, the anatomical, histological, and biochemical similarity of most components of vertebrate blood and circulatory systems was demonstrated. Altogether, this was the overwhelming evidence of common origin of globin genes and their protein products. In this context, sequencing of globins could be perceived more as a confirmation of the phylogenetic hypothesis than a way of proposing their common origin in the first place. Here again, Pauling and Zukerkandl were ahead of their time when they emphasized that sequences of biopolymers are the real foundation for comparing all of their other properties, and that phylogenetic hypotheses may be put forward on the basis of sequence analysis alone, before inferring other shared properties of genes and proteins. This is a dramatic shift in the way we look at genetic information.

Pauling and Zuckerkandl did not discuss at any length how exactly we should compare sequences and how to measure the strength of signals that this comparison may provide. This was an algorithmic problem in the area of pattern matching, and solving it required the help of mathematicians, computational scientists, and statisticians.

Sequence comparison, particularly the crucial role played in it by one class of algorithms, namely dynamic programming, is discussed in almost every book on computational biology and bioinformatics. David Sankoff was one of the most important figures in the field, and reviewed the early work in a short, vivid paper (Sankoff, 2000). Other reviews can be found in Mount (2004), which is also one of the most detailed introductions to the mechanics of database search and sequence alignments, and in Jones and Pevzner (2004). Succinct primers on dynamic programming and other basic elements of sequence analysis (e.g., substitution matrices and hidden Markov models) can be found in notes by Sean Eddy (2004a–2004d), a thorough review of combinatorial and algorithmic aspects of sequence analysis is provided in Gusfield (1997), and the best introduction to the probabilistic aspects of the same is the book by Durbin *et al.* (1998). Finally, the redoubtable family of BLAST programs has been thoroughly covered in a corpus of work by Steven Altschul (Karlin and Altschul, 1990; Altschul, 1991; Altschul and Gish, 1996; Schaffer *et al.*, 2001; Altschul *et al.*, 1990, 2001, 2005). Newer programs suitable for the era of complete genome sequencing, assembly, and multigenome alignment are discussed in Miller (2001), Kent and Haussler (2001), Schwartz *et al.* (2003), Blanchette *et al.* (2004), and Ovcharenko *et al.* (2005).

My goal in this chapter is not to repeat what is written in these excellent books and articles. Rather, I present five challenges of biological sequence analysis that receive relatively little attention but can make a major difference in sequence analysis, and I try to show how some of the well-known sequence comparison approaches address these challenges. In dealing with these concerns, I mostly talk about protein molecules, which, of course, are sequences of amino acid characters drawn, in the first approximation, from the 20-letter alphabet. I only briefly mention comparison of nucleotide sequences, which consist of four nucleotide characters, and other types of comparisons, such as comparison of gene orders in different genomes, when the alphabet may include hundreds or thousands of characters.

Challenge 1. The methods of sequence alignment are often classified into "local" or "global" methods, or, more accurately, into methods that produce local or global alignments. (In a global alignment, each character is forced to be aligned with something, and in a local alignment some characters are not considered. Many special cases of alignment can be given more rigorous definition; Gusfield, 1997.) In one sense, this distinction is important because statistics of local alignments is well-defined, which is not the case for global alignments (Altschul, 2006). In a different sense, this distinction is a red herring because the goal of comparative sequence analysis is really not "to construct an *alignment.*" Rather, the objective is to find evolutionary, functional, and structural signals in biological sense-carrying units—the signals that, as discussed in Chapter 1, are revealed by sequence similarity. Thus, algorithms may be set up to produce either local or global *alignment*, whereas in fact the most important question is whether the *similarity* between sequences is global or local.

Challenge 2. Each method of sequence alignment tries to find an extremum of some value, such as the minimal number of operations required to convert one sequence into another or the maximal matching score (which is most commonly sought and which will mostly concern us in this chapter). This solves an optimization problem but may not do much to solve a biological problem (i.e., to find signals in sense-carrying units). Biological knowledge enters into the picture by way of the scoring function, which is the way of measuring similarities/differences between sequences. For example, if we thought that 4 amino acid residues represented by vowels of the Latin alphabet (A, E, I, and Y) are less important in proteins than the other 16 residues, and decided to only consider matches between the latter

16, any alignment algorithm would work with such a scoring system without complain—even though the idea is absurd on its face. All improvements in sensitivity of sequence analysis are in fact the improvements in measuring similarity between sequences—from less sensitive to more sensitive substitution matrices and then to probabilistic models of multiple sequence alignments. The theory of similarity/distance between sense-carrying units, however, is in its infancy, notwithstanding some important insights (see Altschul, 1991; Zharkikh, 1994).

Challenge 3. Sequence alignment algorithms, even when provided with good scoring schemes, will align any strings of allowed symbols and produce the highest scoring match between any two sequences, whether they contain biological signals or not. But these algorithms will not tell whether this highest match is "high enough" to indicate the presence of a signal we are looking for. To pick out matches that represent biologically important signals, one needs a statistical theory that evaluates alignments and compares them to some kind of a standard. Such theory is available in an exact form for ungapped alignments (Karlin and Altschul, 1990; Altschul, 2006) and in an approximate, yet apparently quite accurate, form for alignments with gaps (Mott, 2000). But even with this theory in hand, and with good scoring schemes, there are many alignments that remain in the "twilight zone" of borderline statistical significance and cannot be directly used to infer the presence of a biological signal. The problem of how to validate (or reject) the alignments in the twilight zone is still not fully solved.

Challenge 4. Related to challenges 2 and 3 is the problem of nontransitivity of sequence similarity scores. The simplest way to state nontransitivity is for the case of three sequences: If sequences A and B can be matched (aligned) with a high score, and sequences B and C can also be matched with a high score, this does not tell us anything about the score between A and C. That score can also be high according to our statistical theory or it can be low—so low as to be indistinguishable from the noise. In the context of the database searches, most matches indistinguishable from the noise are not reported to the investigator, so we may not know about similarity between A and C unless we first know about similarity between A and B. Of course, we can increase sensitivity of sequence comparison, for example, by replacing a single-sequence query by a probabilistic model of a protein family to which this sequence belongs or by aligning two family models instead of two representative sequences. This will pull some of the twilight zone similarities into the high-similarity zone (i.e., some "sequences C" will become directly linked to A), but other sequences and sequence families may remain low scoring with regard to some query A yet pass the significance threshold with a query B that itself is high scoring with regard to A. This nontransitivity problem is not fully solved in any method of sequence comparison.

Challenge 5. Any textbook on bioinformatics will discuss differences between pairwise alignments and multiple alignments. It is important to know what these differences are: For example, some of the theory that is worked out in considerable detail for the case of two sequences cannot be easily generalized to multiple alignments, and some alignment methods that have acceptable speed of execution on two sequences are computationally prohibitive when many sequences are involved. But there is another distinction, which is sometimes overlooked; this distinction is between different types of pairwise alignments. Indeed, we may use methods of pairwise alignment as a tool for discovering similarity that was not known before, but we also can apply alignment methods to study similarity between sequences that are already known to be related. The first type of pairwise alignment, in principle, does not have to be biologically optimal: Arguably, it has to score just high enough to stand out from the background. At the same time, this "type I" alignment has to be arrived at with high efficiency, because discovery of sequence similarity is typically done in the context of database searches, in which a query sequence is matched to all, or at least many, database sequences. The "type II" alignment, on the contrary, has to be accurate, but the program that produces it does not have to be ultrafast. Thus, database search programs, which produce type I pairwise alignments, may sacrifice some accuracy for speed. The relationship between type I and type II alignments, however, is not well understood; perhaps the only thing that can be said with confidence is that, as a rule, type II alignments include larger number of aligned characters than do type I alignments. It is unknown whether there is any other systemic bias between two types of pairwise alignments.

One of the first practical approaches to pairwise sequence comparison, which already wrestled with most of these challenges, was the work of Adrian Gibbs and George McIntyre of CSIRO in Canberra, Australia. They developed what they simply called a "diagram," or a two-dimensional representation of similarities between two sequences (Gibbs and McIntyre, 1970). The sequences of two proteins were written down along the two adjacent sides of a rectangle, and similarities between them were recorded inside the rectangle (Fig. 2.1). The description of the method boils down to a sentence at the beginning of the article: "Within the body of the diagram every match is recorded; a dot is put wherever a row and column with the same amino acid (recorded at the edge of diagram) intersect."

Several properties of a diagram are obvious. If the same string of amino acids is found in both sequences, this is seen as a cluster of dots along a diagonal. If a sequence is compared to

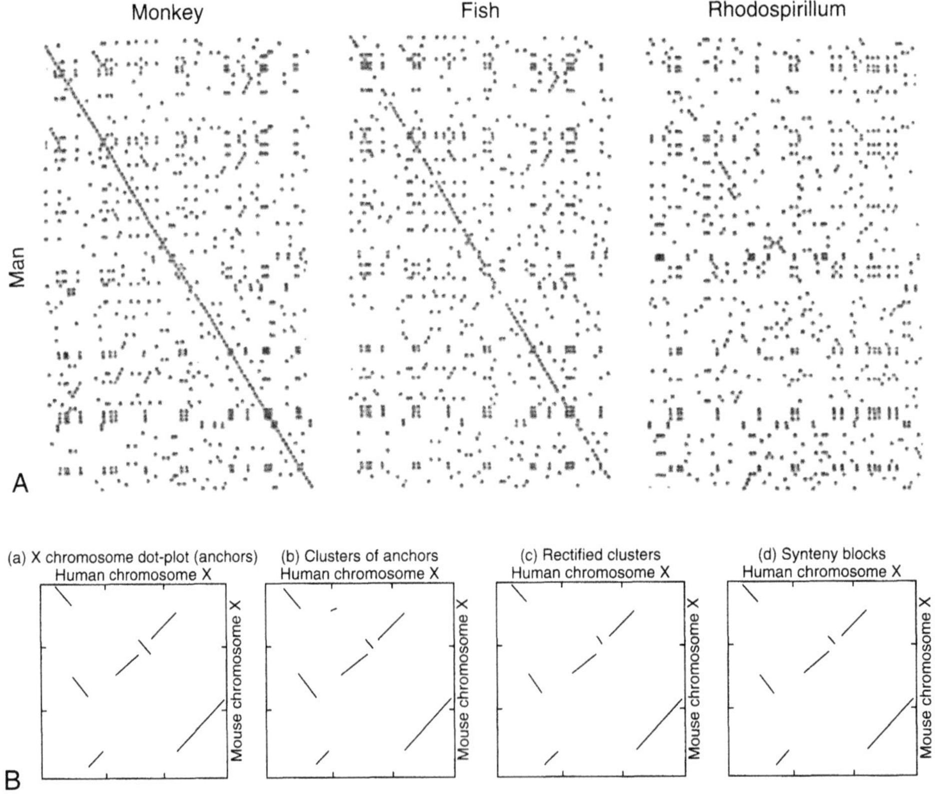

Figure 2.1. (**A**) "Diagram" of pairwise comparison of cytochromes from different species, from vertebrates to bacteria. Reprinted from Gibbs and McEntyre (1970) by permission of Blackwell Publishing. (**B**) Newer use of diagram showing rearrangement of chromosomes in mammalian evolution; here, dots represent shared genes in two genomes instead of identical amino acids in two proteins. Reprinted from Murphy *et al.* (2004) by permission of Elsevier.

itself, the longest (main) diagonal is marked. If a string of amino acids is found more than once within the same protein sequence, segments in other diagonals will also be marked, either symmetrically with regard to the main diagonal in the case of self-comparison or asymmetrically if two different proteins are compared. If two sequences differ from one another by only a few substitutions, the main diagonal will still be seen, although it will break at the substitution sites. If there are insertions and deletions, the highlighted diagonals move away from the main diagonal, and with more changes, the highlighted diagonals become shorter.

When two sequences are not too similar, the diagonals may be more difficult to detect by eye, as in the case of the rightmost panel in Fig. 2.1. How can we automatically improve recognition of these diagonal fragments? Gibbs and McIntyre suggested two measures for doing so. "Run index" computed the distribution of the lengths of all unbroken runs in the diagram, and "diagonal index" computed the distribution of the number of matches in each diagonal (how exactly this was done is not very important at the moment). The two measures are different: The first prefers relatively long matching segments, and the second seeks groups of matches, perhaps even short ones, that belong to the same diagonal. The signal revealed by these two measures is also different: For example, the "diagonal index" suggested that the cytochrome of screwworm (dipteran insect) was most similar to the fish relative, whereas the "run index" indicated that the cytochromes from screwworm and silkworm (lepidopteran insect) were the closest to each other.

Gibbs and McIntyre also proposed a statistics of the observed "runs" of amino acids. They jumbled amino acid letters so that a real sequence could be compared not only to another real sequence but also to its randomized version. For each two sequences, one could now compute a number that quantified the similarity between them, and it was also possible to compute the same number for the comparison of one real sequence and the jumbled other sequence ("similarity expected by chance for the proteins with the same amino acid composition"). One can also produce several jumbled versions of one sequence, compare the other sequence with each of them, compute similarity for each of the comparisons, and calculate the average of these similarities or some other statistical measure. This will be "background," "random," or "chance" similarity, compared to the "real" similarity between two real sequences.

Several decades later, some of these statistics may sound a bit naive (and, indeed, "run index" and "diagonal index" are not commonly used in sequence comparisons anymore). But today, the same as four decades ago, we are concerned with selecting a good measure for assessing similarity between proteins. Note that Gibbs and McIntyre's work mentioned several measures of different nature. In particular, they employed, first, similarity between two real sequences, measured in at least two different ways (run index and diagonal index, which are both derived measures; there were also more direct measures, such as percentage of identical residues, which were quite self-evident even then, and perhaps for that reason not discussed at all); second, similarity between the real and randomized sequences, which, again, can be measured in several ways; and third, the difference between the first and the second measures, which can be expressed, for example, as ratio, or in other ways. Thus, one can come up with many different numbers, and, perhaps, all of them are of interest.

The beginnings of theory of sequence similarity, in fact, predate Gibbs and McIntyre's article. Work done in the 1960s by Walter Fitch, then of the University of Wisconsin and currently at the University of California at Irvine (Fitch, 1966a,b, 1967, 1969, 1970a,b), and by Margaret O. Dayhoff's group at the University of Maryland, and later at Georgetown University (Dayhoff *et al.*, 1965; Dayhoff and Eck, 1968), is especially notable, although some of Fitch's work received criticism from Gibbs and McIntyre for reasons that are no longer obvious, at least to me. In fact, it took researchers another 20 years to develop more sophisticated mathematical foundations of sequence matching. But the rectangular diagram, which was introduced by Gibbs and McIntyre in 1970, stayed around. An example (Fig. 2.1B) shows that it continues to enjoy popularity in modern genomics (Murphy *et al.*, 2004).

Also in 1970, the rectangular diagram of similarity between two biomolecular sequences was put to a different use. Whereas Gibbs and McIntyre were interested in visualizing every kind of similarity between a pair of sequences, Saul Needleman and Christian Wunsch of Northwestern University and VA Research Hospital in Chicago decided to search for what they called "the maximum match"—a correspondence between two sequences in which the maximal number of amino acids in each sequence are aligned to each other, achieving a large, possibly a maximum, score (Needleman and Wunsch, 1970). In order to find such a match, they used an approach that is known in computer science as dynamic programming (which is an algorithmic idea, not programming as in "writing a computer program," and for this reason perhaps should be called something else, for example, "dynamic planning," as in Harel, 1992). Dynamic programming/planning is useful for finding approximate similarities between any strings of symbols—a problem that occurs in many areas of science and technology. Indeed, it appears that the idea of dynamic planning has been proposed independently several times, with perhaps the earliest formal description coming from Richard Bellman of the RAND Corporation (Bellman, 1952; he notes that similar ideas were published in the late 1940s by future Nobel Laureate in Economics Kenneth J. Arrow, then also of the RAND Corporation, and later of Stanford and Harvard, and by statistician Abraham Wald of the University of Chicago). Bellman's work is rather abstract, and he has given only toy examples of possible uses. One of the earliest practical applications of the approach is by Taras Vintsyuk of Ukrainian Academy of Sciences (Vintsyuk, 1968), who worked at the time, as he does now, in the area of speech recognition. The main idea of the algorithm is quite simple and is encountered even in middle-school math.

The problem shown in Fig. 2.2 is from a seventh-grade honors algebra curriculum. We have to travel from the top left corner (Origin) of a grid, say, of 7×8 dimension, to the bottom right corner (Destination) along the gridlines, moving either right or down, but never left or up. How many distinct paths are there? To find the answer, consider the Destination node. Obviously, that node can be reached either from the node immediately to its left or from the node immediately above. If there are a distinct paths to the former and b distinct paths to the Destination, then the number of distinct paths to the Destination is $a+b$. Clearly, the number of paths to any node in the grid is the sum of such two numbers, each of which represents the sum of paths—one to the node on top and the other to the node on the left. We also see that all nodes on the left and the top sides of the large rectangle have just one path leading to them. This is all we need to know in order to count the number of paths to each node in the graph. This is a familiar construct in mathematics—nothing more than a Pascal triangle in disguise.

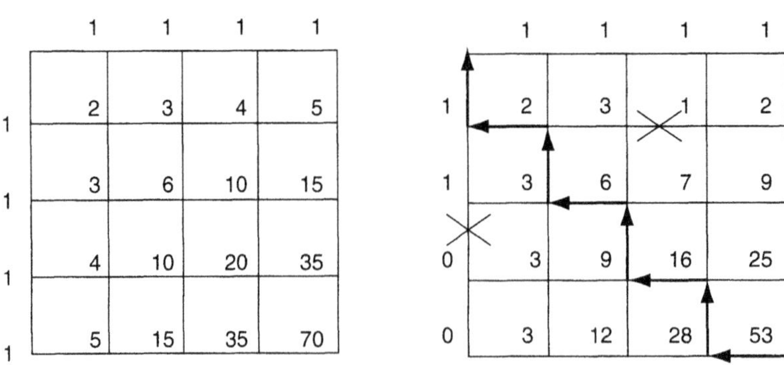

Figure 2.2. Perhaps the simplest application of dynamic planning: counting paths on a rectangular grid.

Suppose now that some of the paths are blocked. It is easy to see how numbers at some of the nodes will become smaller, maybe even zero if a node is obstructed from all sides and becomes unreachable, and how the final number of paths leading to the Destination node will change (Fig. 2.2). Suppose that we are at the Destination node and want to get back to the Origin, using only allowed paths, and to choose such a path that we end up with maximal possible sum of numbers that we passed on the way. To do that, we proceed backwards, at each node selecting the largest number among the two allowed.

This concludes the description of three stages of the dynamic planning process for this task. First, we produce the set of initial conditions—in this case, the size of the grid, the instructions to count the number of paths, and the positions of blocked paths. Second, we set the recursion—that is, the rule determining which number is assigned to each node; in this case, this number is produced by addition of two numbers at the left and on top. Finally, we define the traceback rule.

Let us now write numbers inside the squares and not at the nodes as before. This is very similar to the setting in Needleman and Wunsch's paper. To quote, "In the simplest method, MAT_{ij} [i.e., the value in the *i*th row and *j*th column of the matrix] is assigned the value, one if A_j is the same kind of amino acid as B_i; if they are different amino acids, MAT_{ij} is assigned the value, zero." This is the initialization (Fig. 2.3). Recursion in this case is as follows: For each cell *ij*, instead of its two neighboring cells, we examine the fragments of the $(i-1)$th row and a $(j-1)$th column, find the largest number in any of these cells, and add it to the number in the cell *ij*. Traceback is also self-evident (Fig. 2.3).

Now that the way to obtain the answer is known, let us see what the question was. Needleman and Wunsh were interested in a "maximum match"—the alignment of two sequences that has the highest score. Of course, the value of this score depends on the scoring scheme, but the method of obtaining the maximum match should work the same way for every such scheme. The Needleman–Wunsch approach to approximate matching is able to interrupt one or both sequences, if such interruptions improve the number of matched amino acids. A space between two characters in one sequence, to which one or more characters in another sequence are aligned, is called a "gap."

Gerhard Braunitzer at Max Planck Institute for Biochemistry was one of the first to attract attention to gaps in sequence alignment and may have even coined the name (Braunitzer, 1965). Researchers have been uneasy about "gapping" the sequences in order to improve the

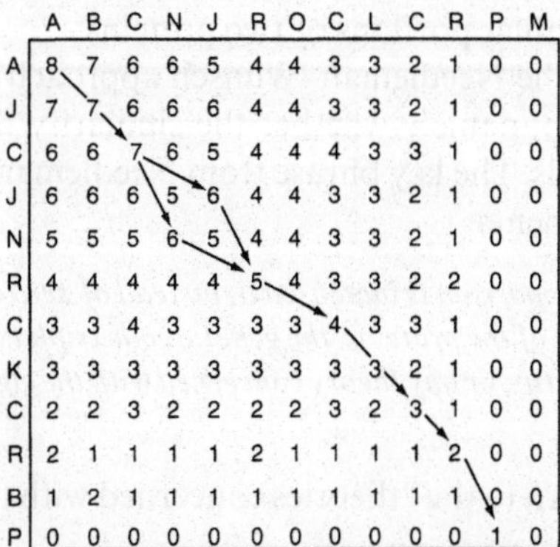

	A	B	C	N	J	R	O	C	L	C	R	P	M
A	8	7	6	6	5	4	4	3	3	2	1	0	0
J	7	7	6	6	6	4	4	3	3	2	1	0	0
C	6	6	7	6	5	4	4	4	3	3	1	0	0
J	6	6	6	5	6	4	4	3	3	2	1	0	0
N	5	5	5	6	5	4	4	3	3	2	1	0	0
R	4	4	4	4	4	5	4	3	3	2	2	0	0
C	3	3	4	3	3	3	3	4	3	3	1	0	0
K	3	3	3	3	3	3	3	3	3	2	1	0	0
C	2	2	3	2	2	2	2	3	2	3	1	0	0
R	2	1	1	1	1	2	1	1	1	1	2	0	0
B	1	2	1	1	1	1	1	1	1	1	1	0	0
P	0	0	0	0	0	0	0	0	0	0	0	1	0

Figure 2.3. Dynamic planning approach and recursion proposed by Needleman and Wunsch (1970). The use of letters J and O as amino acid symbols is now obsolete. Reprinted from *J. Mol. Biol.*, 34(3), Needleman, S. B., and Wunsch, C. D., A general method applicable to the search for similarities in the amino acid sequence of two proteins, pp. 443–453, Copyright 1970, with permission from Elsevier.

alignment: Somehow, the procedure of gap introduction seemed arbitrary. It is true that in the course of alignment gaps are deliberately introduced by a human being (or by a program written by a human being). But it is also important to remember that the processes of DNA replication, recombination, and repair involve occasional insertions and deletions of nucleotides, and some sequences may have added or deleted nucleotides because of that (the reminder emphasized by Doolittle, 1986). Thus, there is a perfectly natural justification for making gaps in the alignments: When introduced at the points of insertions/deletions, gaps record actual events, and therefore they indicate evolutionary and other signals—exactly what we want to study. So it is not true that all gaps in alignments are undesirable; however, it is true that there should be a well-defined procedure to account for them. And if the very purpose of introducing a gap is to improve the quality of the sequence match, it is essential to associate the alignment with a numerical value, so that different alignments can be compared and the effect of gaps on alignment score can be studied. For example, it may be important to compare the alignments before and after introduction of a gap. A related issue, which has to do with the efficiency of computation, is that some gaps do not seem to be worth experimenting with, such as gaps in highly similar regions that align straightforwardly. Therefore, it would be beneficial to limit the number of gaps that are actually examined. These questions were not addressed in Needleman and Wunsch's work and, in a sense, have not been rigorously addressed ever since.

As with many influential papers, Needleman and Wunsch's is widely cited but occasionally misinterpreted. It is often referred to as "Needleman–Wunsch algorithm," whereas Gusfield (1997) and Sankoff (2000) point out that the actual algorithmic implementation proposed by Needleman and Wunsch was not the fastest one: It ran in the cubic time, whereas the quadratic time implementations were already known. Moreover, sometimes it is stated that Needleman and Wunsch gave the solution for the highest scoring alignment with gaps; in fact, even though their approach allows one to introduce some gaps, and not to waste time examining gaps that are far from the alignment path, the proof that the match is indeed maximal was not given.

Another ambiguity is to call Needleman–Wunsch a "global alignment algorithm." In fact, nothing prevents us from using the same algorithm for finding similarities that are local with regard to at least one sequence, for example, by controlling the number and/or length of gaps. Also, of course, the real question is not whether the algorithm we are using is local or global but, rather, whether the similarity between sequences is along their whole length or confined to a shorter segment. If the similarity is local, such as when two protein sequences share one conserved domain but also have other, dissimilar domains, then there is no gain in using a global algorithm. On the other hand, there may be no harm in it either, if the program is able to correctly align the most similar portions of two proteins.

Most important, neither the Needleman–Wunsch approach nor any other algorithm can answer a biological question, namely whether the similarity that we observe contains any biologically important signals. The key phrase from Needleman and Wunsch's paper sets the scene for the future developments:

> *The sophistication of the comparison is increased if, instead of zero or one, each cell value is made a function of the composition of the proteins, the genetic code triplets representing the amino acids, the neighboring cells in the array, or any theory concerned with the significance of pair of the amino acids.*

Later in this chapter, we return to the "theories concerned with the significance of amino acid pairs."

In 1981, Temple Smith, then of Northern Michigan University and currently of Boston University, and Michael Waterman, then of Los Alamos National Laboratory and currently of the University of Southern California, found a way to detect a local match between two proteins that is "the best," meaning that, under a given scoring scheme, this match cannot be improved by either adding or trimming the aligned pairs of characters. Before discussing their work, however, let's ask ourselves, why would one think that such highest scoring match might be local? Indeed, in the Needleman–Wunsch approach, it was usually profitable to include more amino acids into the alignment because the score could only get higher. How might it be possible that the longest alignment is not guaranteed to be the best?

The answer consists of two parts. First, Needleman and Wunsch did not prove that their approach generates the highest scoring alignment (although such proof could be derived from the work of Bellman and other earlier work on dynamic programming, with which most biologists were not familiar at the time). Second, from 1970 to 1981, the approaches to similarity scoring changed, as the negative numbers entered the picture.

It seems natural that biologically significant sequence alignment should have a (high) positive score associated with it. In the examples shown by Needleman and Wunsch, all matches were positive and mismatches were equal to zero, and addition of some such numbers was the only permitted operation. Mismatches, insertions, and deletions were neither rewarded nor penalized, so the score of any alignment of two sequences could only have a positive value. Later, however, scientists decided that it is not good to be so relaxed about gaps. One argument was that most pairs of sequences could be somehow aligned if gaps were allowed, and, therefore, arbitrary gaps might help to "legitimize" a similarity between any two sequences, even if they are unrelated and should not be aligned.

To show that arbitrary gapping is a real concern, let us consider an example from 1991—two decades later than Needleman–Wunsch's work and a decade later than that of Smith–Waterman. Connexins are proteins that are found in animal gap junctions (no connection to gaps in sequences), a specialized type of tight connections between membranes of adjoining animal cells. Plants also have specialized connections between cells, called plasmodesmata, and although they are now known to be morphologically different from animal gap junctions, the cell–cell contacts in plants and animals share some physiological properties (Robards *et al.*, 1990). Thus, it would be extremely interesting to know whether the two types of connections are made up of similar protein components. Meiners *et al.* (1991) used antibodies raised against animal connexin-32 to screen the expression library of the model plant *Arabidopsis thaliana* and found a gene, promptly named CX32, whose product cross-reacted with the anti-connexin antibodies. The same gene product was detected in cell walls by immunohistochemistry, and the authors concluded that they were looking at plant connexin. The purported relationship between connexin-32 and its plant "counterpart" can be seen in Fig. 3 in their paper. Is that alignment good additional evidence that we are looking at plant connexin? Do we see a structural, functional, or evolutionary signal in that alignment?

One view on the role of sequence comparison is that most alignments are at best the auxiliary evidence, except for the cases of extremely high similarity between the aligned sequences. If, however, the similarity is moderate, the alignment does not matter much one way or the other; let us focus on wet-lab experiments, from which the final judgment will come. In the case of the search for plant connexin, serological properties and cellular localization of the proteins appeared to be compatible with its purported role, and one could argue that what is really needed is further physiological and biochemical evidence of its role in cell contacts, not further examination of the alignment.

There are problems with such a view, of course. First, in any inquiry, one should not only consider the observations favoring a theory but also examine all the evidence against the theory. Second, this concerns all kinds of evidence, including both biochemical experimentation and computer-aided sequence analysis. Eugene Koonin and myself reanalyzed sequence relationships between connexin and CX32 by standard (although at the time relatively novel) BLAST search of the complete protein sequence database at the

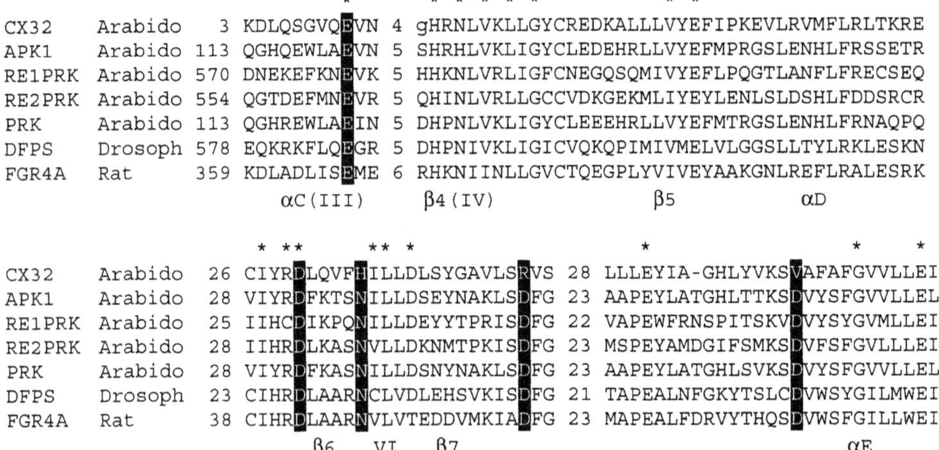

Figure 2.4. The alleged plant connexin is in fact a member of the protein kinase family. Multiple alignment of CX32 with selected protein kinase sequences. Residues shared by CX32 and other sequences (one exception in kinases allowed in the bottom of the figure) are indicated by *asterisks*. Alignment to kinases is supported by BLAST statistics, by the conservation of known or predicted secondary structure elements, and by conservation of residues directly involved in catalysis (indicated by *shading*). The CX32 sequence appeared to have several of these residues mutated, but more recent resequencing has indicated that most of them are in fact conserved in other kinases (see SWISSPROT entry P27450). Modified from Mushegian and Koonin (1993) by permission from the American Society of Plant Biology.

National Center for Biotechnology Information. Connexin sequences matched with high scores only of other animal connexins, but CX32 did not show up in these searches at all. Instead, CX32 turned out to be similar to the eukaryotic protein kinase family (Mushegian and Koonin, 1993). The supporting alignment is shown in Fig. 2.4. Note that this alignment is produced by inserting gaps in the positions different from those in the original alignment, and there are regions of similarity in which gaps are rare. Most important, the residues that are conserved in protein kinases and CX32 are the same as the most conserved residues within the kinase family proper; that is, they indicate that some functional properties conferred by these residues in biochemically characterized protein kinases are most likely also preserved in CX32. This cannot be said about the residues "conserved" between CX32 and connexins. Thus, the indication that CX32 is not connexin is much stronger than the opposite hypothesis, even though the former is provided by computational analysis and the latter derives from the wet-lab experiment.

This example also shows that frivolous gaps can get us in trouble. But this does not mean that gaps should be disallowed or avoided altogether. Insertions and deletions in DNA really happen, and they may be of structural, functional, and evolutionary significance. There is nothing wrong with considering them. The quantitative theory of gaps, however, turned out to be a difficult problem (Storey and Siegmund, 2001; Zachariah *et al.*, 2005). We cannot express it shorter and better than did S. Altschul, the creator of the BLAST suite of programs, in his on-line tutorial (Altschul, 2006):

> *Appropriate gap scores have been selected over the years by trial and error ... and most alignment programs will have a default set of gap scores to go with a default set of substitution scores. If the user wishes to employ a different set of substitution scores, there is no guarantee that the same gap scores will remain appropriate. No clear theoretical guidance can be given, but "affine gap scores"..., with a large penalty for opening a gap and a much smaller one for extending it, have generally proved among the most effective.*

Thus, as a practical solution, it was decided in the late 1970s that gaps should not be scored as zero-value matches but had better be penalized. Gap penalties, therefore, will result in negative values for some cells in the dynamic planning matrix simply because it is possible that a small, slowly increasing score will be offset by a larger gap penalty. The other source of negative numbers in the dynamic programming table came from newer scoring functions, also developed in the 1970s, in which not only gaps but also certain substitutions could be penalized (e.g., between two amino acids that were extremely dissimilar from the chemical point of view or extremely unlikely to mutate in nature; see later).

Returning to the very short and very influential paper by Smith and Waterman (1981), we can say that the essence of their approach is in resetting the negative values in the matrix to zero. For each cell, the Smith–Waterman algorithm examines the values corresponding to the three directions of possible extension of the match. If at least one value is positive, the highest value is selected, and if all three are negative, zero is used instead. Several features of the resulting matrix are of note. First, the traceback is straightforward: Starting from the largest number in the table, select the largest among the numbers in three neighboring cells until these numbers run off to zero. There will be no higher scoring match in the table. Second, the procedure works with any scoring scheme—that is, with any set of values for matches, mismatches, and gaps (although the highest scoring path may change if these values change). Third, the ends of the highest scoring match can be used to define the zones where the next best similarity can be looked for, and there are ways to connect several local similarity runs into a longer (maybe even global) alignment.

The Smith–Waterman algorithm guarantees maximally scoring ("best") alignment. However, as for any other approach to sequence alignment, in itself it is not sufficient to judge about relatedness of two sequences. For that purpose, we need a statistical theory.

Consider a query sequence and one database sequence. A pair of equal-length, ungapped segments, one from each of these two sequences, such that its score cannot be improved by either extension or trimming, is called high-scoring segment pair, or HSP. There may be several such pairs in two sequences that are being compared, often with different scores. The following is a critical question: For any such HSP, how likely are we to observe it by chance alone? If we are not likely to see an HSP with a score that high purely by chance, then perhaps this HSP represents something biologically interesting (e.g., structural, functional, or evolutionary signal).

What is "chance" in this context? To quote Stephen Altschul (2006) again, as far as sequence analysis is concerned, " 'chance' can mean the comparison of (i) real but nonhomologous sequences; (ii) real sequences that are shuffled to preserve compositional properties; or (iii) sequences that are generated randomly based upon DNA or protein sequence model."

We can now use any of these models to estimate the number of HSPs with positive score at least S which are expected to occur by chance. Relevant statistic is called the extreme value statistic, and it is given by the following expression: $Kmne^{-\lambda S}$, where n and m are lengths of two sequences that are compared, K is the parameter adjusting for the search space size, and λ is the scaling parameter for the scoring system. The only other requirement is that the scoring system is such that a score for a random HSP is negative so that two long sequences do not reach a high score simply because they are long.

This statistic is precise only for local alignments that do not have gaps. For gapped alignments, parameters K and λ must be estimated empirically by simulation or a large-scale comparison of unrelated sequences. Besides this statistical theory, the BLAST suite of programs includes many heuristics that are used to achieve the desired speed when searching ever-increasing sequence databases. Those are discussed in much detail in Altschul *et al.* (1990, 1997) and Schaffer *et al.* (2001), and the interested reader is encouraged to read these momentous papers. It is important to remember that all enhancements described in those works are useful only as long as the alignment is based on a good scoring scheme. How to select such a scheme is an extremely important problem.

There are $20 \times 20 = 400$ possible alignments between amino acid pairs, so it is quite reasonable to expect that any scoring system may take the form of a 20×20 table. The question, of course, is how to determine the value in each of the cells in the table. All approaches to scoring amino acid substitutions can be divided into two classes. One class of schemes is of a deductive sort. It relies on some general principles of genetics, on knowledge of DNA mutation mechanisms, or on principles of amino acid biochemistry. The other class can be called inductive; it derives scores from the statistical analysis of amino acid substitutions that are found in nature. (Nucleotide alphabets are smaller, and there are $4 \times 4 = 16$ possible interconversions between nucleotides; many aspects of scoring systems for nucleotides will be easily understood based on richer models for amino acids).

One example of an early "theory concerning the pairs of amino acids" of the deductive type is mentioned by Zuckerkandl and Pauling in "Molecules as Documents of Evolutionary History" (1965). In that model, substitutions are scored using the "mutations required" parameter. Any codon can be converted into any other codon by one, two, or three mutations. Thus, every pair of amino acids can be associated with a value (score) from 0 to 3, and alignment of two proteins can be seen as a sum of scores for each pair of amino acids. If there is more than one way to convert one amino acid to another, involving different numbers of changes, they can be averaged, giving fractional numbers in the same interval. One also has to derive a way to score amino acids aligned with themselves if the codons in two sequences are not the same.

There are several concerns regarding this model. First, in this scoring system, we are measuring distance between amino acids and proteins, rather than similarity, so that the numbers are smaller as the sequences become closer. This does not provide any natural scale for long and short sequences: A pair of identical amino acids has the same score (of zero) as the pair of identical aligned proteins. On the other hand, if a match is associated with a positive number, then the similarity score of a protein with itself will be on the order of hundreds (an average bacterial protein consists of 200–300 amino acids, and an average eukaryotic protein is longer), and we could at least easily distinguish this score from the similarity score of two short peptides. A more serious concern has to do with the assumptions that the "mutation required" approach makes about evolution. If all mutations were independent and equally likely, three mutations in one codon would be indeed less probable than just one. But this is relevant only if the rate of mutation is the limiting factor in protein evolution, which is only true in the absence of selection. Indeed, matrices based on the properties of genetic code are rarely used nowadays (although evolution of genetic code remains an exciting area of computational biology; see Trifonov, 2000, 2004; Knight *et al.*, 2001; Vetsigian *et al.*, 2006).

The other major line of deduction is based on the knowledge of amino acid chemistry. Throughout the years, many properties of amino acids have been determined, including bulkiness, polarity, hydrophobicity, electronegativity, molecular volume, hydropathy index, accessible surface area, and chemical classification of the side chains such as aliphatic, aromatic, or carboxylic. Many classifications can be produced on the basis of these properties or their combinations, and the distance between amino acids can be derived from these classifications.

Several problems have to be addressed here. First, it is desirable that the substitution scores have numeric values, but it is not clear how to quantify different types of physical or chemical similarity. Suppose that aspartic and glutamic acids (D and E), both having carboxylic groups in their side chains, are easily interchangeable. Lysine and arginine (K and R), both with the amino groups in their side chains, may likewise be interchangeable. Surely, the change D-to-K, which fundamentally alters the polarity and size of the side chain, should receive a different score than the conservative D-to-E change. But should the former score be slightly or significantly lower than the latter? Should it perhaps be negative? Should the "conservative" D-to-E change be scored differently from the "also conservative" K-to-R change? How to compare the "nonconservative" D-to-K replacement to the similarly nonconservative E-to-R replacement? What should be done with each of the reverse replacements—are all scores symmetric?

A further problem is that it may be difficult, if not impossible, to partition all amino acids into groups of similar residues without overlaps. This is because a residue has more than one property and may share different properties with different subsets of amino acids. Consider, for example, serine S. It is customary to group it together with threonine T, based on the hydroxyl groups in both side chains. But how do we know that hydroxyl is the most important feature, which explains each occurrence of serine in every protein? Serine also has a small side chain, and in some cases it is this small size, allowing for the greater flexibility of the protein main chain, that matters most; in that respect, serines in certain positions are most similar to glycine G and alanine A. Furthermore, the hydroxyl group of serine often plays a functional role in the active centers of hydrolytic enzymes, where it may participate in nucleophilic attack on a chemical bond that needs to be cleaved. A similar role is often played by another nucleophile, aspartic acid D. Threonine serves as a nucleophile only infrequently, in a limited set of specialized hydrolases. Should we group S and D together, to the exclusion of T? On the other hand, a hydrocarbon moiety of the side chain in threonine is longer than in serine, and this is the likely reason why threonine is often found within the beta strands, where it is interspersed with leucine L, isoleucine I, and valine V. Serine is infrequent in these positions: Does that mean that a group of L, I, V, and T should be created, to the exclusion of S? Perhaps these relationships can be better represented as a Venn's diagram than as a set of non-overlapping groups (Fig. 2.5), but the first problem remains—how to convert these diagrams (or, indeed, any other type of representation) into a 20×20 matrix of numerical values useful for similarity scoring?

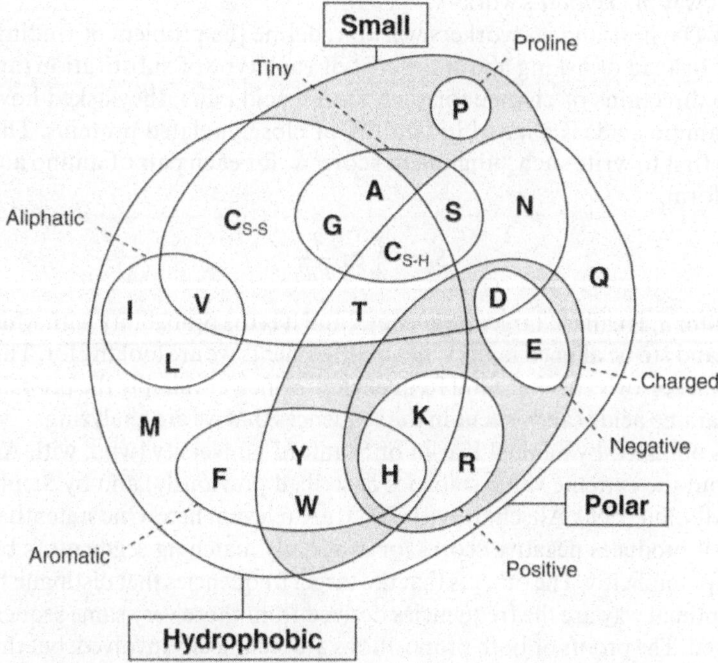

Figure 2.5. Unique, shared, and overlapping properties of amino acids. From www.russell.embl-heidelberg.de/aas/aas.html by permission of Rob Russell.

The solution to the scoring matrix problem has come from protein and gene sequencing. Few publications in the history of computational biology are more important than the "Atlas of Protein Structure and Function" series published by M. O. Dayhoff and co-workers in the 1960s and 1970s (Dayhoff *et al.*, 1965; Dayhoff and Eck, 1968). Many fundamental concepts and data presentations that we now take for granted originated from this work, including the very notion of protein families, the idea that the 20 × 20 matrix should be derived from the observation of frequencies of naturally occurring amino acid substitutions, and the derivation of the evolutionary model from the substitution data.

Dayhoff's atlas included sets of closely related proteins from different species, such as insulins from various mammals or cytochromes c from different mammals, other vertebrates, and even bacteria. The common origin of proteins in each group was not controversial, and construction of alignments was quite straightforward. These alignments were used to study amino acid substitutions and their evolutionary role. The central idea was "accepted point mutation," where "accepted" meant approved by natural selection (Barker and Dayhoff, 1982).

Mutations in DNA may produce different sorts of amino acid changes, but when we look at protein families, we only see those substitutions that have already been chosen by selection. In the Atlas, this was stated as "two distinct processes: the first is the occurrence of the mutation in the gene and the second is its acceptance by natural selection as an improvement" (Barker and Dayhoff, 1982). The latter assumption—that is, "acceptance means improvement"—is actually difficult to test. We usually do not know that the selected substitution is an improvement, and in fact, neutral theory of molecular evolution holds that in most cases it is not even true—the majority of observed nucleotide and amino acids substitutions are thought to be neither useful nor particularly harmful (Kimura, 1983). Fortunately, however, development of substitution scores does not require this assumption of improvement, and it was not used in any significant way in Dayhoff's work.

The idea of Dayhoff and co-workers was to redefine the problem of finding the substitution scores. Instead of asking about the cost of each type of substitution (and having to deal with two directions of change for each amino acid pair), they asked how frequently each pair of amino acids is aligned in families of closely related proteins. They also may have been the first to write such "alignment score" s_{ij} for each pair of amino acids i and j in the log-odds form:

$$S_{ij} = \frac{1}{\lambda} \ln \frac{q_{ij}}{p_i p_j}.$$

The numerator q_{ij} is called "target frequency," and it is the probability with which we expect amino acids i and j to be aligned in the types of alignments we are looking for. The denominator is the product of two "background frequencies," which are simply the probabilities p_i and p_j with which amino acids i and j occur in the sequences that we are analyzing.

Two results obtained by Samuel Karlin of Stanford University (who, with Amir Dembo, also worked out the extreme value statistics described previously) and by Stephen Altschul (Karlin and Altschul, 1990; Altschul, 1991, 2006) are relevant here. One states that every scoring system that produces negative scores for randomly matching segments is based on log-odds, if perhaps implicitly. The other is that the target frequencies that distinguish signal from noise in the optimal way are the frequencies derived from those very same sequences that are being compared. The proofs of both propositions are technically involved, but the conclusion is rewarding: Once you have the matching pair of sequences, you know how to devise a scoring scheme that will best distinguish this match from the background. The problem, of course, is how to devise a scoring function that would be good enough to find HSPs in the first place. This is where PAM and BLOSUM matrices come into the picture.

Margaret Dayhoff and colleagues were the first to explicitly estimate target and background frequencies suitable for sequence comparison in a broad range of situations. They derived their frequencies and scoring functions from the collection of alignments of many closely related proteins. Their PAM ("point accepted mutation"; sometimes also referred to as "percentage accepted mutation") model of protein evolution measures distance between sequences in "PAMs," where 1 PAM corresponds to an average change in 1% of all amino acid positions. After 100 PAMs of evolution, not every residue is changed exactly once; in fact, some (and perhaps many) amino acids will remain the same, and others will have changed more than once, sometimes even returning to their initial state. Thus, at the distance of 100 PAMs or even more, many pairs of proteins will still be sufficiently similar to produce HSPs.

Using this model, the target frequencies and the substitution matrices may be calculated for any evolutionary distance. Alignments that were available to Dayhoff *et al.* included sequences that were much more closely related to each other than 100 PAM (often, on the order of approximately 1 PAM). Such closely related sequences are not very difficult to align with any substitution matrix. However, on the basis of the PAM1 matrix, any larger PAM distances can be obtained by multiplying the matrix by itself, several times if needed (Dayhoff *et al.*, 1965; Dayhoff and Eck, 1968).

The point of all this is that target frequencies and substitution matrices derived from the PAM model for a given distance may be optimal for finding and scoring similarities that are within the same PAM range. Moreover, one does not have to obtain distantly related sequences to deduce the optimal matrix for finding distantly related similarities; extrapolation may be sufficient, assuming, of course, that closely related and distantly related proteins evolve in a substantially similar fashion.

These premises have been debated in the literature, and many modifications of the PAM matrices have been proposed: More sequences were included, different models of evolution were considered, and alternative techniques for extrapolation and transformation were employed (Kawashima and Kanehisa, 2000). However, the real improvement in performance was achieved only with the introduction of the BLOSUM matrices.

Stephen and Jorja Henikoff of the Fred Hutchinson Cancer Research Center in Seattle and Howard Hughes Medical Institute developed BLOSUM matrices as an extension of their studies of the aligned families of related proteins (Henikoff and Henikoff, 1992, 1993). There were several crucial differences from the PAM approach. First, after 20 years of gene sequencing and database searches, there were many more protein families than were available to the PAM project. Each of the families included more proteins, and there was much larger variation in the degree of protein identity within each family. More important, the BLOSUM project focused on selected fragments of multiple alignments, called blocks; these are not arbitrarily chosen fragments, nor are they leftovers of longer alignments. Blocks contain important information about proteins that PAM matrices were not able to reflect. Specifically, blocks tend to correspond to the regions that preserved in protein evolution, presumably because they contain most of the structural and functional signal within the molecule. Initial collection of blocks (the BLOCKS database) from which the BLOSUM matrices were constructed was based on the PROSITE patterns (Bairoch, 1992). In fact, patterns are not required to build blocks: One can extract the regions of high similarity directly from the alignments using the similarity scores or percentage of identity to define these regions. The main reason for using the PROSITE patterns in the first place was the benefit of known biological relevance: If a block is built on a PROSITE pattern, this means that the conserved region has been critically assessed by a human expert, and it usually has a specific, known molecular function.

PAM and BLOSUM matrices are often compared to each other in the literature. Some of the differences are obvious: For example, it is not in doubt that the BLOSUM project used a much larger data set than did the PAM project, or that the evolutionary assumptions used to build PAM matrices were explicitly stated along the way, whereas in the BLOSUM project they were not. However, many other notions are more problematic. For example, it is sometimes said that because the PAM matrices are based on a sound evolutionary model of amino acid changes in proteins, they should be preferentially used if we want to ask evolutionary questions; that because of the way they are designed, PAM matrices are better suited to detect close relationships, whereas BLOSUM matrices are better in detecting more distant relationships; or that PAM is designed to track the ancestral proteins, whereas BLOSUM is set up to detect conserved domains (Mount, 2004).

In my opinion, each of these statements is confusing in its own way. First, an explicit evolutionary model is not necessarily a correct or optimal evolutionary model (although it is true, of course, that when the model is clearly described, it is easier to examine its assumptions and to see how realistic they may be). Second, I do not know what "more successful in detecting close relationships" might mean: Any similarity-scoring approach is initially tested on already discovered, closer relationships, and only in the case of success will it be tried on more distant relationships. It is also unclear in what sense would the BLOSUM matrices underperform on the sequences that are closely related (there is not much evidence that, for example, alignment of closely related sequences is less accurate with BLOSUM matrices than with PAM matrices). Finally, BLOSUM and PAM are collections of matrices, and it is important to specify which matrices from each series are compared. Many would agree, however, that BLOSUM matrices capture more information about distant protein similarities than do PAM matrices because BLOSUM matrices are based on the actual observation of these similarities.

We now come to the Fourth Challenge of sequence comparison. Finding related sequences is not the same as ensuring a complete and correct definition of a gene/protein family. Methods of sequence comparison can (and should) be tested so that when they are applied to database search, they maximize the detection rate of the known members of a family, and they maximize the fraction of these known sequences that are validated by statistical approaches. Moreover, a general-purpose program for database search should work well on a large number of different families. But there is a deep biological problem: Although closely related sequences tend to be close in structure and function, the opposite is not true—distantly related sequences are not necessarily different in structure or function. This asymmetry was discussed in Chapter 1 in the context of the first and second facts of biological sequence analysis, but the current statement is stronger because it applies to the sequences that are *known to be related.* Evolution does not impose a requirement on the related sequences to exceed certain threshold of similarity (or percentage of identity, score, or statistical significance). Furthermore, the picture is not the same in different protein families—some accumulate more changes in the course of evolution, whereas others accumulate less. This results in a complex interplay of sequence, structure, and function conservation and in nontransitivity of database searches. A quite typical example is shown in Fig. 2.6. A database of enzymes involved in carbohydrate metabolism is maintained by the glycobiology unit at AFMB-CNRS in Marseille, France (http://afmb.cnrs-mrs.fr/CAZY/index.html), and the glycosyltransferase (GT) section of the database is one of the largest, containing several thousand sequences and currently organized into 83 families on the basis of high sequence similarity to one or more founding members with experimentally demonstrated GT activity. A natural classification of GTs is nonetheless still unavailable. We have shown (Liu and Mushegian, 2003) that at least 20 CAZy families appear to belong to the large superfamily of GTs, called GT-A. The structure of the similarity network within the superfamily is complex. There are several well-conserved, fully linked protein families, with the average similarity score of $s = 240$ and probability of random match in the context of the NR search ranging from 10^{-15} to 10^{-60}. At the same time, there are groups not directly linked at all. The CAZy family GT-2, which includes bacterial spore coat polysaccharide biosynthesis protein SpsA with known structure, is connected to the largest number of

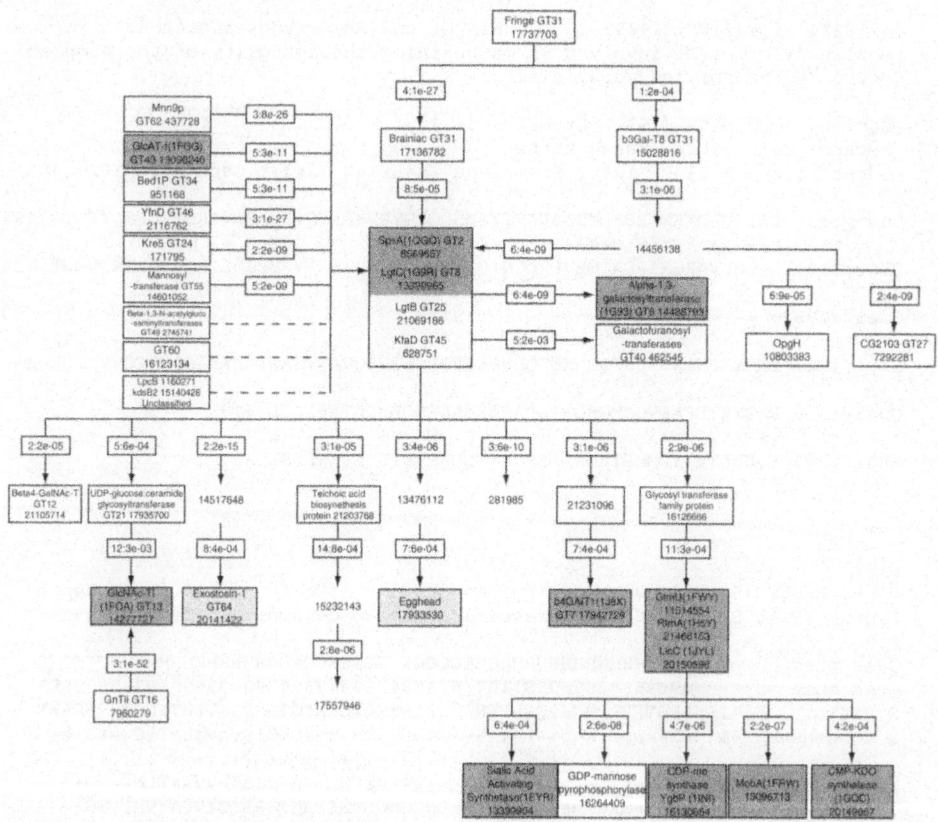

Figure 2.6. Nontransitivity of sequence comparison. Reproduced from Liu and Mushegian (2003) by permission of Cold Spring Harbor Laboratory Press.

other CAZy families. On the other hand, if one starts with a poorly connected sequence, it would be difficult to find other members of the same superfamily.

Our ability to find weak but biologically important sequence similarities improves all the time. In effect, all these improvements have to do with increasingly sophisticated probabilistic models of sequence similarity—from regular expressions to profiles, position-specific scoring matrices, and to hidden Markov models. The succession of these methods is reviewed in Durbin *et al.* (1998), Stormo (2000), Eddy (2004d) and Soding (2005) as well as in almost every textbook on bioinformatics. The improved performance is associated with better sensitivity of each successive method toward distant sequence similarities. And the main thing these methods attempt to do is to overcome the nontransitive property of sequence similarities: When A matches B, and B matches C, but A does not match C, it is hoped that a probabilistic model obtained by alignment of A and B will match C as well as, perhaps, other homologs. (Two examples of observations made in this manner are shown in Fig. 2.7). It remains to be seen, however, whether a general solution of this problem exists. The crucial problem here, perhaps, is to understand the properties of the "special nodes" and to find ways of identifying them in the course of database searches.

I conclude this chapter with two additional remarks. First, most methods of sequence comparison assume that amino acids or nucleotides in each position of the alignment change independently from one another. No one really believes this to be true. Many lines of evidence,

```
>gi|6324424|ref|NP_014493.1| Subunit of the SAGA transcriptional
regulatory complex, involved in maintaining the integrity of the complex;
Spt20p [Saccharomyces cerevisiae]

 Score = 79.7 bits (195), Expect = 1e-12,
 Method: Composition-based stats.
 Identities = 21/162 (12%), Positives = 40/162 (24%), Gaps = 66/162 (40%)

Query:439 LLLQCIDREMLPEFLMDLLVAETVSLSDGEGTRVYAKPSVFYAGCVIAQIRDFRQTFAT- 497
            L+ + R  +P+ +M++L    +                FY G +I Q+ D   T
Sbjct:181 EFLEYVARGRIPDAIMEVLRDCNIQ---------------FYEGNLILQVYDHTNTVDVT 225

Query:498 --------------------------STNI----------------------CDMKHILL 509
                                                                 +   LL
Sbjct:226 PKENKPNLNSSSSPSNNNSTQDNSKIQQPSEPNSGVANTGANTANKKASFKRPRVYRTLL 285

Query:510 RPTNATLFAEVQQMGSQ--LPAEDKLALESQLVLATAEPLCL 549
          +P + T + ++                 ES+++  T   L L
A Sbjct:286 KPNDLTTYYDMMSYADNARFSDSIYQQFESEILTLTKRNLSL 327
```

```
>1yz4_A DUSP15, dual specificity phosphatase-like 15; 2.40A {Homo sapiens}
Probab=90.35 E-value=0.01 Score=43.37 Aligned_columns=76 Identities=17%

Q ss_pred            HHHHHHHHHHHHHHHCCCCCCCCCEEEEECCHHHHHHHHHHHHHHHHCCCCCC
Q ss_conf            99875888899999716766788847997178267899999999999985599987
Q Rit1        416 LRSTFPRIHGEIQSLFTNRDEKIKPMLICCNTGTDMSIGVILSILCTKYTEEWM
Q Consensus   416 LR~~Lp~i~~fi~~~l~~~~~~~~~~~~ilV~CesGkDlSvgvaLaIlC~~fd~e~~
                 +...|+.++.||....    +.+++++|+|..|.+-|+.++.+-|+..++
T Consensus    66 ~~~~~~~~~~~i~~~~~~~~~~~~~g~~VlVHC~~G~sRS~~vv~aYLm~~~~~~~---
T 1yz4_A      66 IKKHFKECINFIHCCR----LNGGNCLVHSFAGISRSTTIVTAYVMTVTG----
T ss_dssp        GGGGHHHHHHHHHHHH----HTTCCEEEEETTSSSHHHHHHHHHHHHHHHC----
T ss_pred         HHHHHHHHHHHHHHHH----HCCCEEEEECCCCCCHHHHHHHHHHHHCC----
T ss_conf         4899999999999875----138707887001126049999999999839----

Q ss_pred         CCCCCCCCCHHHHHHHHHHHHHHHCCCCCCCHHHHHHHHHHH
Q ss_conf         4667788888999999999999872277578872678988875
Q Rit1            LTSELPDISKLIVRKHLTKLISHLKGRNVNPSRATLNSVNSF  511 (513)
Q Consensus       ~~~~~~~~itK~~IR~rL~~I~~~~~~~~~~vNPSRatLqsVNsF  511 (513)
                            ....+-+.+|-+.-|   .+||+++.+++.-.|
T Consensus       ----------~s~~~Ai~~vr~~Rp--~i~pn~~f~~QL~~~  141 (160)
T 1yz4_A        ----------LGWRDVLEAIKATRP--IANPNPGFRQQLEEF  141 (160)
T ss_dssp        ----------CCHHHHHHHHHHTCT--TCCCCHHHHHHHHHH
T ss_pred         ----------CCHHHHHHHHHHHCC--CCCCCHHHHHHHHHH
B T ss_conf       ----------999999999997188--367898689999999
```

Figure 2.7. The power of probabilistic models of protein sequences. (**A**) Significant sequence similarity between uncharacterized protein CG17689 from *Drosophila melanogaster* (query line) and better studied yeast protein Spt20 (subject line). Low percentage of identity and long gaps hide this similarity—it is not observed by either single-pass BLAST or Smith–Waterman comparison of CG17689 to the databases. The match is validated by low E value (obtained at iteration 4 with inclusion threshold 0.002 in iterations 1–3, raised to 0.015 at iteration 4); by reverse PSI-BLAST, in which Spt20 is used as a query and retrieves CG17869; and by the fact that CG17689 is found in the fruit fly chromatin remodeling complex, which has functional similarities with the yeast SAGA complex and shares with it many other protein components (Weake and Workman, personal communication). (**B**) The Rit1 2′-O-ribosyl phosphate transferase family contains a C-terminal domain related to dual-specificity (protein and lipid) phosphatases. Rit1, the enzyme specific to plants and fungi, modifies the initiator methionine tRNA at position 64 to distinguish it from elongator methionine tRNA. This similarity, not reported before and discovered using the profile-to-profile comparison (Soding, 2005), suggests that the C-terminal region of Rit1 may be involved in binding phosphoribosyl pyrophosphate or transferring ribosyl phosphate onto tRNA.

such as theoretical and experimental studies of mutational robustness (Martinez *et al.*, 1996; Wagner, 2005), indicate that some positions in proteins and nucleic acids are more likely to change than others, and that changes in some positions are likely to constrain and "canalize" further acceptable changes. But in most cases, this covariation is extremely difficult to model. Naive proposals, such as considering a dipeptide substitution matrix (400 × 400), have not been any more effective than models with independent substitutions. Thus, radical new approaches might be needed. In what is possibly the only example of such an approach, matching and alignment of RNA molecules that exhibit constrained variation (i.e., many mutations are tolerated, as long as complementary bases co-vary to preserve pairing) is greatly improved using the stochastic context-free grammar formalism (Grate *et al.*, 1994; Rivas and Eddy, 2001; Dowell and Eddy, 2004).

Second, the approaches that were first developed for biological sequences, with their 4-letter or 20-letter alphabets, can be extended to other types of symbolic strings that are found in genomes. For example, versions of dynamic programming have been used to compare the order of genes in genomes (Wolf *et al.*, 2001a) and the order of transcription factor binding sites in the promoters of coregulated genes (Hallikas *et al.*, 2006). In both cases, the alphabets contain several hundred to several thousand symbols (respectively, different genes and distinct types of binding sites). As with sequence alignment, the choice of scoring function appears to be crucial for finding biological signals in these types of strings.

3

Homology: Can We Get It Right?

Thus far, I have discussed sequence similarity without using the word "homology." This cannot last much longer. All organisms on Earth are thought to descend from the common ancestor (discussed further in Chapter 13), and when two organisms have a common trait, this is often because the trait has been inherited by both species from an ancestor—either the distant ancestor of all organisms or, more likely, a more recent ancestor of the two organisms. Walter Fitch (1970b, 2000) defined *homology* as "the relationship of two characters that have descended, usually with divergence, from a common ancestral character." The term is in fact more than 150 years old, believed to be first proposed by Sir Richard Owen, the renowned British anatomist and paleontologist of the 19th century, to describe morphological traits derived from ancestral traits. Homology has separate meanings in other areas of science, such as cytogenetics, organic chemistry, and algebra, with which we will not concern ourselves here.

Characters can be "genic" in the Pauling–Zuckerkandl sense (see Chapter 1)—that is, represent genes or sense-carrying units derived from genes. Characters can also be structural, functional, behavioral, or anything else. Characters have states. Consider the example of the Walker-type NTPases and kinases (Leipe *et al.*, 2003; see also mentions of this famous protein family in Chapters 4 and 6). In the most conserved sequence region within this family, sometimes called the Walker A box, we usually see a tripeptide GKS or GKT. The last amino acid in this tripeptide is a character, whereas S and T are the states of this character. Fitch (2000) asserts that "homology is in the character, not in its state," and therefore, technically speaking, "S and T occupy the homologous position" is more rigorous than "S and T are homologous characters." I think that both usages are acceptable because there is no confusion. We will discuss real terminological confusion soon.

Notice that characters exist on many levels. A string of nucleotide characters forms a gene, which is also a character. Indeed, in Chapters 5–8, we will discuss whole-genome comparisons, in which genes in the genome are matched, sometimes by arranging them in strings much like polynucleotides or polypeptides, and treat gene absences similarly to insertions and deletions of bases or codons in genes (see also Fig. 2.1B).

Furthermore, genes and their products form biological pathways. Pathways in the existing organisms have descended with divergence from ancestral pathways, which existed in the ancestral organisms. Thus, a pathway as a whole may be viewed as a character too. An anatomical organ and a physiological system of organs are produced by many pathways interacting with each other in dividing and differentiating cells. Each organ or system can be treated as a character in itself (again, descent with divergence). Thus, characters are found everywhere in a biological system, from a single nucleotide to a very complex trait. But what counts as a character at one level may be only implicit at another level. For example, we can estimate the evolutionary distance between humans and turnips by finding and counting all genes that we and turnips inherit from our common ancestor; on the other hand, we can estimate human–turnip distance by comparing nucleotides/amino acids at homologous positions in one or more of those homologous genes. In the latter cases, nucleotides or amino acids will be counted as characters, but in the former case, they will only be used to establish the relationships between genes and after that they will not be examined any longer.

Two characters either share a common ancestor or they do not. This all-or-nothingness is a fundamental fact about homology, compared by some authors to pregnancy or death (Petsko, 2001). However, misuse of this term is common. Expressions such as "strong homology" or "two sequences are 75% homologous" persist in the literature, despite powerful arguments against them (Fitch, 1970a,b; Reeck *et al.*, 1987).

Why is this so? Why does the urge to talk about "75% homology" between two sequences appear to be irresistible? Could people be onto something important here? Is there some aspect of homology that asks to be measured?

The most common explanation of this confusion is in my opinion correct. The problem derives from a mix-up of what is measured and what is inferred. The degree of similarity between sequences can be measured: For example, two aligned sequences can be 75% identical. Homology, however, is a statement about the evolutionary history of the characters. That history, with a few exceptions, has not been observed. To make a statement that two characters have descended from the same ancestral character, we analyze the results of our measurements. So, on the one hand, there is the act of computing a number and, on the other hand, the act of using this number to infer a singular event in the past. Thus, the most common misuse of "homology" is to say that "two sequences are 75 % homologous," when the intended meaning is that they are 75% identical or similar and that this observed value is high enough to seriously consider a hypothesis of the common origin of the two sequences. This logic is in most cases straightforward and not controversial. But the measure and the inference from it are not one and the same, and there is no point in convoluting them.

Exceptions to this understanding have been claimed. First, consider the "recombination problem" (Fitch, 2000). A gene may be a recombinational fusion of two unrelated genes. Each of the "parent" genes shares a common ancestor with only a part of the recombinant gene. Fitch stated,

> *If the domain that is homologous to the low-density lipoprotein receptor constitutes 20% of enterokinase, then enterokinase is only 20% homologous to that lipoprotein receptor, irrespective of its percent identity. If, at the same time, this common domain were half of the lipoprotein receptor, the receptor would be 50% homologous to the enterokinase. The homologies are not the same in both directions if the proteins are of unequal length!*

This example, however, shows only that a portion of enterokinase is homologous to a portion of lipoprotein receptor, and both proteins also have additional regions that are not homologous to each other. The homologous region accounts for some percentage of enterokinase length, as well as for some percentage of lipoprotein receptor length, and the nonhomologous regions account for the rest. Homology itself is the inference that helps to explain the evolutionary roots of the observed similarity, but the relative sizes of the aligned fragments have no role in the hypothesis of their common ancestry; for example, if the homologous region accounted for 60% of enterokinase length and for the 34% of lipoprotein receptor length, the arguments in favor of homology of this domain would not change.

A more interesting proposition is that "percentage of homology" should be a legitimate way to express the extent of our belief that the sequences are homologous (Robson, 2001); "more homologous" in this case would be the same as "more likely to be homologous." Robson also mentions that, perhaps, "more homologous" should signify "share a more recent common ancestor." These are two separate ideas, each of which deserves some consideration, but in my opinion both of them should be dismissed. Regardless of what we may think about the role of individual belief in scientific inquiry, imagine some process that goes as follows: (1) Assign the probability for the two characters to share a common ancestor; (2) measure the similarity between characters; (3) use the result of this measurement as a new observation; and (4) update the probability that the characters are homologous, given this observation of similarity. If correctly executed, this approach places the inference of homology in the Bayesian framework. The crucial question here is how to assign the prior probability for the two characters to share a common ancestor.

Most people will agree that the best way to assign prior probabilities is to use a model of the process that generates these probabilities. Suppose that such a model is available. In fact, we are lucky here—evolution of nucleotide and amino acid sequences is relatively well understood from the statistical point of view. (Also note that some hard-core Bayesian statisticians would not see the assignment of priors as a problem at all, even when the process model is unavailable; see Sober, 1991; Felsenstein, 2003; and Eddy, 2004c). But the issue of scale will still remain. That is, when a scientist says "I am 75% confident that two sequences are homologous," she is saying something different than "two sequences are 75% identical." The numbers in these statements are not equivalent; any biologist that has examined homologous and nonhomologous sequences will state that if two (long enough) sequences have 75% identity, their chance of being homologous is close to 100%. On the other hand, two proteins can have extremely low sequence similarity, barely distinguishable from the random level, and yet homology between them, in certain cases, can be inferred with confidence (Kinch and Grishin, 2002). Thus, there remains a clear difference between the measured numeric value and the inference made on its basis. However, there is no use in mingling the two in "percentage of homology."

How about "more homology" as a synonym for "more recent common ancestry"? The time passed from the common ancestor of two characters is itself an inference; some issues related to making such inferences are discussed in Chapters 12 and 13. But determining when the ancestor existed is distinct from determining that such an ancestor did exist in the first place. Here again, I do not know what we would gain by lumping these two assertions together.

Terminological difficulties that involve homology do not stop there. We sometimes read that "two proteins are structurally homologous." If homology is understood as just discussed (i.e., in its true sense), this is the same as stating that "two proteins are structurally derived from the common ancestor." But similarity of three-dimensional shapes of two proteins does not always indicate their common origin, and sensitive and specific discrimination of structure divergence and convergence is still an open problem (see Chapters 9 and 10). If two proteins have similar structures, and additionally there is a statistically significant similarity between their sequences, this joint observation is usually a sound argument for homology. But in such a case, one really infers homology from sequence, not from structure.

Of course, biological characters may evolve toward the same structure or function in the absence of the common ancestor. This is called structural or functional convergence, or, in genome comparisons, "nonorthologous gene displacement" (Koonin *et al.*, 1996). Convergence may involve either homologous characters, which had diverged in the past but then became more similar again, or nonhomologous characters. Nonhomologous characters can also converge, in which case they become more similar but do not become homologous at all.

Two aspects of homology should be emphasized. First, as already discussed at length, there is no constraint on degree of similarity between the orthologs: They may be very similar or quite dissimilar, depending on the time of divergence and rate of evolution. Second, there is no requirement of functional similarity. Homology is the statement about evolutionary events (i.e., the existence of a common ancestor and descent from it), not about function.

There is plenty of confusion about this. Even a popular, if not already standard, introductory textbook, *Bioinformatics* by David Mount (2004), is not much help. For example, page 5 provides the following definition: "Homologous describes genes that have arisen from a common ancestor gene, as evidenced by their having similar sequences." Again, on page 68, the author states, "Homologous sequences refers to two or more sequences that can be quite readily aligned such that they must have originated from a common ancestor sequence in earlier evolutionary time." Page 285 states, "Homologous genes: Genes whose sequences are so similar that they almost certainly arose from a common ancestor gene." Although Mount correctly identifies homology as an evolutionary statement, and is not willing to use "degrees of homology," he seems to insist that sequences are homologous only if we can infer their common ancestor from high sequence similarity or from a "quite ready" alignment. If, on the contrary, two sequences are not similar enough, or not easy to align, Mount denies homology to them. Thus, whether intended or not, homology becomes a statement about our ability to infer common ancestor, not about its existence. I do not think this makes sense, and throughout this book, we will understand homology as the descent from common ancestor, as elaborated by Fitch and others, and not only such descent that can be proved by high sequence similarity.

The notion of homology is quite simple and is important for understanding evolution, so its correct usage is well worth fighting for. However,, there is a need to differentiate between different kinds of homology and different subsets of homologs. This calls for even more specialized terminology, and this is how orthologs and paralogs (Fitch, 1970b, 2000) enter the vocabulary.

The descent from a common ancestor, which produces homologs, is a complex process. First, there is speciation—where there was one species, there become two or more (here and throughout this book, we will consider only the cases in which a lineage is split in two; although real evolutionary histories may also contain multifurcations, those can always be converted into series of bifurcations, for example, by inclusion of branches with zero length). The homologous characters (genes and their products) that are produced by speciation are called orthologs. Recalling Latin, in which *orthos* means "straight" or "exact," *orthology* means "straight homology." Of course, orthologs remain orthologs regardless of the number of the speciation events; we can properly compare orthologs not only from recently diverged species but also from different kingdoms of life or any other taxa. Second, there is gene duplication, which can occur within a lineage, in the absence of any speciation. Homologous genes that are produced in this manner are called paralogs (i.e., "lateral" or "parallel homologs"). As the first approximation, that is all there is to it—there are just two subsets of homology.

One additional, special type of gene relationship, called xenology, describes two orthologs such that one or both have been horizontally transferred [xenology was discussed by Fitch (2000), but this definition is from Jensen (2001)]. Two homologs are either orthologs, in which case they may or may not be xenologs, or they are paralogs. In both cases, as with homologs in general, neither the extent of sequence similarity nor the commonality of biological function matter: Orthologs, as well as paralogs, may be similar or dissimilar in sequence, and they may have either closely related or different functions.

The relationships between different classes of homologs are shown in Fig. 3.1. It becomes immediately obvious that despite simple definitions of orthologs and paralogs, the interplay of orthologous and paralogous relationships between homologous genes in different genomes is a complex affair. The main issue here is that separation of lineages and duplication of genes in evolution may not occur at the same time. Nevertheless, orthology and paralogy can be consistently recognized, as we will now discuss.

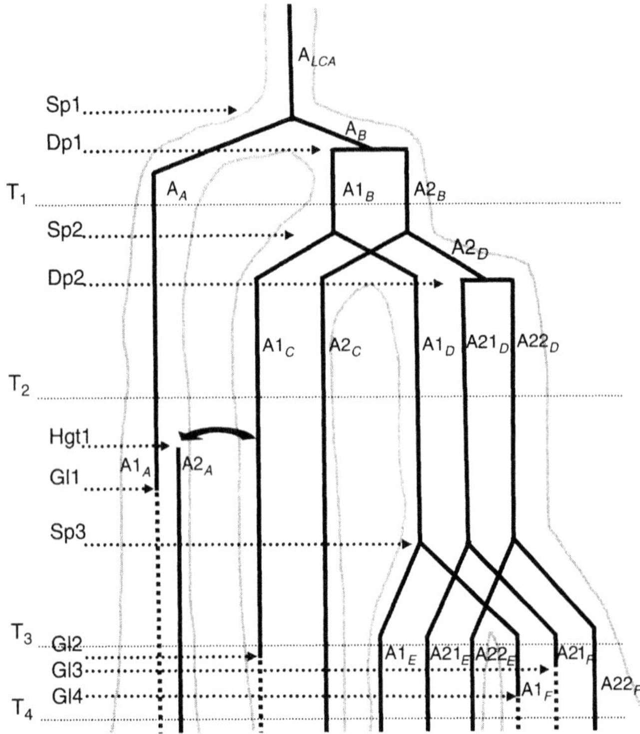

Figure 3.1. Distinguishing between orthologs and paralogs when the complete evolutionary history of species and genes is known. A phylogenetic tree of species is shown by gray outline, and gene phylogeny is shown by solid black lines. Gene copies are indicated by bold type, and species are indicated by italicized subscripts. Thus, $A1_C$ is to be understood as homologous copy 1 of gene A in species *C*. Timelines are indicated by T_1–T_4. Sp, speciation; Dp, duplication; Hgt, horizontal gene transfer; Gl, gene loss.

Let us first sort orthologs from paralogs when the true evolutionary histories of species and genes are known (Fig. 3.1). A straightforward approach in this case is to detect all speciation points as well as all intralineage gene duplication points. A pair of genes produced by the speciation event are each others' orthologs, and a pair of genes produced by the duplication event are each others' paralogs.

Care needs to be taken, however, when more complex combinations of events are considered. There are two things to keep in mind. First, it is not required that every gene has exactly one ortholog in another lineage. Second, paralogs of a gene are not always in the same lineage as this gene—they can be in a different lineage, too. With this in mind, let us start by examining all gene relationships at time T_1. There are three homologous genes at that point: one gene A_A in species *A*, and two genes, $A1_B$ and $A2_B$, in species *B*. Their ancestral gene in the last common ancestor (species *LCA*) is A_{LCA}, and there has been one speciation (Sp1) and one duplication (Dp1) in the history of species *A* and *B*. Genes $A1_B$ and $A2_B$ are paralogs of each other. Each has one and the same orthologous gene, A_A, in lineage *A*.

How should we describe the relationship between A_A and $A1_B$? In order to answer, we have to reverse the arrow of time and move toward the root of the genes' and species' tree until we encounter the event that joins A_A and $A1_B$. If this event is duplication, then genes are all paralogs, and if it is speciation, they are orthologs. The event of interest turns out to be speciation Sp1, so A_A and $A1_B$ are orthologs, and A_A and $A2_B$ are also orthologs. $A1_B$ and $A2_B$ are also called co-orthologs of A_A (see glossary at the end of this chapter).

Consider now the time T_2. We have to examine the relationships between six genes—A_A, $A1_C$, $A2_C$, $A1_D$, $A21_D$, and $A22_D$. The relationships between A_A and the other five genes are exactly the same as that between A_A and the other two genes at time T_1 because the split between A_A and the rest of the tree still maps to Sp1. $A1_C$ and $A2_C$ are paralogs—their split maps to the duplication Dp1. Each pair among $A1_D$, $A21_D$, and $A22_D$ is paralogous too—the split maps to either Dp1 or Dp2.

Let us now examine such pairs of genes where one gene comes from lineage C and another from lineage D. Among those, $A1_C$ and $A1_D$ have been produced by speciation—they are joined at a speciation point Sp2. On the other hand, $A2_C$ and $A1_D$ are paralogs—they are joined at the point Dp1. Finally, to understand the relationships between $A2_C$, $A21_D$, and $A22_D$, note that these are the same as the already discussed relationships of genes A_A, $A1_B$, and $A2_B$. In addition, $A21_D$ and $A22_D$ are in-paralogs, and $A1_D$ is their out-paralog (see Glossary); the difference here is in the order with which we pass the speciation and duplication events as we move to the root. Xenology and gene loss are easily accounted for and do not change the status of orthologs and paralogs, nor the rules we use to distinguish between them.

Fitch (2000) notes that orthology, paralogy, and xenology are all reflexive. This means that for each of the three types of homologous relationships, if A has a certain type of relationship to B, then B has the same relationship to A. None of the three, however, is transitive: If A has a particular type of homologous relationship to B, and B is related to C in the same way, it may not be true that the relationship between A and C is also the same. Homology, of course, is reflexive and transitive, but other relationships mentioned in the glossary are all different: For example, analogy is reflexive but not transitive.

The existence of orthologous and paralogous relationships does not sit well with some authors. According to one strong opinion, orthologs and paralogs are postmodern catchwords lacking real utility (Petsko, 2001). Others, including myself, think that all of the terms discussed in this chapter are proving their usefulness day in and day out, allowing us "to speak more, rather than less, accurately and comprehensibly, about what is really going on in genome evolution" (Koonin, 2001). So what bothers the critics? Why not let "orthology" and "paralogy" be, allowing us to speak comprehensibly about gene gains and losses, about extreme conservation of some genes and extreme divergence of others, and about specialization and takeover of molecular function? All these processes are at the heart of genome evolution and function, and we cannot start making sense of them unless we have ways of describing various kinds of common ancestry (or lack of such ancestry).

I think that the irritation of most critics (Ouzounis, 1999; Varshavsky, 2004) has less to do with the definitions of orthology and paralogy than with our ability to distinguish between orthologs and paralogs in practice. Similar concerns about other words have been raised. For example, inference of any homology between proteins was once thought to be extremely problematic: Early on, Winter *et al.* (1968) argued that because proteins do not leave a fossil record and their evolution cannot be directly observed, the question of their homology is intractable and should not even be asked. They proposed to co-opt "homology" for the nonevolutionary uses, such as the degree of structural similarity.

Not everyone was so pessimistic about the power of inference, and ways of distinguishing homology from analogy were proposed very soon thereafter by Fitch (1970a) and by others. Three and a half decades later, we know that one can infer homology with considerable success, on the basis of an evolutionarily informed measure of similarity between two sequences and a statistical theory that tells us when such sequence similarity is too high to be explained by anything else but common ancestry (see Chapter 2). Thus, establishing homology of two sequences has become quite routine. And so, I believe, it will be with orthologs and paralogs: As the practical approaches for recognizing two types of homology improve, the utility of the words will not be debated anymore.

In fact, computational techniques for inferring orthology and paralogy relationships have been proposed. Let us revisit the events described in Fig. 3.1. When all duplication and speciation events were known to us, our task was only to work out the correct names. Now, suppose that we know the present-day species and the complete set of homologs of a given gene that exist in these species, and we need to sort out orthologs and paralogs among them. The species are A, C, E, and F, and there are six homologous genes, A_A, A_C, A_E', A_E'', A_E''', and A_F. This setup corresponds to time T_4 in Fig. 3.1, but we do not know yet about the speciation and duplication events that have occurred in the past. As in Fig. 3.1, species *D* represents the common ancestor of *E* and *F*, and species *B* represent the common ancestor of *C* and *D*; thus, *B* and *D* are not observed at T_4. We have two evolutionary trees (Fig. 3.2)—one for the homologous genes and the other for the four species. The catch is that the latter tree has to be inferred on the basis of some external evidence, without knowing how gene A and its relatives evolved.

Our goal is to use information represented by the two trees in Fig. 3.2 to infer the scenario of speciation and gene duplication that is shown in Fig. 3.1 (note that by time T_4, four genes have been lost and cannot be observed). More specifically, we need to label each internal branching point in the gene tree as either a duplication or a speciation event.

If two terminal branches are joined in the gene tree, and they represent genes from two different species, intuition tells us that these genes must have been produced by speciation. One such pair of branches in the gene tree is A_E''' and A_F, and the other pair with the same properties is A_A and A_E'. The same probably applies to the relationship between A_C and its tree neighbors: All genes on both sides of A_C are from other species. So perhaps we can manually assign the Sp labels to three nodes in the gene tree, indicated by labels g2, g3, and g5 in Fig. 3.2. The other two nodes should perhaps be labeled as duplications.

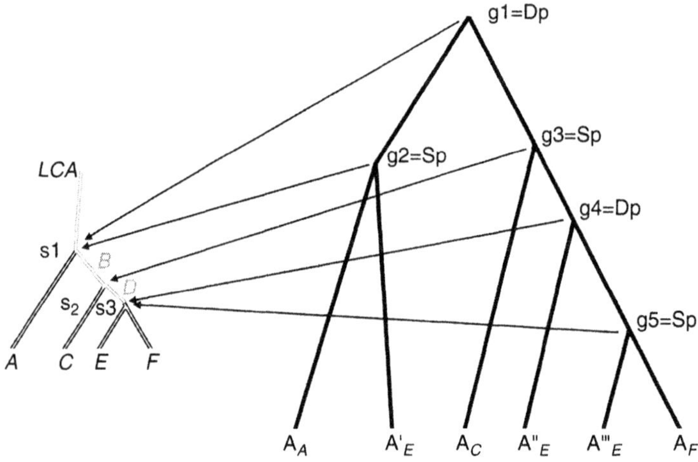

Figure 3.2. Inferring duplication and speciation events by comparing gene tree and species tree. The sets of species and genes are the same as in Fig, 3.1, corresponding to the time point T_4. Species B and D are extinct. Gene A_A in this figure is in fact gene $A2_A$ in Fig. 3.1; A_C is $A2_C$, A_E' is $A1_E$, A_E'' is $A21_E$, A_E''' is $A22_E$, and A_F is $A22_F$.

More formally, let us call the gene tree G and label each internal node in G as g_1, g_2, and so on. Similarly, each internal node in the species tree, S, is designated $s_1, s_2, \ldots, s_n$ (usually, there are more gene duplications than gene losses and, as a result, more genes than species, so $g_n > s_n$). For any g_i, let γ (g) be the set of species in which the descendant genes of g_i occur. For example, in the cases of g_4 and g_5, both γ(g) consist of species E and F. Similarly, for any s_i, let σ(s) be the set of the existing species that descend from s_i. For example, in the case of s_2, σ(s) consists of species C, E, and F.

The mapping function M(g), which relates every g_i in G to a unique s_i in S, is defined as follows: For each g_i, its M(g) is the lowest node in the species tree S, such that γ (g) is included in σ(M(g)). In other words, the species M(g) is the last common ancestor of all species in which the descendants of gene g are found. For example, $M(g_4) = s_3$, because all genes descending from g_4 are found in species E or F, and s_3 is the lowest node from which E and F descend.

The main observation that allows us to identify the duplications is as follows: If, in a gene tree, a node g_i has a direct descendant node, which maps to the same place in the species tree as g_i itself, then g_i is a duplication. In Fig. 3.2, this is the case for nodes g_1 and g_4.

Less formally, if the offspring of a node g_i in a gene tree is distributed among a set of species, and the offspring of a lower node g_j is distributed among the same set of species (or among the subset of that set), this means that no new species were produced between g_i and g_j. That way, every g node that has the same M(g) as its descendant is registered as a duplication, and otherwise it is speciation.

These ideas have been developed in a series of works by many authors [see Zmasek and Eddy (2001) for review, discussion, important considerations of algorithm efficiency, that I omit here, and for a practical algorithm]. This approach really works: In the case shown in Fig. 3.2, it correctly identifies g_1 and g_4 as duplications and g_2, g_3, and g_5 as speciations. The other two types of events, horizontal gene transfer and gene loss, are not explicitly accounted for, although they can be inferred by further analysis.

All this means that given good evolutionary trees for all homologous genes and for species in which these genes reside, we can recover most of duplication and speciation events and thus may distinguish orthologs from paralogs [in my informal discussion, I used a single, although relatively complex, example, but in fact general applicability of the approach has been proven; see Zmasek and Eddy (2001)]. But what this information is good for?

Strictly speaking, the story told by mapped orthologs and paralogs in the tree is mostly about the evolution of the gene family. We learn about the fate of genes and their diverged copies in the history of living organisms, and we can most directly relate this knowledge to trees of other characters in the same set of species. However, there is another reason to be interested in sorting out orthologs and paralogs, and it has to do with predicting protein function. The basic idea is that orthologs are more likely to preserve biological function in the course of evolution, and paralogs are more likely to evolve a new, if usually related, function (discussed in Chapters 7 and 8).

Homology, orthology, and paralogy will be at the heart of almost every theme that we will encounter in the rest of this book.

Glossary

Alloparalogy—same as out-paralogy (Koonin, 2005).

Analogy—any similarity that is not due to the common ancestry.

Co-orthology—relationship in which gene duplication in one species produced a set of genes, each of them orthologous to the single homolog in the other species (Sonnhammer and Koonin, 2002).

Homology—relationship in which two characters have descended from a common ancestor.

In-paralogy—relationship in which paralogs were produced after speciation (Sonnhammer and Koonin, 2002).

Isology—apparently simply a synonym for homology. Modern usage can be tracked to Fields and Adams (1994) and annotations of chromosome 2 of *Arabidopsis thaliana* at TIGR (the Institute for Genome Research). I can make no compelling case for using this term, but it still can be found is some GenBank annotations.
Orthology—relationship in which two homologs in two species (one in each species) are derived by speciation. The common ancestor of these homologs is in the cenancestor of the two species (Fitch, 2000).
Out-paralogy—relationship in which paralogs were produced before speciation (Sonnhammer and Koonin, 2002).
Paralogy—relationship in which two or more homologs are related by gene duplication. When these homologs are in one species, they are always paralogs; when they are in two or more species, they are paralogs, if in each of these species there have been duplications.
Pro-orthology (from Holland, 1999, credited to A. C. Sharman)—relationship of a gene to each of its co-orthologs in another species (i.e., when duplication occurred only in the latter).
Pseudoorthology-relationship between two paralogs after differential paralog loss in two lineages (Koonin, 2005). The problem is that these proteins may be taken for orthologs, unless a genome sequence is encountered in which the two paralogs are still present.
Pseudoparalogy—simultaneous presence of a regular ortholog and a xenolog in the same lineage. There have been no duplications, and yet this pair of proteins can be taken for the paralogs. As with pseudoorthologs, the confusion may be resolved if additional genomes are sequenced.
Sequelogy—a recently invented term (Varshavsky, 2004), apparently the exact synonym for sequence similarity: "A is sequelog of B" means, by definition, that A and B are similar sequences. I am not sure what is gained by using this word.
Semi-orthology (from Holland, 1999, credited to A. C. Sharman)—same as co-orthology.
Spalogy—a recently invented term (Varshavsky, 2004), apparently the exact synonym for spatial structure similarity: "A is spalog of B" means, by definition, that A and B are similar structures. I am not sure what is gained by using this word.
Super-orthology—relationship in which two genes are leaves on a rooted tree with duplications and speciations assigned to each node, and all the nodes in the connecting path between these two genes are speciation events (Zmasek and Eddy, 2002).
Synparalogy—same as in-paralogy (Koonin, 2005).
Synology—relationship between two orthologs after two complete genomes fused, for example, by hybridization, resulting in two orthologs in the same genome (Gogarten, 1994). It appears to be the same as xenology, except that the latter assumes a relatively small fraction of transferred genes compared to the genome as a whole. Synology may also usefully cover orthologs produced by polyploidization. Fitch (2000) notes, however, that if the hybridization of two species is successful, they are effectively the same species, and synology is not different from allele reassortment in a locus. The same word was re-used by Lerat *et al.* (2005) with a different meaning, namely as a union of paralogous and xenologous copies of a gene in the genome (i.e., any intragenomic replicates regardless of their origin).
Trans-homology (from Holland, 1999, credited to A. C. Sharman)—relationship between two sets of postspeciation paralogs; close, if not identical, to out-paralogy.
Ultra-paralogy—relationship in which two genes are leaves on a rooted tree with duplications and speciations assigned to each node, and all the nodes in the smallest subtree containing these two genes are duplication events (Zmasek and Eddy, 2002).
Xenology—relationship between two orthologs, one or both of which have been horizontally transferred (Jensen, 2001).

4

Getting Ready for the Era of Comparative Genomics: The Importance of Viruses

Virologists like to recollect the episodes in the history of science in which viruses played an important role. Max Delbruck and the Phage Group at Cold Spring Harbor Laboratory is the most famous example ("Phage and the Origins of Molecular Biology," 2006). Many other fundamental discoveries in molecular biology were facilitated by the simplicity of viral genetic systems, with their limited number of tractable molecular components. This includes the Hershey and Chase experiment on phage T4 infectivity, which settled the question of whether genes were made of protein or nucleic acid (Hershey and Chase, 1952); Fraenkel-Conrat's demonstration of infectivity of TMV RNA (Fraenkel-Conrat *et al.*, 1957); virus self-assembly studies, also by Fraenkel-Conrat as well as others (reviewed in Fraenkel-Conrat, 1990); understanding of retrovirus genome strategy, which led to refinement and crystallization of the central dogma of molecular biology (Crick, 1958, 1970); and the discovery of RNA splicing (Berget *et al.*, 1977; Chow *et al.*, 1977).

These occurred decades ago. But how about now? Today, we have bacteria, yeasts, the nematode *Caenorhabditis elegans* (which has no common name, although "elegant worm" seems to be gaining popularity), flies, *Arabidopsis thaliana* (too many common names, including mustard weed, Thale's cress, and mouse ear, none of them gaining much traction), sea squirts, mice, mosses, and so on. All of these are immensely useful model systems, some simpler than others, but all with the number of genes in the thousands. The simplicity of genome does not seem to be a major requirement for a model system anymore. So what about viruses—are they still of any use as models of anything? In a world full of interesting living species, who cares about viruses, except for virologists?

There is no doubt that the medical, agricultural, and other social impact of viruses is on the scale from moderate to huge, depending on the disease. But more important to the themes in this book, viruses continue to provide clues to many biological processes that were not known only a few years ago—the phenomena whose significance eclipses host–pathogen interactions. Posttranscriptional gene silencing, RNA interference, and related phenomena—which were all but unknown 10 years ago and are, of course, all the rage of molecular biology now (Zamore and Haley, 2005) and have been recognized by 2006 Nobel Prize—were discovered and understood to be the RNA-level effects by plant virologists. The crucial observation was that RNA produced by virus-derived transgenes in plants is sufficient to shut down the infection by the same virus (Lindbo and Dougherty, 1992a,b, 2005). When the host genes required for this shutdown were isolated, one of them turned out to be eukaryotic RNA-dependent RNA polymerase, which had been cloned originally as the enzyme responsible for viroid replication in tobacco (Schiebel *et al.*, 1998). Most likely, this is the same activity that was discovered by Fraenkel-Conrat decades ago as he was trying to dissect the enzymology of virus replication in plants (Khan *et al.*, 1986). There are multiple evolutionary and functional connections between replication of virus RNA and posttranscriptional gene silencing in eukaryotes (Ahlquist, 2002).

In the future, there will be more discoveries from the observations of viruses. The following is one example of what may be expected: Genomic RNA of brome mosaic virus, a plant virus that replicates in cytoplasm without a DNA intermediate, appears to contain modified bases, pseudouridylate and ribothymidylate (Baumstark and Ahlquist, 2001). We already knew that rRNAs and tRNAs are full of these and other modifications, but to find them in virus mRNA, even though the modified region is between two cistrons and is not translated, is a surprise. In the same vein, several plant viruses encode a domain homologous to the 2-oxoglutarate-dependent dioxygenase/AlkB family, which was discovered by comparative sequence analysis and is predicted to possess demethylase or dealkylase activity (Aravind and Koonin, 2001). Some of the members of this family appear to demethylate RNA (Ougland *et al.*, 2004), although the activity of virus-encoded homologs has not been determined. I believe we are looking at the two facets of the enzymatic modification of mRNA: one equivalent to mutational damage, which needs to be repaired, and the other having a functional role, perhaps in control of mRNA stability, folding, and translation. If, as I expect, we will learn in the near future that cellular mRNAs also contain functionally important enzymatic modifications, perhaps constituting another level of regulation and recoding, the first indications will have come from viruses. (In a sense, discovery of the cap structure on eukaryotic mRNA is part of the same phenomenon, and this observation also came from studying viral transcripts; Wei and Moss, 1974; Wei *et al.*, 1975; Ensinger *et al.*, 1975; Keith and Fraenkel-Conrat, 1975).

But let us return to what this chapter is about—the role of viruses (and virologists) in defining comparative genomics. The reasons why viruses have been so popular as model systems are relatively simple: The number of genes in virus genomes is small, and the amount of genetically homogeneous progeny that can be obtained in the laboratory is large. These were the properties that made it possible to clone and sequence virus DNAs and RNAs early on. Even before the first genome sequence of a cellular life-form, *Haemophilus influenzae*, was determined in 1995, there were approximately 200 complete virus genomes in the databases (the exact number depends on how similar strains and isolates of the same virus are counted). By that time, virologists had already realized that virus genomes should be studied in a systematic way.

In 1971, David Baltimore of the Massachusetts Institute of Technology, currently at Caltech, provided the first system of viruses that was based on the diversity of forms of virus genomes and the modes of their expression (Baltimore, 1971). Whereas genomes of cellular organisms are, in a way, all the same—made of double-stranded DNAs, with all other nucleic acid forms being transient in the life cycle—viruses are all different. Not only do virions of different viruses contain all sorts of genetic material—DNA or RNA, single-stranded or double stranded—but also some virus life cycles do not include double-stranded DNA stage at all. This diversity needed to be rationalized, and Baltimore's system did just that. The fundamental idea was that every virus needs to produce a minimal set of two genetically encoded products ("sense-carrying units"): copies of the genomic nucleic acid and at least one mRNA to express proteins. The pathways sufficient to perform both tasks are schematically represented in Fig. 4.1.

Baltimore noted that "[t]here are many viruses about which so little is known that they cannot be placed in the scheme.... However, as their transcription becomes understood, either

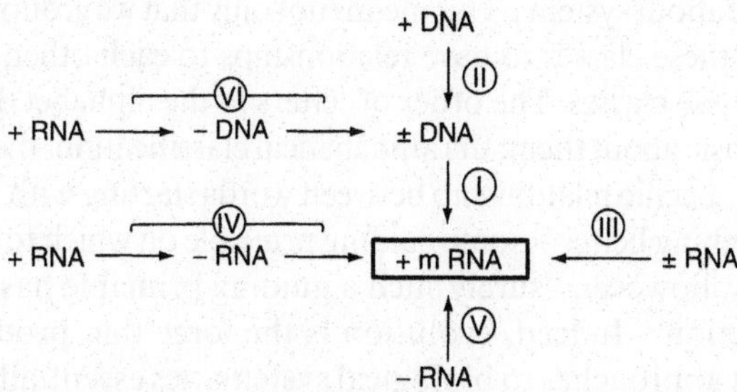

Figure 4.1. David Baltimore's system of viruses. Reproduced from Baltimore (1971) by permission of the American Society for Microbiology.

they should fall into place in a recognized pattern, or new classes will have to be added." What is most amazing is how little of such an addition was in fact required since 1971. In fact, about the only finding that makes it to the classification now, 35 years later, is that some viruses with single-stranded RNA genomes have ambisense segments (i.e., RNAs in which one part is positive-sense and the other is negative-sense). This is a relatively minor adjustment; most groups of DNA viruses also transcribe different portions of genome into separate, less-than-genome-length mRNAs, which Baltimore did not fail to discuss in 1971.

Let us now consider Baltimore's proposal from a broader point of view. Since ancient times, people have been pursuing the quest for the natural system of living forms. We would like to know whether comparison of genomes brings us closer to such a system.

We are looking for ways to rationally organize a large number of diverse objects. As a rule, the objects have complex structure that is not completely known. A sensible start would be to find some number of categories, such that every object could belong to one category. It is good if the number of categories is much smaller that the number of objects. The main way to assign the objects to categories is to find some sort of similarity between these objects and to use the similarities to define groups of related objects, be it genes, viruses, or anything else.

There are different ways to organize things by similarity. To proceed, let us define the meaning of the following words: classification, systematics, taxonomy, and phylogeny. Instead of reviewing the history of (changing) usage of each of the four words, I intend to stay as close to their literal meanings as possible.

Classification should be about classes. Classes, or categories, of objects can be defined in any way we wish—for example, by the shape of the objects, the second letter of their names, or any other properties. Classes do not necessarily uncover any intrinsic properties of the objects that we classify. With the appropriately chosen basis for our classification, however, classes may turn out to be natural categories, representing some objective and essential properties of the entities. Drawing a line between objective and arbitrary classifications is not always easy. For example, grouping words by the first letter or grouping people by the last letter of their last names seem quite arbitrary. Yet, most Russian words that start with "a" are borrowed (often from Greek but sometimes from Arabic by way of other languages), whereas some people may recognize my last name as Armenian. Therefore, at least in a small way, certain patterns reflecting something important about words in a language (or, in this case, about historical trajectories of certain words) can be gleaned merely from grouping them into classes based on a simple alphabetical rule.

Classification is more interesting if it is hierarchical, i.e., some categories are themselves grouped into a category. Again, there is no constraint on the ways in which we build such hierarchies.

Systematics should be about systems. This means not only that we group objects into classes but also that we would like these classes to have relationships to each other, which represent some intrinsic properties of these classes. The order of letters in the alphabet does not seem to reveal anything deep and intrinsic about them; the alphabetical classification, therefore, is not a systematics because there is no specific relationship between words starting with "a" and words starting with "b." We want something better—some guiding principle on which to build the system.

Biologists would say, however, "surely such a guiding principle has been discovered; it is called biological evolution." Indeed, evolution is the force that produced and shaped the observed life, and most approaches to biological systematics essentially amount to uncovering the evolutionary history of the taxa that are examined. But there is no reason why the historical development of organisms should be the only guiding principle for systematics. An analogy from chemistry is the Periodic System of Elements. Conceived in 1869 by Dmitry Ivanovich Mendeleev, the system captures the majority of then-known physical and chemical properties of elements and their similarities and consistent differences between different groups of them (Mendelejeff, 1869). Importantly, many properties of each element are determined by, or at least strongly correlated with, a single parameter. This parameter later turned out to be the positive charge of the nucleus of a chemical element, which, in Mendeleev's times, could be only approximated by atomic weight. Interestingly, the contemporaries of Mendeleev, notably Newlands in England and Meyer in Germany, developed similar systems at the same time. However, a notable distinction of Mendeleev's system is that, unlike the proposals of Newlands or Meyer, it contained empty classes (i.e., places for the elements that remained to be discovered). Therefore, the system is robustly organized by a guiding principle that reflects something profound about the elements and even predicts new things about chemical organization of the universe. It is systematics, under the literal definition proposed here. Yet, it does not tell us much about the natural history of chemical elements.

Biological species are different from chemical elements, of course. The main distinction of biological species from most other things is the foundation of the sense-carrying units, evolving by descent with divergence from the common ancestor. Even so, there is no reason why some important aspects of form and diversity of living things should not be presented as some sort of a "periodic system." Baltimore's system of viruses is a good example of just such a representation.

Taxonomy should be about taxa. That is where evolution starts playing a more prominent role. Taxa are groups nested in an hierarchy. In biological systems, the most common and natural reason for the existence of a hierarchy is evolutionary process. For some thinkers, all taxonomy is by definition evolutionary. On the other hand, biologists often make use of nonevolutionary hierarchies. One example is EC, the enzyme classification employed by the International Union of Pure and Applied Chemistry (Tipton and Boyce, 2000). EC divides all enzymes into six groups—oxidoreductases, transferases, hydrolases, lyases, isomerases, and ligases—each of which is hierarchically divided further. For example, EC 2, transferases, includes group EC 2.7, enzymes that transfer phosphorus-containing group. EC 2.7 includes EC 2.7.4, phosphotransferases with a phosphate group as acceptor, which has a member EC 2.7.4.2, phosphomevalonate kinase. There exist at least two enzymes with phosphomevalonate kinase activity that have different, unrelated sequences and probably do not share any common ancestor (such isofunctional but unrelated enzymes are discussed in Chapter 5). Thus, EC may (although some will say should not) be called a biological taxonomy, but it is not an evolutionary taxonomy.

More recently, the genomics community started putting together Gene Onthology, a knowledge base and controlled vocabulary for annotating gene function. This is also a hierarchy that contains both evolutionary and substantial nonevolutionary components.

Finally, *phylogeny* should be about genesis of phylae. Although *phylae* literally means "races" or "classes" in Greek, the scientific meaning of the word has to do with branches in the trees that depict historic relationships between species. Although phylogenetic trees are familiar to a biologist, trees are also objects of mathematics. They are formally defined in graphs, and there are many different ways to construct tree-like graphs. Not every tree is truly representative of the evolutionary history of the species that we are studying (see Chapter 11). Moreover, even if the objects of interest do not have any evolutionary relationship, we can still construct a tree-like representation of connections between them; this is common, for example, in comparing gene expression patterns and other genomewide numeric data (see Chapter 14). Thus, phylogeny is a tree-like representation of ancestral relationships, but not every tree-like representation of biologically interesting information is a phylogeny.

With this, as literal as possible, understanding of classification, systematics, taxonomy, and phylogeny in hand, let us examine Baltimore's proposal once again. The scheme shown in Fig. 4.1 surely is a classification, and it is also a system: Not only are the objects (viruses) partitioned into classes but also this is done on the basis of a principle. As with any good system, the property chosen as the basis of the system—in this case, the path from genomic nucleic acid to mRNA—allows one to make many predictions of other properties of viruses. One such prediction, discussed by Baltimore, was that virions that do not pack mRNA have to rely on cellular machinery to produce it, or to pack virus-encoded transcription enzyme into virions. Baltimore's scheme, however, is not a taxonomy—there are no nested taxa in it. It is also not a phylogeny. The word "evolution" is not mentioned in the paper at all; even its root ("evolved") is only used once, in passing. I do not suppose this indicates lack of interest in evolution on Baltimore's part; more likely, he did not believe there was enough evidence to suggest a sensible scenario for the evolutionary origin of different virus groups. Indeed, as we will soon see, the evolutionary relationships between Baltimore's groups are not intuitive.

Baltimore also remarked, "Viruses with similar transcriptional systems could have different replicational systems, leading to the necessity to extend the class designations." Such an extended system proposed a few years later by Vadim Agol (1974) of Moscow University and the Institute of Poliomielitis was the next major step in comparative genomics.

Agol defined four types of genetic elements: (+)DNA, (−)DNA, (+)RNA and (−)RNA. Eight "elementary acts of synthesis" are theoretically possible, two for each of these genetics elements; for example, (+)DNA can be copied either into (−)DNA or (−)RNA. There are 44 distinct "full acts of synthesis," or interconversions of single-stranded and double-stranded molecules. If we require that each life cycle contains one act of mRNA synthesis and at least one act of multiplicative synthesis (increase in genome copy number), and only consider graphs with no more than three edges, there are 35 distinct life cycle graphs that can be constructed from these elements (Fig. 4.2).

The constraint on the number of edges was introduced for convenience; no fundamental reason is known as to why the life cycles of viruses should have only two or three edges. In fact, viruses appear to follow this rule: By and large, the life cycles of all known viruses are covered by this set of simple graphs. (A few cases in which four or five edges may give a better explanation of virus reproduction mechanism are discussed in Agol's work, but with reasonable, minor simplifications, they are all reduced to three-edged graphs.)

Agol's scheme highlighted some empty classes and predicted that virus life cycles corresponding to some of these new classes will be discovered. In fact, the class A1 in the DDRD type has since been filled. This class is represented by animal hepadnaviruses and two groups of plant pararetroviruses—caulimoviruses and badnaviruses. Interestingly, this class was identified by the author as one of those for which such future discovery was most likely, based on the observation that all acts of synthesis required for this class were already discovered in nature.

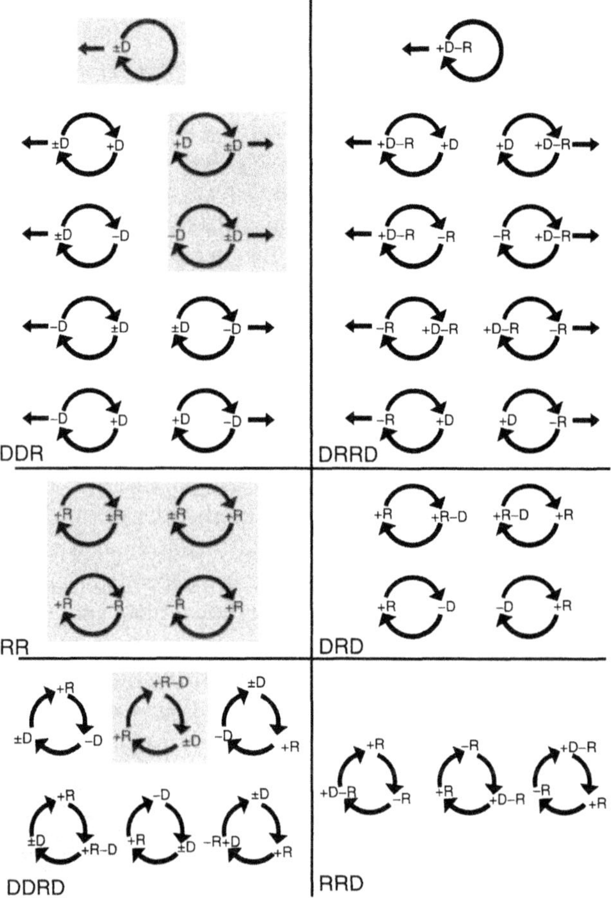

Figure 4.2. Vadim Agol's system of viruses. Polarity of DNA (D) and RNA (R) strands is shown next to each nucleic acid. The nucleic acid that is part of a virion is on the left side of each circular graph. If the cycle does not contain the (+)RNA stage, an mRNA needs to be synthesized in an additional step. Such cases are shown by the linear arrows pointing outside the circles. Shaded graphs are the strategies that had been discovered before Agol's work, and the lightly shaded graph at the bottom left is the life cycle predicted by Agol to exist and indeed discovered later. Modified from *BioSystems*, 6, Agol, V. I., Towards the system of viruses, pp. 113–132, Copyright 1974, with permission from Elsevier.

Agol's scheme is a classification and also a system. It is a taxonomy as well, because it includes a natural hierarchy of classes, superclasses, and types. On the other hand, this hierarchy is not phylogeny: In recent years, it became clear that some of the types, particular RR and DDRD, are evolutionarily related, and that the DDR type may have to be split into evolutionarily independent lineages. Baltimore's classes are distributed among three of Agol's types. For example, Baltimore's classes III, IV, and V belong to Agol's type RR, and Baltimore's class IV is one of the six classes in type DDRD. This is because Baltimore was mostly interested in classifying the existing viruses, whereas Agol was concerned in an exhaustive enumeration of all logical possibilities, and his theoretical system was therefore set up to include "blank" classes. Many graphs remain purely theoretical possibilities 30 years later. The aforementioned class DDRD-A1 is the only novel class discovered since then; three of the highest level groups, types DRRD, DRD, and RRD, remain completely vacant, and approximately half of the classes in other types are also empty.

The large number of "still-empty" cells in Agol's system is interesting because it indicates that classes of biological objects are usually unequally populated. In other words, nothing in biology is purely combinatorial: Some sets of properties characterize a very large number of biological objects, other combinations of properties are found rarely, and there are many seemingly plausible but nonetheless empty classes. It is almost too easy to discover the basic components of living systems and to invent the ways of mixing and matching them. What is much more difficult is to understand the constraints that are imposed on such combinations in the evolution of life. Most of the time, we can only guess about the reasons that "forbid" some of the combinations. In fact, in the case of Agol's empty types, his explanation was very good: The empty superclasses typically involve an mRNA synthesis directed either by a single-stranded DNA or by an RNA–DNA hybrid, and all empty types involve a synthesis of a (+)DNA strand on a (−)RNA strand, or the opposite (+)RNA → (−)DNA reaction. In the environment of the double-stranded DNA genomes, the RNA strands of these duplexes will be prone to destruction, and single-stranded DNAs will be restored to two strands by the DNA repair system. This remains a guess; experimental testing of this suggestion will require the construction of artificial viruses, which appears to be within reach (Cello *et al.*, 2002; Smith *et al.*, 2003).

The systems of viruses described previously mark the beginning of comparative genomics in a most direct sense—that is, the work of comparing different genomes. Thus came to be the idea of virus genomes as complete multigene entities, which are related to each other in specific ways and can be studied as a whole. In the late 1970s and early 1980s, sequences of individual virus genes and of complete virus genomes started to accumulate, and at approximately the same time, biologists started to get better access to computing.

In 1980, David Botstein of MIT (now at Princeton University) presented a metaphor that continues to catch on in molecular evolution and comparative genomics, in a work called "A Theory of Modular Evolution for Bacteriophages." That work was based on the observations of several temperate bacteriophages with double-stranded DNA genomes. Only partial nucleotide sequences were known for these phages, but genetic maps of many of them, particularly the lambda phage and its close relatives, were worked out in great detail. A remarkable feature of these genomes was that the genes involved in one and the same function were, more often than not, positioned close to each other in phage genomes, and arrays of such genes occupied similar positions within the genomes. For example, in several lambdoid phages, the group of genes coding for the phage head component was followed by the group of tail genes. After the tail genes, a stretch of DNA could be found to which no gene functions were mapped at the time, but remarkably, this stretch accounted for roughly the same fraction of each genome. This was followed by genes involved in DNA recombination, then by DNA replication factors, and, finally, by lysis genes (Botstein, 1980).

Why was the conserved gene order remarkable? Colinearity of many genes in cellular lifeforms had already been demonstrated by genetic mapping of related enterobacteria, such as *Escherichia* and *Salmonella*. In phages, however, there was a crucial difference: In some cases, only the locations of the isofunctional genes were conserved, whereas the gene sequences were quite different. One example was phage tails that look completely unlike one another. Lambda's tail is long and flexible and the tail of P22 is short and rigid, yet both are encoded by tail modules located at identical positions in these phage genomes. Another example is lack of similarity between biochemical activities of lysis genes in phages 80 and P22, respectively represented by glycosidase and endopeptidase, which target different chemical bonds within peptidoglycan of the cell wall. In order to explain this contrast between gene colinearity and lack of molecular similarity between functional groups of genes, Botstein proposed that these groups can be transferred between genomes independently of one another, which was in agreement with the observation that different lambdoid phages could recombine in multiple spots along the genome. This ability, then, provided the mechanism for exchanging genes or, more important, groups of genes or "modules"—for example, the complete set of a half dozen genes required for tail assembly.

Thus, the "module" idea has been released into comparative virology, from where it passed to other areas of biology and, in due course, to the emerging discipline of comparative genomics. It is worth remembering, however, that as early as the late 1950s, the idea of bacterial operons (i.e., groups of functionally linked genes in bacteria that are also adjacent on a chromosome, ensuring ease of regulation by way of a single control element) had been proposed (Jacob *et al.*, 1960). Dozens of operons in bacteria and in bacteriophages were known by 1980. In fact, the choice between lysis and lysogeny pathways in temperate bacteriophages was one of the favorite models to study operon organization of genes and transcription regulation in prokaryotic cells. So, what was the difference between operons and Botstein's modules?

One distinction is that operon theory focused mostly on the functional aspects of gene clustering on a chromosome—adjoining genes in bacteria are easier to coregulate, using elaborations of the basic scheme that involves a single activator of transcription (operator) and expressing multiple genes by internal initiation of translation on polycistronic mRNA. The theory of modules considers the same gene clusters and emphasizes the ease with which they can be transferred between genomes by DNA recombination. Although evolutionary mobility of operons had been discussed before Botstein's work, it had been viewed as a relatively rare event, occurring in the background of much larger bacterial genome, which was believed to evolve mostly by mutational divergence and by recombination that involves long regions of high sequence similarity. In contrast, Botstein's proposal decomposed the entire phage genome into a moderate number of parts that could be inherited independently of one another—there was almost no "stable background." Botstein (1980) stated, "A rather large and apparently diverse group of temperate bacteriophages are related in ways not easily accounted for by standard ideas of evolution along branching trees of linear descents." Thus, evolution of viruses may have to be depicted as something different than a tree. Notably, this idea derived from the empirical observation of molecular–genetic characters in completely mapped, if not yet sequenced, genomes.

In Botstein's proposal, the relatively short regions of high nucleotide sequence similarity, located at the junctions between different modules, were thought to be necessary for the mechanism of module exchange. Nowadays, many DNA recombination pathways are known that do not require extended complementarity: When DNA homology is present, the recombination machinery will take advantage of it, but when there is no lengthy stretches of identical DNA, recombination may occur anyway. Recombination between distantly related DNA genomes may have played an exceptional role in evolution of life, especially early on (see Chapter 11). Genomes of the RNA viruses also contain plenty of evidence of gene exchange.

Modular evolution theory fits well with Pauling and Zuckerkandl's idea that a molecular function can be performed in several mechanistically and evolutionarily unrelated ways. Functionally analogous proteins do indeed exist, as the example with lysis genes of temperate phages shows. Functional convergence at the molecular level will be discussed in much more detail in Chapter 6.

Finally, the concept of modular evolution indicates that morphology may not always be a reliable guide in evolutionary studies. The type of phage tail has been often used as a phylogenetic marker, but if tail genes constitute just one module among many, accounting for only a fraction of all phage genes, and if this module is free to "mix and match" with other modules, then two phages with similar sets of tail genes may be placed into the same taxonomic group regardless of the other genes they have. On the other hand, phages that share most of the genome but have different sets of tail genes may fall into separate groups.

One could argue that the situation is perhaps different when we deal with morphology of cellular (especially multicellular) organisms. Here, the hope may be that morphology is determined by interaction of many genes and reflects their joint presence and coevolution—in contrast to viruses, in which morphology may be determined by too few genes and be less representative of the rest of the genome. The International Committee on Taxonomy of Viruses, nonetheless, still relies heavily on its virion morphology for defining at least the higher order virus groups. This may be satisfactory for the purposes of classification and taxonomy, but it would hardly be enough if we want our taxonomy to be informed by evolution and to represent phylogenetic relationships. The impact of genome modularity on phylogeny is discussed in Chapters 11 and 12.

These are just some of the implications of the "modular theory": Chapter 14 provides a brief discussion of "modular biology," a new proposal for organizing and studying genomewide data. But, perhaps surprisingly, the beginnings of this project are in sequencing virus genomes.

Another nonphylogenetic approach to virus classification was put forward by Eugene Koonin in 1991 when he was still at the Institute of Microbiology, Academy of Sciences of the USSR (Koonin, 1991). That work, dealing only with positive-stranded RNA viruses, had already benefited from sequencing of virus genomes, but it examined not so much the individual sequence relationships but, rather, similarities in genome strategies (i.e., the molecular mechanisms employed by viruses to express their genes and replicate their genomes). The topic was thoroughly familiar to Baltimore in 1971 and Agol in 1974, but they were mostly interested in the strategies of viral replication and transcription. However, for the largest group of viruses, those of the RR type, the diversification of molecular strategies of genome expression takes place at the level of mRNA translation into proteins.

The main outcome of virus genome expression is the production of individual virus proteins. All viruses usurp ribosomal machinery of the host cell in order to express their proteins, and almost all viruses encode more than one mature protein. Therefore, the main task of a virus expression strategy is to produce several protein species from their mRNAs.

In bacteria, all virus genes can be translated into separate proteins from one polycistronic RNA, which is the same as genomic RNA (the RNA that is translated and the RNA that is packaged into virions may not be physically the same molecule, but, as a first approximation, they have the same nucleotide sequence). Translation of many open reading frames (ORFs) from the same polycistronic mRNA is the first expression strategy.

Many viruses use another mechanism to produce individual virus proteins: They code for a large precursor protein that can be proteolytically processed into fragments with distinct roles in the virus life cycle. Usually, the proteases required for such processing are included in the large protein precursor and are able to release themselves as well as process the rest of the protein (Bazan and Fletterick, 1989; Gorbalenya *et al.*, 1988, 1989, 1991).

The third strategy of expressing individual proteins is to produce several different mRNAs by transcribing the minus strand of the polycistronic RNA genome. The full-length mRNAs will serve as the genomic RNA that can be encapsidated in the progeny virions and possibly also as the messenger for translation of the 5′-proximal ORF. But (–)RNA may also be transcribed, starting at some internal position, into a (+)RNA that has the same 3′ terminus as the genomic RNA but is less than genome length. The role of such subgenomic RNAs is to direct translation of the downstream genes in the virus genome.

Finally, and quite trivially, a virus can possess a fragmented genome, in which each RNA fragment encodes exactly one protein that does not need to be processed further. All possible combinations of the four mechanisms (polycistronic expression, processing of a polyprotein precursor, synthesis of subgenomic RNAs, and genome fragmentation) give 15 possible strategies (Table 4.1).

Koonin also examined, but decided not to include into his classification, such translation-level events as frameshift, readthrough of leaky termination codon, and other recoding mechanisms. This "ribosome gymnastics" is popular among positive-stranded RNA viruses and also among distantly related retroviruses and some viruses with double-stranded RNA genomes. Another discovery in recent years is the discontinuous synthesis of 3′-coterminal RNAs in some virus groups (e.g., coronaviruses); it may be sufficiently different from other mechanisms of subgenomic RNA formation and could be placed into a separate column.

In truth, only 9, rather than 15, expression strategies were proposed in Koonin's 1991 paper, and accordingly there were 18 classes. All those classes that involve polycistronic translation were collapsed into just 2, with and without VPg. The reason was that for a long time, most people thought that polycistronic translation was unavailable in eukaryotic cells. There,

Table 4.1. Classification of Genome Replication and Expression Strategies in Viruses with Positive-Stranded RNA Genome[a]

	Genome expression mechanisms	Genome segmentation	NonVPg utilizing replication	VPg-utilizing replication
1	Polycistronic	Segmented		
2	expression (1)	Nonsegmented	ssRNA phages	
3	Polyprotein processing (2)	Segmented		Comoviruses Nepoviruses Bymoviruses
4		Nonsegmented	Flaviviruses Pestiviruses	Picornaviruses Potyviruses
5	Subgenomic RNA formation (3)	Segmented	Dianthoviruses Tobraviruses Tricornaviruses Hordeiviruses	
6		Nonsegmented	Carmoviruses Tombusviruses Potexviruses Carlaviruses (Capilloviruses)	Luteoviruses (BYDV)
7	(1)+(2)	Segmented		
8		Nonsegmented	Engineered flavivirus derivatives[b]	
9	(1)+(3)	Segmented		
10		Nonsegmented		
11	(2)+(3)	Segmented	Nodaviruses	
12		Nonsegmented	Alphaviruses Rubiviruses Coronaviruses Tymoviruses	Sobemoviruses Luteoviruses (BWYV, PLRV)
13	(1)+(2)+(3)	Segmented		
14		Nonsegmented		
15	Genome segmentation			

[a]Modified from Koonin, E. (1991). Genome replication/expression strategies of positive strand RNA viruses: A simple version of a combinatorial classification and prediction of new strategies. *Virus Genes* **5**, 273–282. With kind permission of Springer Science and Business Media.

[b]Engineered construct corresponding to one predicted class of viruses.

translation initiation most often proceeds by ribosome scanning from the 5′ end, which provides for efficient translation only of the ORF closest to the 5′ end of mRNA. Or so it was believed for several decades, until evidence of many exceptions to this rule was discovered; most notably, several viral and some cellular mRNAs have sequence segments to which ribosome can bind directly, without scanning from the 5′ end, and to initiate translation on an ORF placed downstream of such element (called IRES). In fairness to the scanning mechanism, it has to be noted that these control elements seem to be used in nature mostly for scanning-independent expression of the 5′-terminal ORF and not for polycistronic expression. However, artificial bicistronic and polycistronic eukaryotic messengers have been constructed, in which each ORF is placed under the control of its own IRES element and all ORFs in such constructs can be expressed. Therefore, it seems possible that sooner or later we will find a eukaryotic virus that also uses this expression mechanism.

The mechanisms of RNA replication are less well studied (although significant progress has been made in this field since 1991; see Ahlquist *et al.*, 2005). Koonin sensibly chose the mechanism of initiation of RNA synthesis as a character that can have two states, namely use of a protein (called VPg) with a covalently attached nucleotide as the primer of RNA synthesis or VPg-free initiation. Combination of these two modes with 15 modes of individual protein generation gives 30 strategies (Table 4.1).

Fifteen years after Koonin's study, the state of the affairs has not changed much. Despite ongoing sequencing of virus genomes and the discovery of several novel virus taxa, empty cells mostly remain empty: Replication/expression strategies of newly described virus groups tend to fall into already occupied classes. However, the principal possibility for the existence of some currently empty classes has been demonstrated by genetic engineering (Geigenmuller-Gnirke *et al.*, 1991). But correlations observed by Koonin in 1991 mostly retain their status: Prokaryotic-style polycistronic expression is actualized only by RNA phages and remains a theoretical possibility for eukaryotic viruses; RNA phages remain the smallest and most homogeneous group, not known to rely on VPg-primed RNA replication; and, for reasons that are unclear, VPg-dependent replication, which is common among eukaryotic viruses, is closely associated with the polyprotein processing strategy.

In summary, Koonin's combinatorial scheme is a classification and a system but not really a taxonomy. As indicated by Koonin, it also does not capture phylogenetic signal very well. Although the same class often contains viruses that are evolutionarily closely related, the opposite is not true: Some groups of viruses that are closely related at the sequence level may have quite different strategies of replication/expression.

A more general conclusion is again the same as before: The combinatorial approach to classification (at least to virus classification) appears to produce a larger number of possibilities than is actually employed by nature. This results in many empty classes in Agol's scheme and in Koonin's classification. Apparently, the evolutionary process operates under constraints, so its results do not look like the product of indiscriminate mixing and matching. Perhaps every combinatorial classification should be expected to contain many empty classes (more examples of this will be provided when the patterns of the presence and absence of genes in genomes are discussed in Chapters 6 and 11).

It is now time to examine another line of comparative virus genomics, namely the study of evolution of individual protein sequences. In the 1980s, these studies began producing important, if not quite expected, results.

Three biochemical activities play a major role in genome replication and expression of positive-stranded RNA viruses (Baltimore's class IV; note that some of the comparisons that we are about to discuss will show that Baltimore's classes III and VI share a common ancestor with class IV). First, RNA strand synthesis on an RNA template requires the processive nucleotidyltransferase activity provided by the enzyme RNA-dependent RNA polymerase.

Second, many positive-stranded RNA genomes encode another enzyme, RNA-dependent ATPase, which plays a nucleic acid remodeling role (e.g., removal of secondary structure from self-paired regions of RNA or, conversely, RNA duplex formation), and may also control association and dissociation of proteins with RNA. Viral (and homologous cellular) proteins capable of doing all this are known under a slightly imprecise name—helicases. Third, cleavage of virus polyproteins requires virus-specific proteases, which belong to several unrelated classes. In addition to these enzymes, viruses code for capsid proteins. In the case of positive-stranded RNA viruses, there is often just one such protein, or there may be several, either expressed independently or produced by proteolysis of a larger precursor, sometimes while or after the capsid shell is assembled. (Other classes of virus-specific proteins—some possessing interesting enzymatic activities and others playing regulatory roles in virus life cycle and host evasion—are less widespread and are not examined here).

The relationships between various virus proteins, and between them and their cellular homologs, began to come to light in the early 1980s. Everything started falling into place in 1984. One the most important activities, template-dependent RNA synthesis, has been genetically mapped to a specific translation product, or at least to a specific RNA segment, in genomes of several viruses. Kamer, Argos, and co-workers (Kamer and Argos, 1984; Argos *et al.*, 1984) reported statistically significant sequence similarity between what was (correctly) inferred to be the main virus RNA replication enzyme RNA-dependent RNA polymerase (RdRp). At approximately the same time, Haseloff and co-workers (1984) compared a subset of these proteins and noticed the same similarities but also additional relationships in two other protein domains.

The region of the highest sequence conservation in RdRp enzymes contained a string of approximately 10 residues, with the tripeptide GDD in the middle (Fig. 4.3), often preceded by Y or another bulky hydrophobic residue. Some positive-stranded RNA viruses (e.g., coronaviruses sequenced later) have SDD instead of GDD. RNA viruses from other groups, those with double-stranded and negative-stranded RNA genomes as well as those that employ reverse transcription, all possess related enzymes, now known to perform processive synthesis of virus nucleic acid. The most conserved site in these proteins is always a variation on the "GDD" theme, typically taking the form of YxDD in retroid viruses and UUDD (where U stands for a bulky hydrophobic residue) in negative-stranded RNA viruses. The active center of RdRp contains two magnesium ions that play a direct role in catalysis, and the first of the two aspartic acid residue as well as some of the preceding residues (tyrosine in the best studied reverse transcriptases) play a direct role in interaction with these metal cofactors.

Common features of these enzymes are not limited to the 10-residue region: There are several other areas with especially high similarity, including some positions where the residues were the same in all sequences. In addition, the approximate distances between these conserved sites were similar in all sequences, and the positions of the RdRp-related regions as a whole within each virus genome were partially conserved. This alignment is quite different from what was considered, at the time, to be typical sequence conservation in a protein family.

The existence of families that consist of homologous proteins (orthologs in different species or paralogs in the same species; see Chapter 3) was well established during the 1960s and 1970s. However, the methods of finding homologs were less sensitive than what we have today, and they tended to recover proteins that were globally similar to one another. Despite many point mutations, and occasional insertions/deletions, each sequence included in these alignments mostly consists of segments that were clearly aligned to their counterparts in all other sequences, and the percentage of residues that fall into conserved regions was quite high.

This does not mean, of course, that the existence of more remote relationships between proteins was completely unsuspected. Not only did theoretical consideration of descent with divergence suggest that such relationships would exist but there was also plenty of empirical

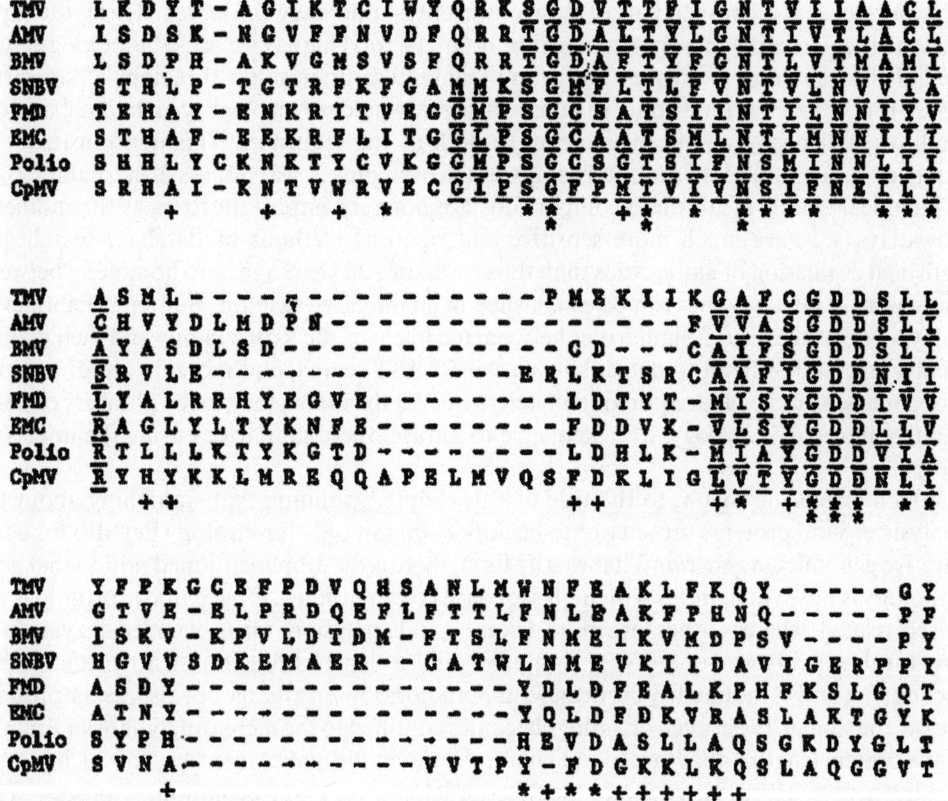

Figure 4.3. Conserved motifs in virus RNA-dependent RNA polymerases. Reprinted from Kamer, G., and Argos, P. (1984). Primary structural comparison of RNA-dependent polymerases from plant, animal and bacterial viruses. *Nucleic Acids Res.* **12,** 7269–7282, by permission of Oxford University Press.

evidence—morphological, cytological, and biochemical—for the common ancestors of various taxa and, most likely, of all living things. In more doubt, however, was the sensitivity of our methods and whether we would ever learn to distinguish extreme evolutionary divergence from convergence or from random coincidence.

In the meantime, however, unexpected sequence relationships were observed increasingly more often. In the early 1980s, genomes of cancer-causing retroviruses were sequenced, and when sequences of their oncogenes were compared to sequence databases, they turned out to have homologs with known functions encoded by cellular genomes (Barker and Dayhoff, 1982; Doolittle *et al.*, 1983). However, even though these similarities were unexpected, they were not remote but, in fact, quite high—so high that no statistical theory was needed to convince the audience that they were real indications of evolutionary relationship and similar molecular function of viral oncogenes and their cellular homologs. Thus, we were already doing well with "unexpected but high" sequence similarities; the problem was validating similarities that were plausibly expected and yet quite low.

The auxiliary evidence, however, strongly supported sequence similarities between proteins conserved in RNA viruses. First, there was plenty of genetic data mapping replication ability to the RdRp domain. Second, conservation of gene order in virus genomes, pointed out by Botstein in 1980, was likewise observed in RNA viruses. By gene order, we understand both the colinearity of genetically mapped functions and the conservation of several different domain sequences, even if each such pair of domains is only moderately conserved. Third, the rapid growth of sequence databases became an important factor: One day we could be looking at two, possibly homologous, sequences, trying and failing to distinguish short regions of similarity signal from the background noise, and the next day we would see a new entry in the database that produces statistically significant matches to both of the sequences. This nontransitivity of sequence searching and scoring, and some ways to overcome it, was discussed in Chapter 2.

Technical details of all these comparisons are not very important to us at the moment. Nowadays, we have much more sensitive and rigorous methods of database search and statistical evaluation of similarities than those available 20 years ago, and homology between many virus RdRp enzymes can be established in an unbiased fashion. But remarkably, even today, some of the distant similarities between members of the RdRp family and their distant relatives are still difficult to detect by a casual BLAST search, and more involved, iterative probabilistic searches are required to match them. All the more glory to the pioneers of comparative molecular virology, who were able to unravel these similarities using the imperfect tools.

Which brings me, again, to the title of this chapter, claiming that something about the analysis of viral proteins turned out to be important not only for virology but also for comparative genomics in general. What was it? First, there is the aforementioned notion that gene orthology is not synonymous with high sequence conservation—some pairs of orthologs are closely related, whereas others are more distant—and analysis of virus-encoded enzymes and structural proteins was one of the first case studies that has drawn our attention to this. Second, not only is the identity between virus homologs low on average but also, as the proteins evolve, the similarity between orthologs becomes confined to the increasingly shorter fraction of the protein length—that turned out later to represent the mainstream way in which protein families evolve. Third, viruses were the first model of a very long evolutionary process. This obviously, does not have to be long in absolute time—more important is the rate of change per generation, and in the case of viruses, generations are short and the number of changes per genome per generation is higher than in cellular genomes—again, with obvious parallels to subsequent analysis of cellular DNA genomes.

Rapid evolution of viruses with RNA genomes is sometimes attributed to a high rate of nucleotide misincorporation in reactions catalyzed by RNA-dependent polymerases (although, of course, high mutation rate will not automatically translate into high evolution rate; see Koonin and Gorbalenya, 1989). But despite this handy molecular explanation, the pattern of conservation in virus-encoded proteins could not be easily dismissed as yet another unusual, extreme adaptation to intercellular parasitism. On the contrary, further sequencing of viral and nonviral genomes alike provided more of the same, and a portrait of a "typical" protein family started to emerge, looking like an assembly of sequence motifs separated by noisy linker regions. A previously more familiar pattern of high sequence conservation along the full length of the aligned proteins may in fact be a special, extreme case of sequence similarity, whereas the "virus-like" conservation, centered on most important sequence motifs, became a new null hypothesis of evolution within the protein family.

At approximately the same time as the observations of the unity of distantly related RdRp enzymes, two other protein families encoded by RNA viruses came to light. One such family consisted of cysteine proteases encoded by genomes of animal picornaviruses and plant comoviruses (Argos *et al.*, 1984). The other family included RNA helicase, at that time known as a putative protein that was commonly found next to the RdRp domains and most likely was involved in replication. Both families displayed the same picture of significant sequence divergence as RdRp, with several relatively short conserved regions interspersed with longer regions where similarity could not be easily detected. Importantly, within a few years, each of these two families was connected to a particular family of cellular enzymes, which themselves showed extreme sequence divergence.

The relationship between cellular helicases and the second most conserved protein of RNA viruses was understood not all at once. In fact, the report of a large ATPase superfamily, of which helicases are a part, was first published in 1982 (Walker *et al.*, 1982). In 1988, Hodgman published the alignment of helicases in the form of a series of conserved blocks along the sequence, removing from consideration the portions of proteins that were not amenable to proper alignment (Fig. 4.4; this was one of the first printings of sequence conservation in such condensed format, which remains popular today). So much for the idea that low conservation and short motifs are the property of virus proteins only.

The other family also extended beyond viruses. The title of an article, "Poliovirus-Encoded Proteinase 3C: A Possible Evolutionary Link between Cellular Serine and Cysteine Proteinase Families" (Gorbalenya *et al.*, 1986), speaks to the significance of sequence comparisons of virus enzymes: Not only are viral and cellular enzymes distantly related but also their similarities illuminate the evolutionary relationships between different classes of cellular enzymes. In this case, again, the most pertinent information was presented as two short blocks of local sequence conservation.

These reports were followed throughout the next two decades by increasingly sensitive sequence analysis, resulting in many other conserved virus domains shared with prokaryotes and eukaryotes (Aravind and Koonin, 2001; Putics *et al.*, 2005). In almost every case, the extent of sequence variation among virus homologs was comparable to variation in the members of the same family encoded by cellular organisms, confirming that the extreme divergence between viral proteins is not a fringe phenomenon.

```
Motif              I                II              III              IV

uvrD     26 VLAGAGSGKTRVLV_174_NILVDEFQNTN_16_VMIVGDDDQSIY_26_QNYRSTSNI
rop      19 VLAGAGSGKTRVIT_175_YLLVDEYQDTN_16_FTVVGDDDQSIY_26_QNYTSSGRI
recB     20 IEASAGTGKTFTIA_345_VAMIDEFQDTD_18_LLLIGDPKQAIY_24_TNWRSAPGM
recD    164 ISGGPGTGKTTTVA_ 82_VLVVDEASMID_16_VIFLGDRDQLAS_24_QLSRLTGTH

EBV      69 ITGTAGAGKSTSVS_113_VIVVDEAGTLS_26_IVCVGSPTQTDA_44_NNKRCTDVQ
HCMV    117 VTGTAGAGKTSSIQ_125_IIVIDECGLML_26_IICVGSPTQTEA_44_HNKRCTDLD
HSV      94 ITGNAGSGKSTCVQ_136_VIVIDEAGLLG_26_LVCVGSPTQTAS_44_HNKRCVEHE
VZV      87 ISGNAGSGKSTCID_135_VIVIDEAGLLG_26_IVCVGSPTQTDS_44_NNKRCQEDD

PIF     255 YTGSAGTGKSILLR_ 46_ALVVDEISMLD_25_LIFCGDFFQLPP_29_KVFRQRGDV

AlMV    821 VDGVAGCGKTTNIK_ 55_RLIFDECFLQH_15_VIGFGDTEQIPF_22_ITWRSPADA
BMV     687 VDGVAGCGKTTAIK_ 54_RLLVDEAGLIH_15_VLAFGDTEQISF_22_KTYRCPQDV
CMV     709 VDGVAGCGKTTAIK_ 54_RVLVDEVVLLH_15_ALCFGDSEQIAF_22_TTFRSFQDV
TMV     829 VDGVPGCGKTKEIL_ 57_RLFIDEGLMLH_15_AYVYGDTQQIPY_24_TTLRCPADV
TRV     901 VDGVPGCGKSTMIV_ 56_VLHFDEALMAH_15_CICQGDQNQISF_24_ETYRSPADV
SFV     183 VEGVPGSGKSAIIK_ 50_ILYVDEAFACH_16_VVLCGDPKQCGF_21_ISRRCTRPV
SV      183 VIGTPGSGKSAIIK_ 50_VLYVDEAFACH_16_VVLCGDPMQCGF_58_ISRRCTQPV

IBV    1209 VQGPPGSGKSHFAY_ 50_ILLVDEVSMLT_15_VVYVGDPAQLPA_30_KCYRCPKEI

BNYVV1  893 VKGGPGTGKSFLIR_ 48_IIFVDEFTAYD_11_IYLVGDEQQTGI_25_MNFRNPVHD
BNYVV2  121 VLGAPCVGKSTSIK_ 49_TMLVDEVTRVH_11_VICPGDPAQGLN_19_ASRRFGKAT
BSMV2   267 ISGVPGSGKSTIVR_ 41_LLIIDEYTLAE_11_VLLVGDVAQGKA_18_TTYRLGQET
```

Figure 4.4. Fragment of multiple sequence alignment showing the most conserved motifs in putative viral replication enzymes with helicase or DNA/RNA-dependent ATPase activity and in their cellular homologs. Modified from Hodgman (1988) by permission of Nature Publications, Inc.

For the most important viral enzyme, RNA-dependent RNA polymerase, the discovery of true cellular homologs has taken the longest time. It was known since the early 1980s that the most conserved regions of virus RNA-dependent polymerases contain the universally preserved diaspartate DD. Even though many other polymerases and nucleotidyltransferases also contain aspartic acid residues in their active centers, the sequence comparisons have unequivocally supported only the relationship between RdRp and RNA-dependent DNA polymerase, the replication enzyme of retroviruses and related retroelements. The latter, of course, are found in abundance in eukaryotic genomes and, to lesser extent, in prokaryotes. So the homologs of RdRp encoded by cellular genomes, or at least the prime suspects for this role, were known for a while. When x-ray structures of RdRp and reverse transcriptases became available, their close similarity and equivalent positions of the most conserved residues left little doubt that the two classes of polymerases are homologous. The shape of reverse transcriptase resembled the right hand, with "palm," "fingers," and "thumb" domains; the palm domain contained the residues required for nucleotidyltransferase activity. However, the cell-encoded reverse transcriptases were all associated with integrated proviruses, retrotransposons, and other such elements that seemed to be genomic parasites (the only homolog that appears to have entered the mainstream of cellular function is telomerase, a distinct eukaryote-specific reverse transcriptase involved in maintaining the integrity of chromosome ends). Then, finally, two connections were made (Fig. 4.5). First, the structure of the catalytic domain of one type of eukaryotic adenylate cyclase was solved, and it had the same palm topology as the RNA-dependent polymerases. In fact, the reaction of nucleic acid polymerization is mechanistically similar to the formation of a cyclic nucleotide: In both cases, the 5′ phosphate is attached to the 3′ hydroxyl, and the difference is whether these two groups are in two different molecules, as in polymerization reaction, or in the same molecule, as in cyclic nucleotide synthesis. Second, sequence similarity between the same adenylate cyclase and a large, mysterious family of bacterial proteins, known as the GGDEF family after the most conserved peptide, was shown (Pei and Grishin, 2001), suggesting a palmlike fold and a role in nucleotide conversion for the GGDEF family. Both predictions were confirmed; the fold of GGDEF proteins is similar to the palm domain (and GDE tripeptide is homologous to xDD in polymerases), and some of the GGDEFs are diguanylate cyclases. In this case, however, GGDEFs and adenylate cyclases, although distantly related to each other, are still closer than the most dissimilar virus polymerases.

I described some notable work on comparative virology from 1971 to the early 1990s. Even before virus nucleic acids were completely sequenced, they seemed small enough to be amenable to genome-level analysis. The ideas and approaches first brought up in connection with virus genome comparisons would reach the full bloom in the second half of the 1990s, when, finally, complete genomes of cellular life-forms were sequenced. However, before the completion of the first bacterial genome of *H. influenzae*, comparative genomics of RNA viruses produced a crescendo: in 1993, when Eugene Koonin at the National Center for Biotechnology Information, and Valerian Dolja, then at Texas A & M University (currently at Oregon State University), published a long article titled "Evolution and Taxonomy of Positive-Strand RNA Viruses: Implications of Comparative Analysis of Amino Acid Sequences" (Koonin and Dolja, 1993). Not only does this work demonstrate the power of comparative analysis for understanding ancient events in virus evolution but also it previews the developments in computational genomics of the cellular organisms, which will be examined in the rest of this book. Here, I list several themes that take us from here to there:

1. Weak similarities between viral proteins are important; they take the form of conserved sequence motifs, which can be validated by comparison of sequences with known properties and by other auxiliary information, such as similar genomic layout. If analyzed correctly, motifs reveal the mode of sequence evolution, where signals indicative of homology, common

```
DNA polymerase I (phages, eukaryotes, and subset of bacteria)
DPOL1_Taqu    601 EEGWLLVALDYSQIELRVLAHL 125 AFNMPVQGTAADLMKLAMVKLFPRL-  0 EEMGARMLLQVHDELVLEAPKE------RAEAVARLAKEVMEGVYP-  0 -LAVPLEVEVGIGE  825
DPOL1_Bste    348 ESDWLIFAADYSQIELRVLAHI 125 AMNTPIQGSAADIIKKAMIDLNARLK  1 ERLQAHLLLQVHDELILEAPKE------EMERLCRLVPEVMEQAVT-  0 -LRVPLKVDYHYGS  574
DPOL1_Ecol    373 PEDYVIVSADYSQIELRIMAHL 125 AINAPMQGTAADIIKRAMIAVDAWLQ  1 EQPRVRMIMQVHDELVFEVHKD------DVDAVAKQIHQLMENCTR-  0 -LDVPLLVEVGSGE  599
DPOL_T7       460 GKPWVQAGIDASGLELRCLAHF 110 ALNTLLQSAGALICKLWIIKTEEMLV  7 WDGDFAYMAWVHDEIQVGCRTE-----EIAQVVIETAQEAMR-----  0 --WVGDHWNFRCLL  684
DPOL_Tgor     395 GLWENIVYLDFRSLYPSIIITH  87 WYCKECAESVTAWGRQYIETTIREI-  0 -EEKFGFKVLYADTDGFFATIPGADAETVKKKAKEFLDYINA-----  0 -KLPGLLELEYEGF  582
DPOLD_Scer    599 YYDVPIATLDFNSLYPSIMMAH  97 LPCLAISSSVTAYGRTMILKTKTAVQ  6 NGYKHDAVVVYGDTDSVMVKFGTTDLKEAMDLGTEAAKYVSTLF---  0 -KHPINLEFEKAYF  806
DPOLA_Scer    855 LHKNYVLVMDFNSLYPSIIQEF  84 FYAKPLAMLVTNKGREILMNTRQ---  0 LAESMNLLVVYGDTDSVMIDTG---CDNYADAIKIGLGFKRLVN---  0 -ERYRLLEIDIDNV 1037
DPOLZ_Scer    966 FYKSPLIVLDFQSLYPSIMIGY 116 MPCSDLADSIVQTGRETLEKAIDIIE  0 KDETWNAKVVYGDTDSLFVYL-----PGKTAIEAFSIGHAMAERVTQ  0 NNPKPIFLKFEKVY 1185
POLII_Ecol    410 GLYDSVLVLDYKSLYPSIIRTF  78 FFDPRLVSSITMRGHQIMRQTKA---  0 LIEAQGYDVIYGDTDSTFVWLK----GAHSEEEATKIGRALVQHVNV  8 QQLTSALELEYETH  597

Adenylate cyclase, soluble (bacteria)
ACYC_Tbru     897 TDPVTLIFTDIESSTALWAAHP   0 ----DLMPDAVAAHHRMVRSLI----  0 GRYKCYEVKTVGDSFMIASKS-----PFAAVQLAQELQLCFLHHDWG 41 RVRVGIHTGLCDIR 1032
CC1415_Ccre   405 RAMKAMLFADIQGFGALRDDQI   0 ---PVFVDGVMGTLARAIEAL-----  0 -AAPPIHVETWGDGLFLVFDE-----PIDAALAALALLEAHRAQDLR  8 GLRIGGHYGPVHLR  507
ACYC9_Hsap    390 IEEVSILFADIVGFTKMSANKS   0 ---AHALVGLLNDLFGRFDRLC----  0 EETKCEKISTLGDCYYCVAGCPE--PRADHAYCCIEMGLGMIKAIEQ  9 NMRVGVHTGTVLCG  498
CyaC_Anab     960 PRLITVLFSDIVGFTQLANTLR   0 ---SRRVAELLNEYLEFMTKAV----  0 FDNGGTVDKFMGDAILALYGAPEELTPNEQVRRAVNTARAMHSSLAQ 19 QFRCGIHQGTAVVG 1080
mll0576_Mlot   35 RRVLTALCYDLVASTELLGLLG   0 ---IEDFEELILAFQMAAKEAI----  0 VSCSGTVRVEVGDGGVAVFPV----DLGAKDAASLTISAGLEIVRAC 12 HVRVGVATSMTLVG  143
OGcyc_Rnor    889 FDQVTIYFSDIVGFTTISALSE   0 ---PIEVVGFLNDLYTMFDAVL----  0 DSHDVYKVETIGDAYMVASGLPRRNGNRHAAEIANMALEILSYAGNF 10 RVRAGLHSGPCVAG 1000

GGDEF family of diguanlyate cyclases (bacteria)
sll0821_     1129 REPLALLLCDVDFFKGFNDN--   0 -YGHPAGDRCLKKIADAMAKVAK---  0 -RPTDLVARYGGEEFAIILSET---SLEGAINVTEALQVEVANLAIP  9 TLSIGIAVYTPERH 1236
YHCK_Bsub     222 HFQFALIYMDIDHFKTINDQ--   0 -YGHHEGDQVLKELGLRLKQTI----  0 -RNTDPAARIGGEEFAVLLPNC---SLDKAARIAERIRSTVSDAPIV  5 ELSVTISLGAAHYP  324
YddV_Ecol     324 GTPLSVLIIDVDKFKEINDT--   0 -WGHNTGDEILRKVSQAFYDNV----  0 -RSSDYVFRYGGDEFIIVLTEA---SENETLRTAERIRSRVEKTKLK  9 SLSIGAAMFNGHPD  430

Predicted polymerases involved in DNA repair or in small RNA pathways (bacteria and archaea)
PH0162_Phor   430 AKLVGVIKGDVDHLGLFF--S-   4 ISEYATASRFMDYFFKFYLKQIIR-- 17 ERPNVVVVYAGGDDFFIVGAWNE--IFELAFRVRNAFSSYTGNNL--  0 TISMALGYFHPKTP  550
MJ1672_Mjan   577 TRKIGILKMDVDNLGEIFTTGL   5 ISRMSTLSSMLTLFFTGYIPHLIK-- 12 FKDNIYLVYAGGDDTLIVGAWDA--VWELAKRIRGDFKKFVCYNPYI  0 TLSAGIVFVNPKFE  698
TM1811_Tmar   488 GKKIASLLVDVDNLGKIFLKGL   4 LSRYSTLSRLMSFFFKERVESIVE--  0 -GKNVMVIYSGGDDLYLVGGWND--VLDVAKELREAFGRFTTNDFM-  0 TFSAGYVITDEKTS  594
TM1794_Tmar   494 NGYIAVLLMDGDRMGDWML-GE  35 PAYHRGVSRTLGIFSQLVGKIV----  0 DRHNGMLVYSGGDDVLALLPADS--VLECANDIRKFFSGHLEYEIEI 31 TMSAGIAIVHHKFP  663
AF1867_Aful   697 PKYYAILMMDGDEMGKLLS-GE  33 PAAHSSISRALKNFSVNHVPDVV---  0 RKGNGTLIYSGGDDVLVLLPVDT--AFDVATELAMTFSTSWNGWEML  3 KLSAGLLIVHYKHP  835
BH0328_Bhal   326 TPYYAFLVCDGDQMGKALR---   4 IEDHQAFSKKLSEFAAKARKIVTT--  1 KRDEGELVYAGGDDVMAYLPLHR--CLDVAAKLQQLFGELMNEALPK  5 TLSVGIVIAHMMEP  437
aq_387_Aaeo   339 NSYFSILMADGDEMGKWL--GL  10 ENFHKKFSEALFKFAQKITKIE----  0 DNICLKFVYLGGDDVLAVAHPSV--ILKAAKIIRKRFSEILKKELKP  7 TMSAGLVIAHEKEN  456
SSO1429_Ssol   13 SRYIALIKADGNNAGKIF--G-   4 FSEYVDKSFRLDFGVKKMFYDTLL-- 16 SRILLGVLYLGGDDIMLLSPSAI--AVPFAVKMFKRSLEYTGFTFKV 21 -MEESKIHTGEKSS  154

Viral RNA-dependent RNA polymerases
RDRP_HC      2630 KKNPMGFSYDTRCFDSTVTEND  51 VLTTSCGNTLTCYLKASAACRAA---  0 -KLQDCTMLVNGDDLVVICESAG---TQEDAASLRVFTEAMT-----  0 --RYSAPPGDPPQP 2775
RDRP_PV      1972 LMEEKLFAFDYTGYDASLSPAW  23 NSGTSIFNSMINNLIIRTLLLKTYKG  0 IDLDHLKMIAYGDDVIASYP---------HEVDASLLAQSGK-----  0 --DYGLTMTPADKS 2108
RDRP_Phi6     315 KEWSLCVATDVSDHDTFWPGWL  58 QGATDLMGTLLMSITYLVMQLDHTAP 20 QGHEEIRQISKSDDAMLGWTKGR--ALVGGHRLFEMLKEGKVN----  0 -PSPYMKISEHGGA  495

Reverse transcriptases - viral and cellular (telomerase)
RT_MMLV       261 PSHQWYTVLDLKDAFFCLRLHP  26 QQGFKNSPTLFDEALHRDLADFRI--  0 -QHPDLILLQYVDDLLLAATSEL-----DCQQGTRALLQTLG-----  0 --NLGYRASAKKAQ  380
RT_HIV        101 KKKKSVTVLDVGDAYFSVPLDE  27 PQGWKGSPAIFQSSMTKILEPFRK--  0 -QNPDIVIYQYMDDLYVGSDLEI----GQHRTKIEELRQHLL-----  0 --RWGLTTPDKKHQ  222
TERT_Hsap     703 PPELYFVKVDVTGAYDTIPQDR 107 PQGSILSTLLCSLCYGDMENKLFAG-  0 IRRDG-LLLRLVDDFLLVTPHL----THAKTFLRTLVRGVP------  0 --EYGCVVNLRKTV  904
TERT_Spom     581 GRKKYFVRIDIKSCYDRIKQDL 102 PQGSILSSFLCHFYMEDLIDEYLSF-  0 TKKKGSVLLRVVDDFLFITVNK----KDAKKFLNLSLRGFE------  0 --KHNFSTSLEKTV  778
```

Figure 4.5. Conserved sequence motifs in viral RNA-dependent RNA polymerases, viral and cellular reverse transcriptases, and their cellular homologs with polymerase and nucleotide cyclase activities. Modified from Makarova et al., Nucleic Acids Research Vol. 30, 2002 by permission of Oxford University Press].

function, and similar structure persist in the form of short regions of conservation interspersed with nonconserved regions of variable length. This mode of evolution is relevant not only to virus proteins but also to any large and diverse superfamily of proteins in prokaryotes and eukaryotes.

2. Although the diversity of viruses may be mind-boggling, the number of building blocks—in this case, conserved virus proteins—is limited. Only one protein, RNA-dependent RNA polymerase, is found in all nondefective RNA viruses; nevertheless, the majority of virus genes are conserved in several virus groups, and none of the viruses contain more than 20 discrete genes. In Chapter 5, we will see that the number of gene/domain blocks that make up the cellular organisms is also finite, although of course much larger than in viruses, and is amenable to analysis by present-day computer technologies.

3. Clustering by similarity produces a limited, tractable number of related protein groups (families or superfamilies; there is no real difference between the two). For example, all RdRp enzymes in RNA viruses fall into one of three conserved superfamilies. Likewise, all RNA helicases encoded by positive-strand RNA viruses belong to one of the three helicase families. Chapters 5 and 10 discuss how proteins encoded by complete genomes can be sorted into a finite number of protein families, greatly facilitating the analysis of the molecular setup of a cell.

4. Combinations of genes are not random. Three types of RdRps and three types of helicases could give rise to nine polymerase–helicase combinations; however, in nature, only three such combinations are widely distributed. Chapter 7 discusses the tendency of some genes to occur in conserved arrays.

5. Conserved arrays of genes or protein domains are not merely the auxiliary factor in determining the identity of protein sequences. The order of genes in virus genomes is a molecular trait in its own right, as indeed is the information on gene co-occurrences mentioned previously. These traits can be used for function prediction and in evolutionary reconstruction.

6. There is abundant evidence for recombination between different virus genomes. Extended regions of nucleotide identity are apparently not required for such recombination. Recombination complicates reconstruction of phylogeny, but evolutionary and taxonomic chaos does not ensue, as discussed in Chapters 11 and 12.

5

The First Fact of Comparative Genomics: Protein Sequences are Remarkably Resilient in Evolution

In Chapter 4, I discussed the early years of comparative genomics, when completely sequenced virus DNAs and RNAs helped us to sharpen our analytical tools and to define what is important in comparative genomics. It was in the 1970s, 1980s, and early 1990s that we learned to pay attention to subtle sequence similarities to genome context and non-random combinations of molecular characters, some of which turned out to be more common than others. But the time of full assault in genomics came with the advent of high-throughput sequencing of DNA of cellular organisms. The first cellular genome, *Haemophilus influenzae*, a gram-negative proteobacterium of gamma subdivision and a close relative of well-studied *Escherichia coli*, was completely sequenced in 1995 (Fleischmann *et al.*, 1995). Even earlier, individual chromosomes of various organisms and organelles had been sequenced; many genome sequences from mitochondria and chloroplasts were available already in the 1980s and early 1990s, and the two smallest chromosomes of yeast *Saccharomyces cerevisiae* were finished by the same time as the first bacterial genome. Approximately 75% of the *E. coli* chromosome was also finished by 1995, including one long contiguous DNA segment that covered more than one-fourth of the genome. Finally, GenBank contained gene sequences from many major groups of organisms, and it could be argued that the diversity of genomes on Earth has been quite extensively sampled. So, when did studies of genes and genome sequences finally become "genomics"?

An enormous number of new genes are sequenced every day, whether in the context of complete genome sequencing or not. This increase in density of sequence space coverage is a benefit in and of itself, regardless of which sequence comes from what species. But one theme to which we will return many times in this book, and which I believe marks "genomics" in the most proper sense, is that complete genomes are a special kind of sequence information, qualitatively different from just a very large collection of sequences. Importantly, if the genome of a life-form is available in its entirety, then the facts about this life-form have to be explained using information about the known and finite set of genes (I leave out the problems of verification of genome completeness and accurate prediction of all genes encoded by the DNA sequence; these, nonetheless, are important tasks, also facilitated by comparative genomics approaches). This completeness of the data changes the way we formulate and test hypotheses. One example of this new thinking was discussed in Chapter 3, in which it was stated that the correct assignment of orthologs, paralogs, and evolutionary events can only be achieved when all homologous genes are known.

Because of this special role of genome completeness, a good date for the beginning of the genomics era is July 28, 1995, when *H. influenzae* genome information, obtained by the group headed by Craig Venter and Hamilton Smith at the Institute for Genome Research, was published in the journal *Science* (Fleischmann *et al.*, 1995). Even more important for genomics was the rapid accumulation of other completely sequenced genomes, sampling the diversity of taxonomic positions, habitats, genome sizes, lifestyles, complexities, and almost every other imaginable trait. The growth of the genome division of GenBank in the years since the *H. influenzae* genome information was published is shown in Fig. 5.1.The numbers of complete genome sequences of bacteria are exploding, no doubt because of their small size that makes them easier to sequence than eukaryotes and their well-recognized importance in society, and the numbers of archaeal and eukaryotic genomes are also growing, slowly but steadily.

With these data in hand, and remembering the two "facts of sequence analysis" from Chapter 1, I state the following as the "first fact of comparative genomics":

> *Protein sequences and proteomes are well conserved in evolution; most proteins encoded by completely sequenced genomes contain at least one region of significant similarity to sequences in distantly related species.*

Proteome means the complete set of proteins predicted to be encoded by the genome. In the literature, "proteome" also has a slightly different meaning—that is, proteins actually encountered in a sample, for example, "yeast membrane proteome" or "proteome of snake venomous glands" (irresistibly called "venome" by Fry, 2005). "Distantly related" also needs some elaboration.

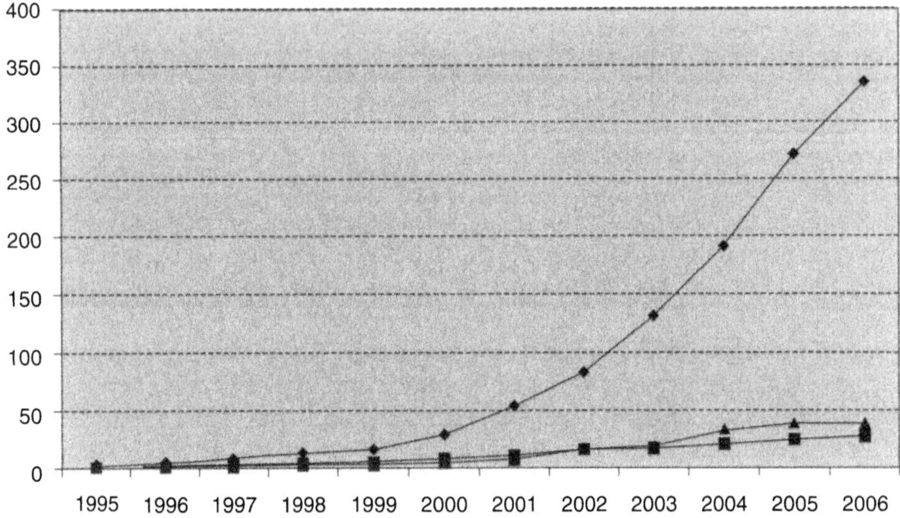

Figure 5.1. Complete genome sequences in public databases. The 2006 data are for the first 6 months of the year. *Diamonds*, Bacteria; *squares*, Archaea; *triangles*, Eukarya. Data are from the NCBI genome division and Nikos Kyrpides' GOLD database (www.genomesonline.org).

Figure 5.2 shows some of the more reliably reconstructed portions of the Tree of Life. Some of the numbers are based on direct evidence, such as dating the fossils. Other are the result of extrapolation on the basis of sequence comparison and other data. Major animal clades have separated nearly 600 million years ago. Regarding prokaryotes, it appears that deep clades of Bacteria, such as proteobacteria and actinomycetes, or proteobacteria and spirochaetes, are separated by more than 1 billion years, and the cyanobacterial clade is perhaps more than 3 billion years old. Bacteria and Archaea may have split at approximately the same time if not earlier (although in this case, the exact date is still not settled; see Chapter 12). In any case, when comparing bacteria, archaea, and eukaryotes, we are dealing with enormous time spans and a very large number of generations.

In the first fact of comparative genomics, distances between these extremely ancient lineages is exactly what is meant by "distantly related." A random protein from a gram-positive bacterium has a better than 50% chance of sharing a related sequence fragment with at least one protein from a cyanobacterium. Note that the existence of a high-scoring fragment is not the same as a reliable global alignment. Some proteins in distantly related genomes may

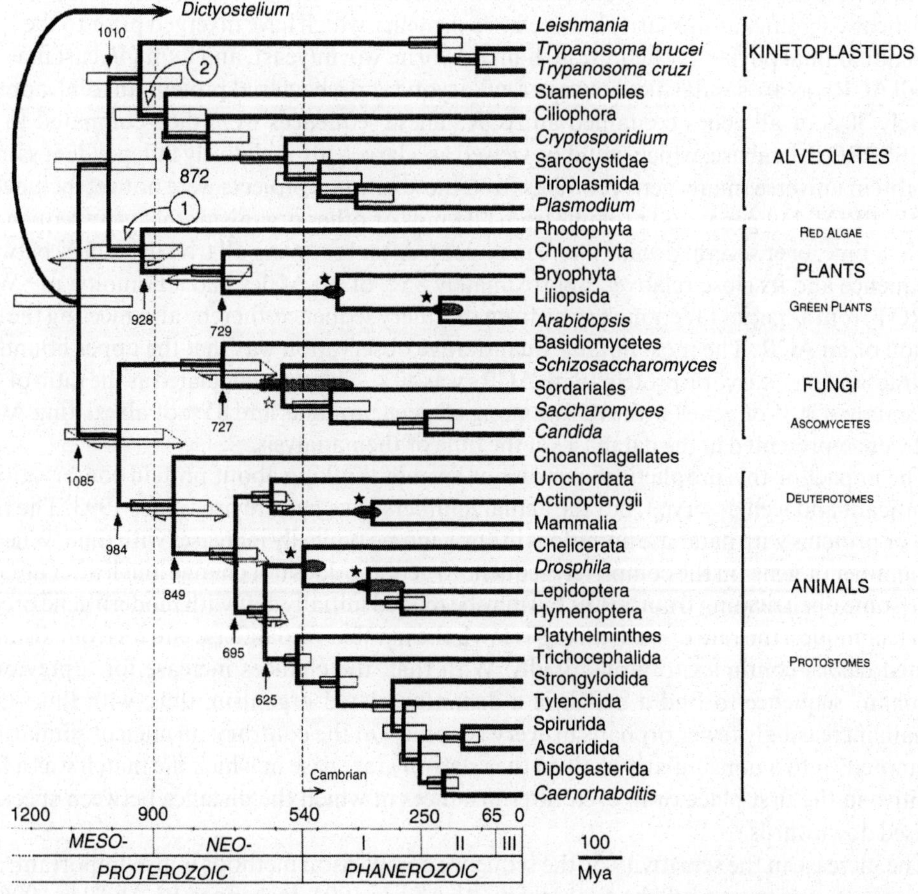

Figure 5.2. Phylogenetic tree of Eukarya with the current estimates of divergence times for various clades. Reproduced from Douzery, E. J., Snell, E. A., Bapteste, E., Delsuc, F., and Philippe, H. (2004). The timing of eukaryotic evolution: Does a relaxed molecular clock reconcile proteins and fossils? Proc. Natl. Acad. Sci. USA **101,** 15386–15391. Copyright (2004) National Academy of Sciences, U.S.A.

contain both conserved domains and nonconserved regions, but the fraction of proteins in a proteome of any cellular species that are at least partially conserved in extremely diverse organisms is high. In other words, proteins truly unique to a narrow phylogenetic group and lacking homologs outside that group are in the minority in every proteome.

This first fact of comparative genomics may seem self-evident today, but it was not that obvious as recently as 10–12 years ago. There was much evidence that similarities between proteins in some families can persist despite large evolutionary distances, but the total number of these conserved regions, the number of proteins that contain these regions, and the proportion of such proteins in each proteome were all unknown.

However, as genome sequencing projects started generating data, Philip Green, then of Washington University (currently at the University of Washington), with colleagues from the same university and from the National Center for Biotechnology Information (NCBI), made the first attempt to estimate these numbers (Green *et al.*, 1993). They used five sets of sequences that were thought to be more or less random samples of protein sets encoded by several genomes: 2644 expressed sequence tags (ESTs) from human brain; 1472 ESTs of a nematode *Caenorhabditis elegans*; 234 genes predicted in the sequenced portion of *C. elegans* genome; 182 yeast genes, also predicted from the genome sequence; and the 1916 genes of *E. coli* that were sequenced at the time. The authors defined ancient conserved region (ACR) as statistically significant similarity between two proteins, which have diverged prior to the splits of major animal phylae. By definition, human/worm, worm/yeast, and human/yeast matches are all ACRs, as are similarities between a eukaryotic and a bacterial protein. In total, approximately 30% of all genes contained an ACR. These sequences were then compared to the SWISSPROT database, which could be viewed as a larger, also relatively independent sample of protein universe (many gene products from these genome projects were not yet included in SWISSPROT, but genes and proteins from all kinds of other organisms that were sequenced, one at a time, over several decades were included). After removing all trivial matches between a sequence and its close relatives, approximately 85% of all ACRs had a homolog in SWISSPROT from a species far enough away from the query sequence, thereby also meeting the definition of an ACR. The most notable quantitative observation was that the upper bound for the fraction of eukaryotic proteins with ACRs was 40%. This was calculated as the ratio of two percentages: 30% of genes with ACRs among all gene products and 85% of all existing ACRs that were represented in the databases at the time of their analysis.

The impact of this prophetic paper on our current thinking about protein conservation is significant and well deserved, but the actual numbers have been revised since 1993. The fraction of proteins with database homologs in any genome typically increases with time: Whereas the number of genes in the completely sequenced genome does not change much after annotation [some open reading frames (ORFs) may be missed initially, but with modern gene prediction techniques, the rate of such misses is low], the number of database entries from distantly related clades continues to grow rapidly. With that, the chances increase for a previously "orphan" sequence to find a match in a distantly related organism; thus, with time, there remain increasingly fewer orphans in every genome. On the contrary, an ancient similarity is "demoted" into a nonsimilarity only in the relatively rare case in which the match was a false positive in the first place or in even rarer instances in which the distance between species is revised downwards.

The increase in the sensitivity of the sequence comparison methods is also important: The same set of sequences could be analyzed by BLAST in 1995, by gapped BLAST2 in 1996, by PSI-BLAST in 1997, and by even more sensitive probabilistic searches later on—each time reducing the number of orphan sequences. For example, in 1996 *Methanococcus janaschii* proteome was analyzed using side-by-side searches by BLAST and BLAST2 against the NR database at NCBI, and BLAST2 produced statistically significant sequence similarity for a much larger number of proteins, indicating that a more sensitive method may change our understanding of the genome dramatically, even when the sequences are compared to the same reference set (Koonin *et al.*, 1997; this is further discussed later in this chapter and in Chapter 11).

Returning to the estimates of Green and coauthors, we now see that among the ACRs shown in their Table 3 and believed to be eukaryote specific at the time, approximately two-thirds are now known to also be present in Bacteria. For some of them, the sequence information was simply not available, as in the case of Ser/Thr protein kinase catalytic domain, which later was detected in many Bacteria and Archaea (Leonard *et al.*, 1998). For other proteins, the sequence was known but the fact of similarity could not be established until the advent of more sensitive methods, as in the case of bacterial homologs of cytoskeletal proteins tubulin and actin.

Three more remarks are in order. First, definitions of "ancient" in the first fact of comparative genomics and in the study of Green *et al.* are not the same, nor are they the only two possibilities. Separation of metazoan clades, used as the milestone by Green *et al.*, is surely very old, but other time points, either more or less recent, may be more appropriate in other contexts. Second, it is common for the large-scale sequence analysis projects to use an operational definition of similarity that involves some sort of across-the-board threshold (e.g., Green *et al.* did not examine matches with BLAST scores less than 90). In an effort to minimize the rates of false-positive matches, such thresholds are sometimes set "conservatively," but the definition of conservative is itself arbitrary. The percentage of conserved proteins in the same genome on the same day will not be the same when the threshold of BLASTP E-value is set at 10^{-10} and 10^{-4}. Thus, without knowing the details of computational protocol, one cannot be sure what has actually been measured. Yet, we have seen in Chapters 2–4 that the theory of homology, as well as the practice of detecting homologous genes and proteins, is not restricted to sequences with very high similarity—a low-scoring match may still be indicative of common ancestry. The most sensitive analysis does not use any arbitrary cutoff but instead evaluates the significance of both high- and low-scoring matches, as we will soon discuss. Third, in order to answer biological questions about gene and protein evolution and function, we not only have to detect all sequence similarities but also have to define all homologs and sort them into orthologs and paralogs (see Chapter 3). However, in a confluence of confusions, some authors define orthologs as "highly similar homologs," or all matches that pass a certain score threshold. This is misleading because different pairs of orthologs may have different degrees of similarity.

In 1995, the analysis of a partial protein list of *E. coli*, by Eugene Koonin, Roman Tatusov, and Kenn Rudd of NCBI, appeared (Koonin *et al.*, 1995; the first and second completely sequenced genomes, of *H. influenzae* and *Mycoplasma genitalium*, were already available, but not in time to be considered in that work). The 2328 putative proteins of *E. coli* known at that time, which would turn out to represent approximately 55% of the *E. coli* proteome, were compared to sequence databases using BLASTP and analysis of conserved sequence motifs, and approximately 70% of all proteins were shown to have homologs in distantly related bacteria. The results of this comparison are shown in Fig. 5.3. Even more convincing is the comparison of the proteomes obtained by the subsequent genome projects involving *H. influenzae*, *M. genitalium*, *Methanococcus jannaschii*, and *Synechocystis* sp. (Fig. 5.4).

The trend is the same in every genome: Most proteins—approximately 60–80% of them—have homologs in distantly related species, separated by hundreds of millions of years. The high fraction of broadly conserved proteins holds despite the difference in the number of proteins in the proteomes—an order of magnitude within bacteria and even more between bacteria and eukaryotes (and within eukaryotes). Differences in lifestyle also seem to have relatively little impact; parasites, symbionts, and free-living species, phototrophs, chemotrophs, and heterotrophs, simple and complex cells, fungi, and animals are all within

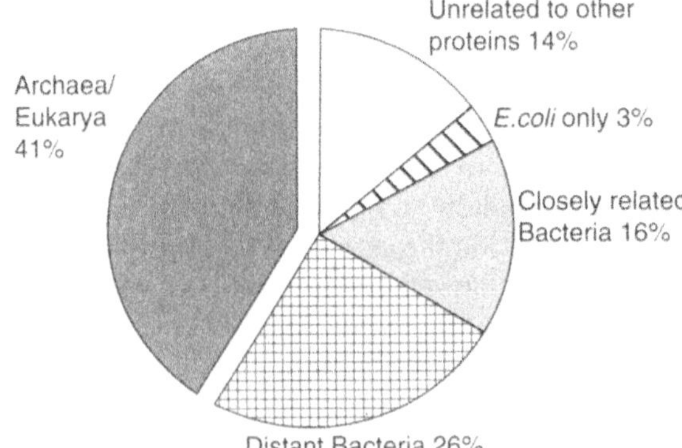

Figure 5.3. Overview of sequence similarity detected in a genome-scale sequence analysis of the extensively sampled genome of *E. coli*. Percentage points indicate the fractions of proteome that are most closely related the homologs in the indicated taxa. Reproduced from Koonin, E. V., Tatusov, R. L., and Rudd, K. E. (1995). Sequence similarity analysis of *Escherichia coli* proteins: Functional and evolutionary implication. *Proc. Natl. Acad. Sci. USA* **92,** 11921–11925. Copyright (1995) National Academy of Sciences, U.S.A.

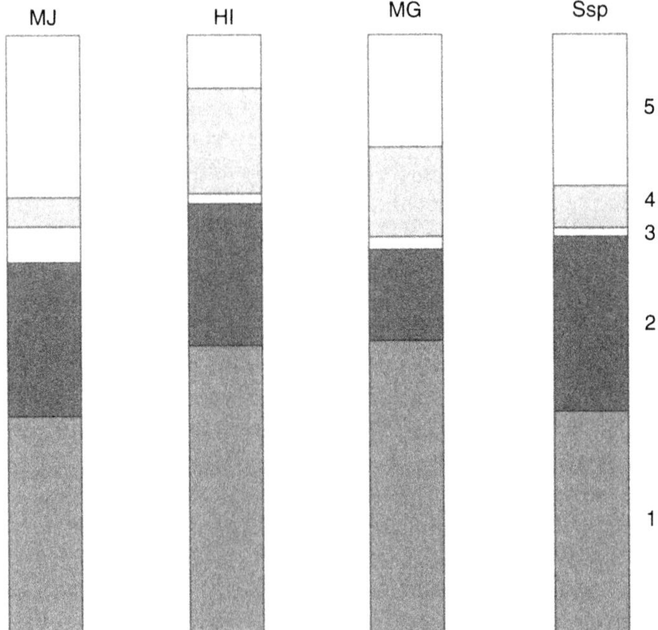

Figure 5.4. Overview of sequence similarity detected in a genome-scale sequence analysis of four completely sequenced genomes. MJ, *Methanococcus janaschii* (current name: *Methanocaldococcus jannaschii*), a methanogenic archaeon; HI, *Haemophilus influenzae*, a gammaproteobacterium; MG, *Mycoplasma genitalium*, a gram-positive-related mollicute; Ssp, *Synechocystis* sp., a blue-green bacterium. Reprinted from Koonin *et al.* (1997) by permission of Blackwell Publishing.

the range. Let us call this number the percentage of proteins with homologs in distant organisms (PHIDO).

At the time of sequencing of the first archaeal genome, *Methanococcus janaschii* (Bult *et al.*, 1996), much has been said about its uniqueness. The original report stated that most proteins in this species did not look like any other proteins, resulting in a low PHIDO of 44%; almost every observed similarity in that case was, by definition, to distantly related species because not much sequence was available from archaea. Soon, however, we found that this low PHIDO value resulted from the fact that the version of similarity search program used by the TIGR group did not work very well at relatively large evolutionary distances, like those separating archaeal proteins from their then available bacterial and eukaryotic database homologs. Only the more advanced database search programs, such as gapped BLAST, could detect these similarities. When we increased search sensitivity, many matches that were missed by the original analysis could now be examined; most of them had strong statistical support and could be further validated by analysis of the conserved sequence motifs (Koonin *et al.*, 1997). All similarities considered, PHIDO for *Methanocccous janaschii*, as of the beginning of 1997, was almost 70%. This percentage continues to increase as the diversity of prokaryotes becomes better sampled by genome projects. So *Methanocccous janaschii* was not as unique as originally believed: Perhaps as much of a surprise was the nonuniqueness of the majority of archaeal genes. (In Chapter 11, we will talk about another evolutionary surprise offered by archaeal genome sequences.)

One archaeal genome, *Aeropyrum pernix*, seems to be the only exception from the 60–80% PHIDO rule; it has been noted, however, that the "unique" ORFs in its genome seem to be significantly shorter than is typical of other archaea. It cannot be excluded that many of those ORFs are spurious predictions rather than real genes.

It is important to note that PHIDO is about sequence similarity and not about what proteins do. Many proteins belong to conserved families for which we cannot predict a biological function. On the other hand, the existence of such conserved uncharacterized protein families is in itself an important result of complete genome sequencing: If we know that they exist, we can direct our experiments toward identifying their function and structure.

The only group of DNA genomes that appears in violation of the first fact of comparative genomics is viruses and bacteriophages. Most of these genomes are smaller than genomes of cellular organisms, although a recently discovered mimivirus that infects amoebas appears to code for approximately 1200 genes, which is more than some bacteria and archaea possess, and the rumors about bacteriophage G suggest that its gene set size may also bridge the gap between viruses and cellular organisms. However, regardless of the proteome size, virus proteins are indeed likely not to match anything in the databases (except for trivially similar sequences of different strains/isolates of the same virus). The PHIDO of DNA virus genomes rarely exceeds 50% and commonly is closer to 20–30% (Liu *et al.*, 2006).

The short generation time of most viruses presents more opportunity for divergence, so it is possible that a fraction of "orphan" virus gene products are proteins that in fact do have database homologs, but the divergence has changed them beyond recognition. It is also likely that the diversity of viruses and their hosts has not been well sampled by sequencing projects, and the homologs of the orphan virus proteins are waiting to be discovered. There is little doubt that virus PHIDO will increase as we sequence more genes and improve sequence comparison approaches; the only question is by how much. On the other hand, it cannot be excluded that some aspect of virus lifestyle requires a truly high proportion of unique proteins (as opposed to copies of homologs in other virus genomes that are changed beyond recognition); this would involve some process of frequent gene innovation.

Thus far, we have examined conservation of protein sequences between completely sequenced, distantly related genomes. A different, but equally important, aspect of evolutionary conservation in proteins is similarity between proteins encoded by the same genome (i.e., the occurrence of paralogs). Every protein either has paralogs in the same proteome or is a "singleton" in this genome. In practice, some paralogs may be easier to recognize than others, but in principle every proteome can be represented as a distribution of protein families by the number of paralogs.

Some paralogs are highly similar, whereas others are extremely distantly related. Often, but not always, the former corresponds to evolutionarily recent family expansions, whereas the latter represents ancient divergence. How should these different levels of similarity be represented, and which of them are of interest? We discussed the ways to detect the homologs and build their hierarchy in Chapters 2 and 3, and these will be discussed further in this chapter. Now, however, let us set the technicalities aside and examine some of the conclusions made from the observations of paralogy in many completely sequenced genomes.

First and most important, the distribution of families by the number of paralogs in each genome has one long tail, i.e., there are many proteins without paralogs and many families with only two or three paralogs, but a small number of families consist of many paralogs. It can be shown that the largest 5–10 families account for a significant proportion of proteins in every proteome.

Second, the largest families of paralogs seem to be domain specific. In almost every prokaryotic (bacterial or archaeal) proteome, the 10 largest families include Walker-type ATPases/GTPases, permeases, helix–turn–helix transcription factors, and Rossmann-fold enzymes that bind nucleotide cofactors (including NAD/NADP-dependent and FAD-dependent oxidoreductases, SAM-dependent methyltransferases, nucleoside-diphosphosugar transferases, and a few other families). On the other hand, the top 10 list in eukaryotes includes the serine/threonine/tyrosine/lipid kinase superfamily, various classes of regulatory proteins with cysteine finger motifs, and, in multicellular species, Ig-like domains and other modules involved in protein–protein interactions—none of which are prominent in prokaryotes. There are also species-specific expansions of individual families, such as Fe–S oxidoreductases in *Methanococcus* or proteinases of the S1 clan in *Drosophila*. Other families may experience lineage-specific reduction; for example, odorant receptor families are large in species that use olfactory communication often, such as *C. elegans* or mouse, but in humans most of the paralogs are functionally inactivated (Zozulia *et al.*, 2001).

Third, we can calculate a summary number for each genome—the percentage of proteins with homologs in the same organism (PHISO). Unlike PHIDO, which is relatively stable from species to species, PHISO displays large variation between species and seems to be correlated with genome size.

PHISO is also sometimes called "paralogy level" and "extent of paralogy," which, unfortunately, sound similar to "degree of homology" and other ambiguous words that we decided to avoid (see Chapter 3). We have to remember that "paralogy level" should not be taken as an indication that paralogs are somehow distinguished by the level of sequence similarity.

In 1995, while analyzing the extent of paralogy in completely sequenced *H. influenzae* and substantially covered *E. coli*, we noticed that the ratio of gene numbers in these two closely related gammaproteobacteria was approximately 2.5 (~4200 in *E. coli* and ~1700 in *H. influenzae*), whereas the ratio of PHISO was 1.45 (50% in *E. coli* and 35% in *H. influenzae*). In a later work, I noted that in several other cases the ratio of PHISO in two genomes was also roughly equal to the square root of the ratio of genes in the same two genomes (Mushegian, 1999). Is this a general trend? What drives evolution of PHISO and how it is related to the evolution of gene number in the genome?

Let us first ask what could have been going on with the total number of genes in *E. coli*, *H. influenzae*, and the common ancestor of these two relatively closely related gammaproteobacteria.

One possibility is that the ancestor was more like *E. coli*, with a large genome and the ability to utilize a wide range of substrates for growth and survival, not restricted to the habitat of the human gut. Under this hypothesis, after divergence, the *Haemophilus* lineage experienced dramatic gene loss, most likely while becoming a parasite of the nutrient-rich cavities in the human body. The other hypothesis is that the ancestor had a smaller number of genes, like *Haemophilus* today, and acquired genes in the course of evolution, adapting to the lifestyle that involved many habitats with different resources. A middle-of-the-road hypothesis is that the ancestor had a medium-sized genome, which split into two lineages, one of which experienced more gene losses and led to *H. influenzae* and the other had more gene gains and led to *E. coli*.

I will not reconstruct gammaproteobacterial evolution in any detail here (for a thorough examination of gammaproteobacterial gene content, see Lerat *et al.*, 2005), but two observations are worth mentioning. First, it is very likely that many gene losses have indeed occurred in *H. influenzae* lineage. Analysis of completely sequenced genome—which genes are in it and which are not—indicates that many metabolic pathways have missing components in *H. influenzae* (Tatusov *et al.*, 1996). Interestingly, all known nutritional requirements of the bacterium can be explained by the absence of specific metabolic enzymes. On the other hand, it appears that every genetic lineage is experiencing gene gains and gene losses simultaneously, and the observed number of genes is the balance of the two processes (Snel *et al.*, 2005). Thus, individual genes, or even complete pathways, may be lost even as the total number of genes in a genome increases. On the contrary, a genome may experience reduction while at the same time some genes may be gained.

What are the molecular mechanisms of all these changes? Gene losses are easier to understand because mutations and deletions in DNA are relatively well studied from both molecular and evolutionary standpoints. Regarding gene gains, three main types of mechanisms can be postulated for protein coding genes: (1) *de novo* generation of a coding ORF when a previously untranscribed/untranslated nucleic acid acquires signals that facilitate synthesis and translation of an mRNA (or where coding potential of a preexisting mRNA is changed by frameshift or another RNA recoding event); (2) horizontal gene transfer, which is a distinct mechanism of gene gain as far as a particular lineage is concerned and will be discussed in that context in Chapters 6 and 11; note, however, that when we discuss gene gain by life as a whole, horizontal gene transfer is not very helpful because we still need to explain how this gene emerged in the first place; and (3) duplication of a preexisting gene, followed by sequence divergence and perhaps change of function.

Let us now examine the effect of gene gains and losses on the evolution of PHISO. The picture here is complicated. For example, if a single-copy, no-paralogs gene is lost from the genome, the total number of genes will decrease, and so will the fraction of unique genes: But then PHISO will increase, if only slightly. The general theory of these processes has been developed (Karev *et al.*, 2002, 2004, 2005), but evolutionary events that have taken place in the actual lineages leading to the existing species remain to be fully understood. However, analysis of paralogy in many bacterial genomes (Pushker *et al.*, 2004) has confirmed our previous observations about the proportionality between the total number of genes in the genome and PHISO value (Fig. 5.5). Moreover, the same trend seems to hold in eukaryotes. A biologically plausible explanation for the wide applicability of this rule remains to be discovered.

It has been reported that if "essential families" are excluded from the genomes, then PHISO is not dependent of genome size but is constant (Enright *et al.*, 2003). In contrast, it is the number of protein families (or TRIBES in Enright *et al.*'s framework) that displays linear dependency of the number of genes in the genome. It is difficult to evaluate the significance of these observations for two reasons. First, Enright *et al.* used a stringent similarity threshold (BLASTP E-value 10^{-10}) in order to determine whether a protein has any homologs. In contrast, Pushker *et al.* (2004) used a more inclusive 10^{-5} cutoff. As argued many times in this book,

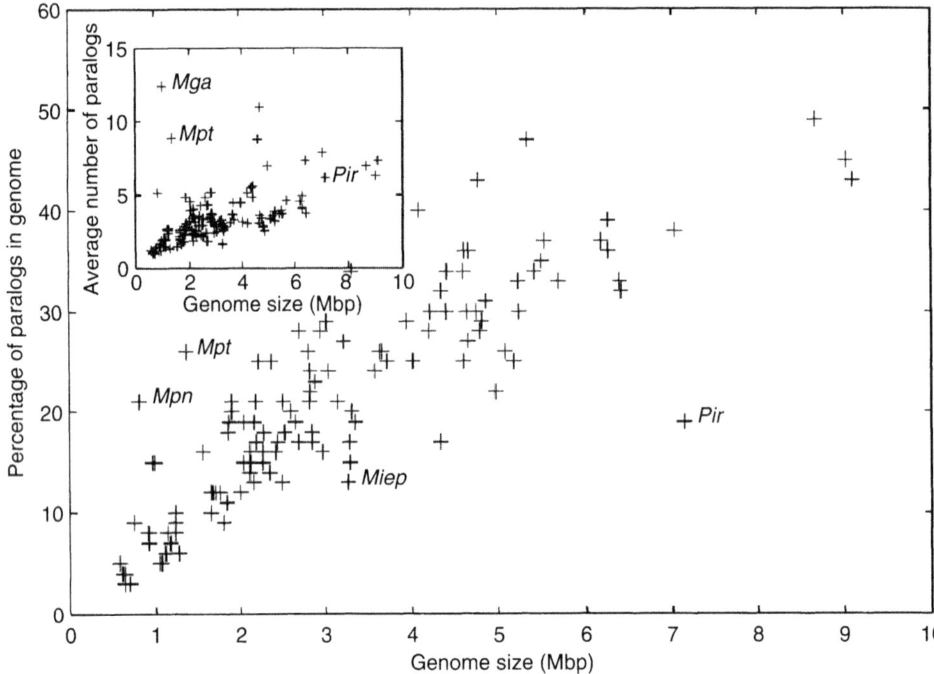

Figure 5.5. Proportional relationship between genome size and the extent of paralogy. From Pushker *et al.* (2004), reprinted under BioMed Central Open Access license agreement.

both cutoffs are arbitrary; they generally underestimate the number of homologs, with faster evolving families suffering from this underestimation more than slow evolvers, but the sample obtained with the use of E-value 10^{-10}, obviously, misses more homologs than the set with the cutoff of 10^{-5}. Second, the essentiality of an individual protein can be defined experimentally—for example, if a null mutant of the corresponding gene is nonviable or otherwise reproductively unsuccessful—but it is unclear to me how to define an essential *family*. Enright *et al.* seem to use the conservation of families in different species as a proxy for their essential role, and they excluded from their counts the families that are found in all domains of life. However, even if some members of such families are essential, it is not clear that every member of each such family is essential (in fact, there is evidence that this is not the case; Hutchison *et al.*, 1999; Glass *et al.*, 2006). Thus, it is not clear how many proteins were removed from the examination and, most important, why they were removed. TRIBES may be an interesting construct to study, but it remains to be seen what it really tells us about paralogy and PHISO.

Interestingly, although PHISO is intended to characterize just one genome, in practice the determination of the state of paralogy can be done in earnest only if we consider multiple genomes. For example, if we want to study the closest similarities, as with in-paralogs (lineage-specific gene multiplications), we need to know the speciation events to determine which paralogs have duplicated before and after each speciation (see Chapter 3). On the other hand, suppose that we want to find all homologs of each protein in a given genome, regardless of the degree of similarity. If some of the paralogs have been duplicated in a very distant ancestor, they may be extremely diverse. If we match each protein to every other protein in the same genome, the similarity between such paralogs may not stand out from the background. For example, when annotating the proteomes of *H. influenzae* and *M. genitalium* in 1995 and 1996, we studied two proteins with lipoate–protein ligase activity, LplA and LipB. They have the same molecular function, yet we and everyone else could not detect significant sequence similarity between them and thought that they were unrelated to each other (Mushegian and Koonin, 1996a). *Escherichia coli* has both proteins, and both *H. influenzae* and *M. genitalium* have one lipoate–protein ligase each and of different type—LplA in *M. genitalium* and LipB in *H. influenzae*—but direct comparison of these proteins to each other is not very illuminating. If, however, we search databases of all known sequences using either LplA or LipB as a query, we will find many sequences with significant similarity to the query—some closely and others not so closely related. This would allow us to construct a multiple alignment and a sensitive probabilistic model and find more remote homologs. Indeed, in 2000 these more sensitive methods of analysis and more dense coverage of the sequence space proved that LplA and LipB and, for good measure, biotin–protein ligases are all homologous to each other (Reche, 2000). Thus, even to establish homology of sequences within one genome, we may need sequences from many genomes.

To sum up the list of main challenges is to reiterate what was said in Chapter 3: There are homologs that are closely related to each other, and there are other homologs that are very distant; some homologs are orthologs, and some are paralogs; some paralogs are ancient duplications, and others are lineage-specific gene expansions. Add to this the co-orthology problem: Because duplications can occur before and after speciations, there may be cases in which gene A in genome *A* has no single ortholog in genome *B* but instead has a set of co-orthologs. Thus, the relationships between some homologous genes and proteins cannot be defined if we only consider one-to-one mapping; one-to-many and many-to-many mappings may be more appropriate.

An innovative resolution of many of these issues was proposed by Roman Tatusov, Eugene Koonin, and David Lipman at NCBI in 1997. The task of organizing orthologs in many genomes was redefined as delineation of Clusters of Orthologous Groups (COGs; Tatusov *et al.*, 1997).

A simple COG, by definition, is a set of orthologs in three or more genomes that belong to three or more phylogenetically distant lineages, one ortholog in each genome. One has to define, of course, which clades are close enough to be treated as one lineage. In the latest version of the COG project, such discrete lineages include, for example, proteobacteria subdivisions alpha, gamma, delta, and epsilon; low-GC gram-positive bacteria; actinomycetes; spirochetes; Euryarchaeota; Crenarchaeota; and some others. Thus, if three orthologous genes are found, one each, in *E. coli*, *H. influenzae*, and *Pasteurella sp*. and not found anywhere else than in these three gammaproteobacteria, this will not satisfy the definition of a simple COG—the clades in question are too close to each other. However, three orthologous genes, one each in *Bacillus halodurans*, *Thermotoga maritima*, and *Caulobacter crescentus*, which represent three distantly related lineages—respectively, gram-positive Bacteria, deep clade of Bacteria with uncertain affinities, and alphaproteobacteria—are a proper simple COG. Incidentally, there is exactly one such triplet—COG03661, an alpha-glucuronidase.

Lineages can be lumped and split depending on the event horizon of interest. For example, if we are examining evolution of only gammaproteobacteria and want to organize their proteins in naturally defined groups, then *Escherichia*, *Haemophilus*, and *Pasteurella* may be legitimate clades, and the approach will work as well.

Simple COG is the most basic unit of protein classification. But why three and not two or five? The answer is linked to the method of COG construction. The most important notion here is the BeT, which is a best hit in a similarity search (by the way, "BLAST hit" is a widely used jargon, which I am trying to avoid in this book and replace with a "match" or, where possible, by a longer explanation of similarity type and of the protocol with which it was detected). BeTs are produced as follows: Suppose that we have *N* genomes and *N* lists of complete proteomes encoded by each of these genomes. Each of the *N* lists is turned into a sequence database. Each protein is used as a query to search every other database. This produces *N*-1 BeTs for each protein. For a moment, we will ignore the possibility of ties, when two proteins $A1_B$ and $A2_B$ in genome *B* have the same level of similarity to the query protein A_A in genome A.

Let us consider query protein A_A in genome *A*, and its BeT in genome *B*, protein A_B. What can be said about the BeTs for protein A_B? There are two possibilities regarding its own BeT in genome *A*: It is either protein A_A or some other protein. The latter is not of interest, but the former represents a special type of relationship: If A_A and A_B are each other's BeTs, this is called symmetric, or reciprocal, BeT (sometimes called SymBeTs).

Now consider A_A and its SymBeT in the third genome *C*, protein A_C. What can be said about the relationship between A_B and A_C? Note that both A_B and A_C have selected A_A as their BeTs in genome *A*. Setting aside some minor details, such as the dependence of the similarity score of length and amino acid composition of the queries, we can say that the probability of two proteins to randomly choose, with the highest rank, the same protein from the database is the reverse of the number of proteins in the database (i.e., in genome *A*). For most genomes, the latter value is between 0.00001 and 0.002, and the chance probability of two proteins that are SymBeTs to share a SymBeT in a third genome is approximately the square of that (although not exactly, because the trials are not completely independent; the exact statistics for COGs remain to be developed and would be of great interest).

In this construct, two important objectives are achieved. First, there is a clearly defined way to select triplets of proteins, one per genome, that are highly unlikely to be related by chance only and therefore are most likely to be homologs (and, moreover, orthologs). Second, all this is done without regard to the absolute level of sequence similarity: A weaker similarity will be recorded the same as a stronger one because the top rank, not the threshold score, is used. The criterion of homology used in this framework (i.e., symmetry of two or more proteins in being each other's highest ranking matches) and the use of triangles of symmetric BeTs are two main novel ideas of Tatusov *et al.*'s paper and of the immensely useful COG framework.

Consider an arbitrary (but finite) number of completely sequenced genomes, each representing a distinct lineage of interest. (Technically speaking, we can register BeTs for any pair of genomes, but in accord with our definition of phylogenetically distant lineages, we may examine only BeTs between proteins that belong only to such lineages. Following the previous example, BeTs between *Haemophilus* and *Pasteurella* will not be used for simple COG construction, but BeTs between *Pasteurella* and spirochetes will be used.) Obviously, triangles of BeT can be constructed for any three lineages. Suppose that A_B and A_C connect, by way of symmetric BeTs, not only to A_A in genome *A* but also to A_D in genome *D* (Fig. 5.6). We then will be looking at two simple COGs that share one side (the one defined by genes A_A and A_C). Merging all such triangles with shared sides, we can produce polygons representing groups of homologous proteins from evolutionarily distant clades.

The hypothesis of homology between BeTs agrees with intuition and is supported by calculations. Most important, the empirical testing indicates high precision of the method: The fraction of triangles that consist of unrelated proteins is negligible. However, we also want some evidence that the homologs recovered by this method are indeed orthologs rather than paralogs.

Suppose that two orthologs A_A and A_B in genomes *A* and *B* are not a symmetric BeT. Then, there are three possibilities: (1) Protein A_A has a paralog (out-paralog; see Chapter 3) $A1_B$ in genome *B*, such that similarity between A_A and A_{1B} is higher than similarity between A_A and A_B; (2) the same situation is true for A_B with regard to genome *A*; or (3) both A_A and A_B have "false orthologs" in the other genome. However, there is no mechanism by which an out-paralog could preserve higher evolutionary conservation than the ortholog; the only conceivable possibility is sequence drift, which could result in occasional rank swaps between the best and second best scoring homologs. Note, however, that in order for all this to happen, the effect of random

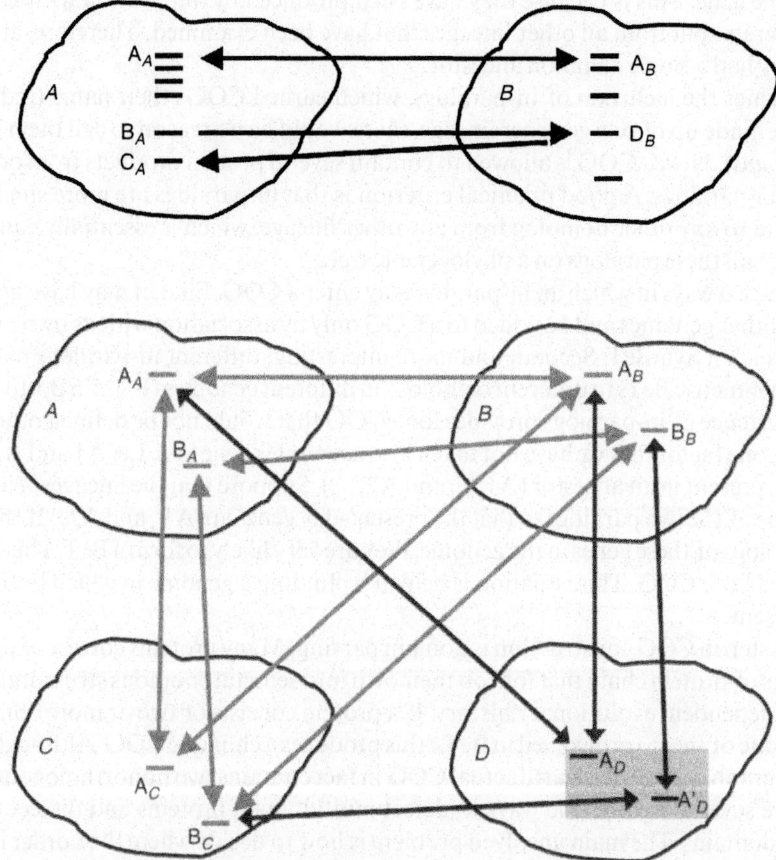

Figure 5.6. Best hits (BeTs), symmetric best hits (SymBeTs), and construction of COGs. **(Top)** A_A and A_B are SymBeTs. D_B is a BeT of B_A, but they are not SymBeTs because the BeT of D_B in genome *A* is gene C_A, not B_A. **(Bottom)** Consider three species, *A–C*. Genes A_A, A_B, and A_C form one triangle of SymBeTs (COG1), and genes B_A, B_B, and B_C form another such triangle (COG2). Both of these COGs, indicated by gray lines with arrows at both ends, are simple COGs. Consider a newly added genome *D* and gene A'_D in it. This gene forms SymBeTs with genes B_B and B_C and thus completes the triangle $B_CB_BA'_D$ that shares a side with COG2. Gene A'_D therefore joins COG2. Gene A_D forms SymBeTs with genes A_A and A_B and therefore joins COG1. Suppose now that A_D and A'_D are in-paralogs: In this case, COG1 and COG2 are merged into one complex COG, the evidence for which was not available before addition of genome *D*.

processes has to be comparable, or slightly higher, than the difference between scores of the first and second highest matches. This means that the difference between these two matches is very small (i.e., they are closely related). Hence, the main reason why the BeTs would not be defining the true sets of orthologs is that the picture of orthology is complicated by co-orthologs and in-paralogs—that is, by lineage-specific duplications.

There may be several approaches to solving this problem. One way out is to turn tables on the notions of orthology and paralogy and to work out some other way of arranging homologs into groups. In practice, this means either applying a more or less arbitrary similarity threshold (i.e., favoring higher similarities and losing information about more subtle ones) or simply declaring the orthology assignment too difficult (at least, too difficult for full automatization). But the authors of the COG framework had a more interesting proposition, namely that, from the evolutionary point of view, in-paralogs could be appropriately treated as essentially one and the same gene. This is because they have been produced by duplication within their own lineage, after its split from all other lineages that have been examined. Therefore, at the point of split, they had a single common ancestor.

Along comes the inclusion of in-paralogs, which earned COGs their name (indeed, if all COGs were made up of one gene per lineage, there would be no reason to call them both *clusters* and *groups*). Now a COG is allowed to contain several protein products from one species, if they are in-paralogs. A good practical criterion is that in-paralogs are more similar to one another than to any other homolog from any other lineage, which is essentially equivalent to clustering of all these paralogs on a phylogenetic tree.

There are two ways in which an in-paralog may enter a COG. First, it may have no symmetric BeTs in other genomes and be added to a COG only by association with its own in-paralog, which has such a SymBeT. Second, and more interesting, different in-paralogs may be connected by symmetric BeTs to different orthologs in different genomes (Fig. 5.6 Bottom). In this case, the existence of in-paralogs provides for a COG that would not be defined otherwise.

Thus far, one factor that we have not taken into account is gene loss. Let A1 and A2 be a pair of paralogs present in an ancestor ($A1_{LCA}$ and $A2_{LCA}$). Suppose that the lineages *A* and *B* have each lost one of the two paralogs so that the present-day genes are $A1_A$ and $A2_B$. If there are no other homologs of these genes in the genome, they are very likely to form a BeT. This is one way to produce a false COG. This situation is helped by finding a genome in which both paralogs are still present.

An extra step in COG construction is domain parsing. Many proteins consist of domains—that is, parts of protein chain that fold on their own into semiautonomous structural units and can have independent evolutionary history. If a protein consists of two or more domains, and more than one of them are involved in BeTs, this produces a chimeric COG. Although all edges (BeTs) in a graph are real, such artifactual COG in fact contains two nonorthologous proteins.

There are several algorithmic ways to detect multidomain proteins and dissect them into individual domains. The main unsolved problem is how to decide where the border is between two domains, especially when sequence similarity within each domain is low. In the COG project, dissection was done mostly manually, but with the growth of the numbers of complete genomes and phylogenetically distant lineages, some automation becomes imminent.

The first version of the COG database was built on the basis of just 7 genomes, which represented five lineages—three bacteria (gammaproteobacteria, gram-positive-like mycoplasmas, and blue-green algae), one archaeon (*Methanococcus janaschii*), and one eukaryote (yeast)—and there were 720 COGs altogether. The latest release utilizes information about 120 genomes and contains almost 14,000 COGs.

The growth of the COG database with the addition of new genomes is an interesting process. Suppose that we have COGs constructed before ("old COGs") and also some proteins from the old genomes that do not belong to any COGs. Some pairs of these proteins, however, form SymBeTs among two lineages; the only problem is that they lack the SymBeTs in any third lineage. These pairs are called TWOGs or "pre-COGs." Some pre-COGs may also contain more than two proteins because of in-paralogs.

When a new genome is added, several changes to the old version of the COG database may happen. First, old COGs get new members. The relationships between proteins that are already components of the old COGs will not change, but some of these proteins may form triangles with orthologs from a new genome (see Fig. 5.6 Bottom). Second, some of the pre-COGs may become COGs. Third, a new gene may link two preexisting COGs into one. Fourth, the information from a new genome may indicate that some COGs are chimeric and need to be split. The two main reasons to split a COG were discussed previously: One is pseudo-orthology, which can be uncovered by a genome that has retained both paralogs, and the other is domain fusion, which can be uncovered by a novel combination of domains in a new genome.

In-paralogs in COGs behave in several ways. Some of them are connected to orthologs in other genomes by regular symmetric BeTs, and others may be involved only in asymmetric BeTs or in a mix of different types of BeTs; some paralogs may even be connected only within their own genome. This differential connectivity of paralogs (and of orthologs, for that matter) has not been well studied.

In addition to the graph of BeTs, several other types of information are associated with each COG. Although COG construction relies on ranks instead of similarity scores, those scores are also available and can be used, for example, for building approximate distance-based phylogenetic trees of proteins included in each COG.

Another important piece of information associated with each COG is phyletic pattern, or the set of species in which members of each COG reside (Fig. 5.7). Phyletic patterns are most appropriately encoded by binary vectors (i.e., strings of ones and zeroes). On the other hand, "ones" can be converted to actual counts of in-paralogs in each species. A binary vector thus becomes an interval vector, with coordinate values represented by real numbers from zero to perhaps some large number. This is an interesting object that needs to be studied further.

Finally, COGs have been functionally annotated by careful analysis of sequence similarity to all functionally characterized homologs, including the remote ones (which do not have to belong to the same COG or to any COG at all). The genome context of COGs is also examined, sometimes providing additional clues to protein function (see Chapter 8).

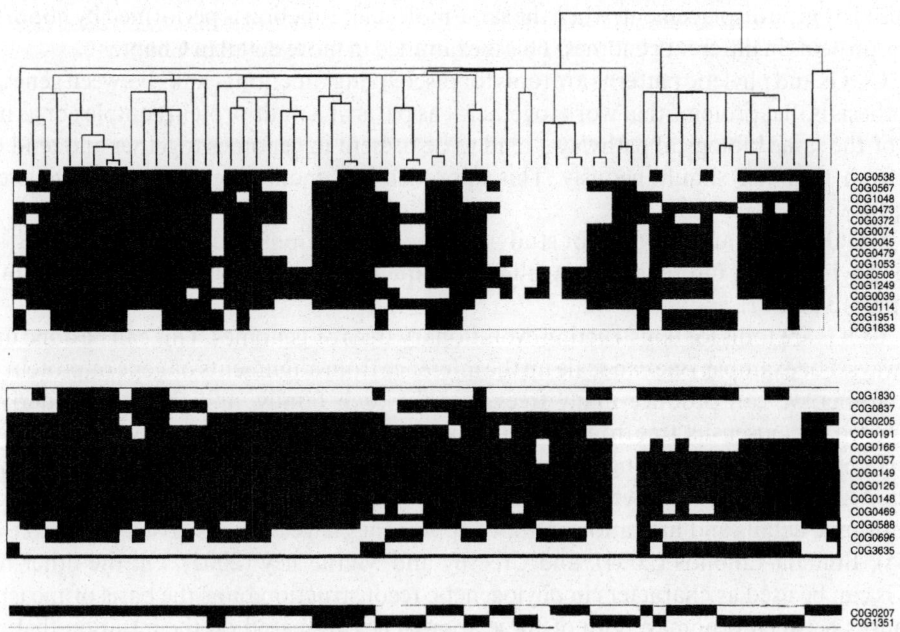

Figure 5.7. Phyletic patterns. Each row represents a COG, and each column represents a completely sequenced genome. Genomes are sorted in the approximate order of traversing the cladogram shown at the top. The presence of a COG in a given genome ("1" in a phyletic vector) is indicated by a black square. (**Top**) enzymes of the TCA cycle; (**middle**) enzymes of glycolysis; (**bottom**) two types of thymidylate synthases. The patchiness of phyletic patterns contains important information about biological function and evolution of metabolic pathways, as discussed in more detail in Chapters 6, 8, and 11–14. Reprinted from Glazko and Mushegian (2004) under BioMed Central Open Access license agreement.

Let us now examine various uses of COGs in comparative genomics:

1. COGs are tools for genome annotation. A list of proteins from a newly sequenced genome can be quickly compared to the COG database, resulting in instantaneous functional prediction of a significant fraction of putative proteins. Family relationships can be determined for even larger proportion of proteins. This is because there are many functionally uncharacterized but conserved families of proteins, in which we can assign orthologs and paralogs by comparing two phylogenetic trees—one for the gene family and another for the species in which these genes are found (see Chapters 3 and 7). Borrowing the functional annotation from a COG to which a new protein is similar represents perhaps the most common practical application of COGs.

2. COGs are indicators of potential errors in gene prediction. A "patchy" phyletic pattern, when an orthologous gene is found in most lineages but is missing from some lineages, can prompt a reinvestigation of sequence similarity and a search for remote homologs if there is evidence that it should be present in genomes in which it has not yet been found. Sometimes, especially when the sought ORF is short, it may be recovered from what was thought to be an intergenic portion of the genome or as an inconspicuous appendage to a longer protein. For example, this is how the smallest component of the tRNA-glutamate aminotransferase complex in *M. genitalium* (Koonin and Aravind, 1998; Mushegian, 1999) and the gamma subunit of 2-ketoglutarate ferredoxin reductase in *Pyrococcus horikoshii* (Huynen *et al.*, 1999) were identified.

3. COGs and phyletic patterns, on the other hand, may be indicators of genuine cases of missing orthologs. Sometimes, this is explained by gene/pathway loss, for example, in the course of genome simplification in parasitic species. In other cases, the patchy phyletic pattern is caused by gene displacement, when the same molecular function is performed by nonorthologous proteins in different genomes. This is examined in more detail in Chapter 6.

4. COGs and phyletic patterns are tools for discovering functional links between genes. The hypothesis is that proteins that work together—as parts of a multiprotein complex or as members of the same biological pathway—tend to be present in genomes together and tend to be lost from genomes simultaneously. This approach to function prediction is discussed in Chapter 8.

5. COGs are a starting point for studying gene duplications and in-paralogy using standard phylogenetic inference from aligned sequences, as discussed in more detail in Chapters 3 and 11.

6. COGs are tools for constructing evolutionary trees of genomes. This can be done in several ways. For example, one can scale up the inference from alignments of gene or protein families: Either we can produce many trees, one for each family, and then try to derived a reconciled, or consensus, tree from them (the "supertree" approach), or we can join alignments of all families into one very long alignment and build the tree from that mega-alignment (if the method of tree building involves a distance matrix, this is called the "supermatrix" approach). Algorithmic details and limitations of both approaches have been discussed by Semple *et al.* (2004), Bininda-Emonds (2004), and Creevey and McInerney (2005). On the other hand, COGs can be used as characters in phylogenetic reconstruction or as the basis of measuring distances between genomes; both of these approaches are based on the intuition that more closely related genomes have more COGs in common than more distantly related ones (see Chapters 11–13 for details).

7. COGs and phyletic patterns are tools to study gene histories when they are different or nonrepresentative of the species' histories. In particular, such events as xenology (horizontal gene transfer) and gene loss can be inferred from the analysis of the discordance between species' history and phyletic pattern (see Chapters 6 and 11).

8. COGs and phyletic patterns are tools for ancestor inference. Given COGs phyletic pattern, the species' tree, and the evolutionary model, one can infer, with various degrees of certainty, the presence or absence of this COG in various ancestral species (see Chapter 13).

From these examples, it is quite clear that there are close connections between COGs, phylogenies of genes, and phylogenies of species in which the COG members are found. Many uses of COGs and trees overlap, whereas other uses are complementary to each other. In fact, this list of COGs uses was inspired by a remarkable article, "Uses of Evolutionary Trees" (Fitch, 1995), which will be examined again in Chapter 11.

The COG framework is used as a resource by a growing number of scientists. COGs are even mentioned in textbooks, including David Mount's *Bioinformatics* (2004). Mount describes the COG framework as follows (pp. 524–525):

> *When entire proteomes of the two organisms are available, orthologs may be identified as the most-alike sequences in reciprocal proteome similarity searches.... Using the protein from one of the organisms to search the proteome of the other for high-scoring matches should identify the ortholog as the highest-scoring match, or best hit. However, in many cases, each of the orthologs belongs to a family composed of paralogous sequences related to each other by gene duplication events. Hence, in the above database search, the ortholog will match not only the orthologous sequence in the second proteome, but also these other paralogous sequences. The objective of the clusters of orthologous groups (COGs) approach is to identify all matching proteins in the organisms, defined as an orthologous group related by both speciation and duplication events. Related orthologous groups in different organisms are clustered together to form a COG that includes both orthologs and paralogs. These clusters correspond to classes of metabolic functions. A database produced by analysis of the available microbial genomes and part of the yeast genome has been made, and a newly identified microbial protein may be used as a query to search this database. Any significant matches will produce an indication to the metabolic function of the query protein (Tatusov et al., 1997).*
>
> *To produce COGs, similarity searches were performed among the proteomes of phylogenetically distant clades of prokaryotes. Orthologous pairs were first defined by the best hits in reciprocal searches. A cluster of three orthologs in three different species was then represented as a triangle on a diagram. Some triangles included a common side, representing the presence of the same orthologous pair in a comparison of four or more organisms. Triangles with this feature were merged into a cluster similar in appearance to Figure 11.6C, part i. Paralogs defined by sets of three matching sequences in the selected organisms were also added to these clusters. The proteins encoded by many prokaryotic organisms have been analyzed for COG relationships (Koonin et al., 1997). A COG analysis provides an initial assessment of the genome composition of prokaryotic organisms and should be followed by a more detailed analysis as described above for the worm and yeast proteomes.*

Descriptions of the COG framework, such as the one just cited, should be taken with caution. The objective of the COGs approach is resoundingly not "to identify all matching proteins in the organisms" (i.e., all homologs). Neither can it be "to identify all matching proteins in the organisms, defined as an orthologous group related by both speciation and duplication events." Any set of homologs is related by some combination of duplication and speciation events, and it is not clear from that description what is the specific definition of the orthologous group. But we now know that the goal of the COG approach is to select, among all homologs, only the union of orthologs and in-paralogs, if the latter exist. Furthermore, some COGs correspond to classes of metabolic functions (these are mostly large COGs, which indeed may include many proteins with similar function, such as transporters for various divalent metal ions or large superfamilies of class I and class II DNA and RNA helicases). However, the majority of COGs correspond to exactly one molecular/metabolic function, not the whole class of them. Such functions, of course, can be hierarchically organized into classes;

for example, several nucleotidyltransferases can form a class of ferredoxin-fold nucleotidyltransferases or, using a different principle, we can assign some of them to the class of DNA replication and repair enzymes and others into the class of proteins involved in posttranscriptional modification of RNA. But formation of classes is external to construction of COGs. Finally, not all paralogs, but only in-paralogs, are sought in COG building.

Previously in this chapter, we discussed PHIDO and PHISO. It is now time to introduce PICO—percentage of proteins in COGs. In agreement with the first fact of comparative genomics, the PICO value is high for most genomes. In prokaryotes PICO is close to 80%, with only a few exceptions: The lime disease pathogen *Borrelia burgdorferii* is an extreme case of low PICO, approximately 43 %. It has been noted that among many replicons that constitute the *Borrelia* genome, there are several plasmids encoding many short, poorly conserved ORFs. Perhaps these plasmids are more similar to DNA viruses, which also have lower COG coverage. Genomes of eukaryotes have lower PICO too, approximately 50%, but the expectation is that their PICO will increase when more eukaryotic lineages are completely sequenced.

It is worth remembering that high percentages represented by PHIDO, PHISO, and PICO, however obvious they may seem now, were not expected by prior scientific experience. Indeed, as recently as 1993 and 1994, nothing of the sort was evident. The fraction of seemingly unique proteins in the database was thought to be high, the limited ability of sequence comparison programs to detect weak sequence similarities had been pointed out many times, and even pilot projects on analysis of partially sequenced genomes of model organisms did not seem to be very encouraging. It was not until the detailed examination of partially sequenced bacterial genomes, *E. coli* (Koonin *et al.*, 1995) and *M. capricolum* (Bork *et al.*, 1995), that the indications for high fraction of conserved proteins started to accumulate. Only after completion of several genome projects, and following the development of rapid and sensitive programs for database searches (notably PSI-BLAST), was the question settled.

One special class of paralogs that is of considerable interest comprises the so-called lineage-specific duplications. Let us call it PIPO—percentage of in-paralogs in the organism. Here, we are interested only in those paralogs that have evolved by duplication in a given lineage after its separation from all the other lineages that we are simultaneously considering. PIPO may be determined by finding all pairs of paralogs in the same genome, such that they are closer to each other than to any other protein in this or any other lineage under consideration. Every such pair is part of a lineage-specific expansion, and in fact at least two-thirds of all lineage-specific gene expansions in bacteria and archaea contain just two genes (Jordan *et al.*, 2001; Lespinet *et al.*, 2002). But there is also a relatively small number of very large groups of in-paralogs. Two of the largest known bacterial expansions are the PPE and PE families of surface proteins involved in interactions with the host cells in *M. tuberculosis*; they consist of 90 and 67 proteins, respectively.

As with many other studies of homologs, it is not very useful to define in-paralogs on the basis of absolute degree of similarity; any predetermined cutoff may turn out to not be sufficiently accurate in distinguishing between in-paralogs and other homologs. For example, average score density (i.e., similarity score per unit length of the alignments, averaged over all aligned pairs) of yet another lineage-specific family of surface proteins in *M. tuberculosis* (Mce1-like proteins, which have 24 family members altogether) is 0.21, whereas the average score density of one 25-member expansion in *Mycoplasma pneumoniae* is 0.62. Neither value is good for predicting the scores or score densities of other in-paralogs or, for that matter, any homologs. What is of ultimate importance is the relative similarity: In-paralogs are closer to each other in the evolutionary hierarchy than to any other homologs, whereas the absolute scores and score densities may vary widely in different expanded families.

The PIPO values are different from all the other measures of inter- and intragenomic protein similarity discussed in this chapter. Whereas PHIDO, PHISO, and PICO each cover more than half of proteins in every completely sequenced genome (exceptions are insignificant in prokaryotes, and even eukaryotic proteomes seem to be within the range), the PIPO value is lower and more widely varied—from 5 to 33% among prokaryotes (Jordan *et al.*, 2001). In many cases, this can be directly explained by a specific adaptive role of the proteins within the expansion: Several large families in human pathogens are cell surface proteins involved in host cell adhesion, and the largest lineage-specific expansion in *E. coli* is a 31-member set of in-paralogs that encode LysR-like transcriptional activators consisting of HTH DNA-binding domains fused to solute-binding regulatory domains. The relationship between this expansion and the ability of *E. coli* to metabolize a huge variety of small organic molecules is obvious.

So, the first fact of comparative genomics tells us that most proteins belong to conserved sequence families. What it does not tell is how many such families exist. This question will be discussed in Chapter 10.

In practice, completely sequenced genomes are easily and routinely annotated every day with the aid of COGs and other databases of conserved protein families (Daraselia *et al.*, 2003; Maheswari *et al.*, 2005; Powell and Hutchison, 2006; Grzymski *et al.*, 2006). This, perhaps, is the most important practical application of the first fact of comparative genomics.

6

The Second Fact of Comparative Genomics: Functional Convergence at the Molecular Level

The main conclusion of Chapter 5 is that the majority of genes have relatives in the same genome, in other genomes, and in the databases. As we improve our ability to detect sequence similarities and to evaluate them using statistical criteria, we detect increasingly larger numbers of such relationships between genes and connect increasingly more proteins into families of homologous sequences. "Superfamilies" and sometimes even "hyperfamilies" are used to characterize the unions of more distantly related families. For now, we will use these words as intuition suggests: Members of a family are on average more similar to each other than members of a superfamily to which this family belongs. The most important point to remember is that sequences within a family, as well as within a super-/hyperfamily, are homologous: They share a common ancestor. The support for the hypothesis that such ancestor existed comes from the statistics of database search or, as with the COG project, the statistics of symmetric BeTs across several genomes. However, the degree of sequence similarity, be it percentage of identity, similarity score, or score density, may be different from family to family.

The evolutionary connections between different proteins, as well as between whole families and superfamilies of proteins, are discovered constantly. There are many ways in which this can lead to new structural, functional, and evolutionary inferences. Let us examine several types of such inferences.

Proteins with Same or Similar Functions

We have two or more proteins, which are known to have similar functions. The immediate question is whether they are evolutionarily related. Sometimes, the similarity between protein sequences is high and easy to discover; other times, the similarity is so low that its significance is difficult to evaluate. Such was the case, for example, with two classes of DNA ligases—enzymes that close gaps and nicks in DNA and are therefore essential for genome replication and repair. For decades, enzymologists knew that bacterial DNA ligases require NAD for activity, and ligases encoded by eukaryotes, viruses, bacteriophages, and archaea depend on ATP. The two classes of enzymes were thought to be unrelated. There are other key components of DNA replication that are different in Bacteria and Archaea/Eukarya, including initiator ATPase, main replicative helicase, DNA primase, and, most important, replicative

DNA polymerase. The most important evolutionary implication of these differences is that DNA replication machinery may have evolved twice independently (Leipe *et al.*, 2001); we will discuss these matters in more detail in Chapter 12. DNA ligase, however, is not as good a confirmation of this theory as was once thought. When the sensitive and specific PSI-BLAST program became available, it was shown that iterative searches with probabilistic models of both families indicate that the main catalytic (nucleotidyltransferase) domains of NAD-dependent and ATP-dependent DNA ligases display statistically significant sequence similarities (Aravind and Koonin, 1999), not obvious in pairwise sequence comparisons. The three-dimensional structure of a representative ATP-dependent ligase was known at the time, and it was clear that the short sequence motifs conserved between NAD-dependent and ATP-dependent ligases correspond to well-defined elements of secondary structure in the latter family, suggesting similar fold in two classes of ligases and mechanistic parallels in reaction mechanism (i.e., transfer of nucleoside monophosphate group onto the 5′ end of the ligated DNA fragment). At approximately the same time, the three-dimensional structure of the catalytic domain of NAD-dependent ligase was published, and its structure turned out to be very close to that of the ATP-dependent enzyme, just as predicted by Aravind and Koonin. Curiously, the authors of that study unequivocally stated that there was no way to notice that similarity by sequence analysis alone, without seeing the three-dimensional structures.

In a more recent example, we (Liu and Mushegian, 2004) and others (Cheng *et al.*, 2004) studied a protease that cleaves capsid protein of DNA bacteriophages. This cleavage occurs simultaneously with the capsid assembly and is important for correct formation of phage head, in which genomic DNA is packaged. The protease is encoded by phage genomes but is not found in bacterial genomes, except for integrated prophages. Database searches, using sensitive probabilistic approaches, showed that phage proteases are related to better studied proteases encoded by herpesviruses of eukaryotes. This is of particular interest because the capsid formation in herpesviruses is mechanistically quite similar to phage capsid assembly. Moreover, formation of capsid and DNA packaging into it in both virus groups requires terminase—the enzyme with ATPase activity that is orthologous in herpesviruses and in phages but has only paralogs in other genomes. Added to all this evidence, the evolutionary relationship between phage and herpesvirus head protease helps to build a stronger case for common ancestry of capsids in viruses of bacteria and eukaryotes.

Proteins with Superficially Different Functions

We have two or more proteins with diverse functions, and use sequence comparison to establish evolutionary connections between them. When the evolutionary relationship is established, the common molecular details of different functions may come to light. For example, years ago we noticed that MutL, a protein that is involved in mismatch repair in bacteria and has orthologs in humans that are mutated in many patients with colon cancer, displays sequence similarity to a large family of bacterial signal transduction histidine kinases. More detailed analysis with PSI-BLAST proved that two additional groups of proteins, chaperones of the HSP90 family and one family of topoisomerases, also had a related sequence domain (Mushegian *et al.*, 1997). Although four classes of proteins may seem to have different functions (mismatch repair in MutL, signal transduction in histidine kinases, protein folding and other chaperone-like activities in HSP90, and DNA unwinding and rewinding in topoisomerase I), in fact they all bind and hydrolyze ATP. Indeed, the wealth of biochemical, pharmacological, and structural evidence indicated that the conserved region corresponded to the ATPase domain. At the time of these observations of sequence similarity, only one representative structure, that of topoisomerase, was known. However, several months later, structures of the sequences from the other three groups were published. Again, structural biologists were explicit in their conclusion that there was no way to make the connection at the sequence level without seeing the structure (Stebbins *et al.*, 1997; Bilwes *et al.*, 1999). Granted, at the time, the sequence similarity was not very easy to observe; it required new methods and careful analysis of all sequence similarities, not only those between proteins with the known structure. But I think that this also illustrates lack of interest in sequence analysis among many structural biologists. One team of crystallographers, however, was not so oblivious and noted that their structure determination confirmed our prediction (Ban and Yang, 1998). The landscape of structural biology, however, changes very fast; now, 10 years later, analysis of sequence families and superfamilies is an accepted, necessary prerequisite to structure determination, at least in the high-throughput, structural genomics approaches (discussed further in Chapters 9 and 10).

Completely Uncharacterized Proteins

This is, obviously, one of the most important applications of sequence similarity in the era of complete genomes: We establish sequence similarity, infer homology, and use information about well-studied homologs to infer the functions of the uncharacterized ones. Enough said; most chapters in this book deal with this matter in one way or another.

The enterprise of finding remote sequence similarities between proteins and making biological inferences from these similarities has been a resounding success. Genomic biology is shaped by these approaches, and they are possible because we became exceedingly sophisticated in detecting evolutionary signal in protein sequences, despite high divergence in many protein families.

One may wonder how far our ability to infer the presence of a common ancestor goes, and how far divergent evolution goes in proteins. Can it be that all proteins have a common ancestor, and can we hope to determine what it was? Or, perhaps, if this cenancestor did not exist, or if it is intractable, have there been a relatively small number of ancestral proteins that we can track down? For example, can it be that each discrete molecular function has a common ancestor?

The rest of this chapter deals with the answer to this latter question. The answer, by and large, appears to be "no"; indeed, the following can be stated as the "second fact of comparative genomics":

> *A molecular function does not require homologous genes; one and the same function can be performed by several different gene products, which give no evidence of their common ancestry.*

As discussed in Chapter 3, similarity in the absence of common ancestry is called analogy. So, the other way of stating the second fact is to say that functional analogy at the molecular level does exist. It is still not known how common functional analogy is, but the diversity of examples that I discuss later seems to indicate that it is common enough—in fact, too common to be ignored. Analogy of form and function, of course, has been discussed in evolutionary literature for several centuries. But I believe that study of analogy at the molecular level is of particular importance for understanding of the other levels of organization of living matter. And I will argue that it is only in the context of complete genomes that analogy at the molecular level can be studied in a definitive way.

Often, analogy is said to be the exact opposite of homology. But looking closer, we see that this is not the case: Homology is the relationship by descent from a common ancestor, whereas analogy is not just the absence of a common ancestor but, rather, *similarity that exists despite the lack of such an ancestor*. Thus, two characters may be homologous even if they are not similar—for example, when similarity between them becomes so low in the course of divergent evolution that it can no longer be distinguished from random background. However, two characters may be called analogous only if they have some kind of similarity in the first place. Thus, homology and analogy, when applied to characters that change in evolution, can be treated as opposite hypotheses only after similarity between the characters has already been established.

When two characters are analogous, sometimes it is said that they have evolved by convergence. But many authors, from morphologists of old to our contemporary Walter Fitch, who discussed molecular homology and analogy in an important paper (Fitch, 2000), noted that there are actually two distinct evolutionary situations. In one of them, the characters in the past were less similar than they are now, and in the other, the degree of similarity did not change. The first scenario can be called convergence, but the second should not be—as Fitch stated, "Why call convergence what fails to converge?"—and may be called parallelism instead. Also note that the same analogous trait may have history of both convergence and parallelism. Here, as almost always in biology, it is important to decide what kind of evolutionary times are of interest to us: It may be analogous and parallel at a more recent time horizon, but it may be analogous and convergent if there has been a more ancient convergence that then stopped. I use "analogy" whenever possible and use "convergence" where the case can be made that the ancestral traits were indeed less similar to each other than the contemporary ones.

Thus, what can be analogous at the molecular level? A short paper by Doolittle (1994) is illuminating. The beginning of that article states,

> *One of the most frequently asked questions after any lecture on the phylogenetic analysis of amino acid sequences is "What about convergence?" ... The term "convergence" is used in many different contexts, however, and much confusion can occur when the subject is raised. As in all matters, a little care taken to define just what is meant can eliminate needless controversy.*

Replacing "convergence" with "analogy," but otherwise following the logic of Doolittle's paper, we can see several types of analogy.

Analogous Function

The function of two proteins is the same, but the proteins themselves are not similar in sequence or in three-dimensional structure. Doolittle discussed hydrolysis of peptide bond: This function can be performed by cysteine proteases, serine proteases, aspartyl proteases, and metalloproteases (threonine proteases, which were not characterized until later, should be added to the list; Lowe et al., 1995). Many of these proteases are completely unrelated to each other; even serine proteases are not monolithic, and they are thought to have emerged several (at least three) times. Proteases as a whole may be too broad a class to expect homology: Even much lower levels of their functional hierarchy contain analogous enzymes. But consider a relatively narrow functional class of signal peptidases—for example, proteases that remove and/or degrade a leader peptide of secreted proteins as they leave the cytoplasm. In this functional class, at least four activities are known from unrelated clans of metalloproteases, aspartic proteases, and two clans of serine proteases unrelated to the first two and only distantly related to each other (Rawlings et al., 2006).

Analogous Mechanism

Spatial arrangement of a small number of functionally important amino acids is similar in two proteins, but the proteins themselves are not similar in sequence and in three-dimensional structure. Doolittle discussed the "catalytic triads"—that is, similarly arranged triplets of residues in the active centers of two structurally different proteases, chymotrypsin and subtilisin. Catalytic triads of similar configuration, usually containing at least one histidine and often also a D/E/N residue in one position and S/T residue in another position, are found in a variety of proteases and also in other hydrolases. The arrangement is so distinctive that a hydrolase activity can be predicted for a protein when nothing is known about it except its sequence and three-dimensional structure. However, there is no evidence that all such proteins are homologous, and dissimilarities in both sequence and structure are in some cases significant enough to conclude that they most likely are not (Fig. 6.1).

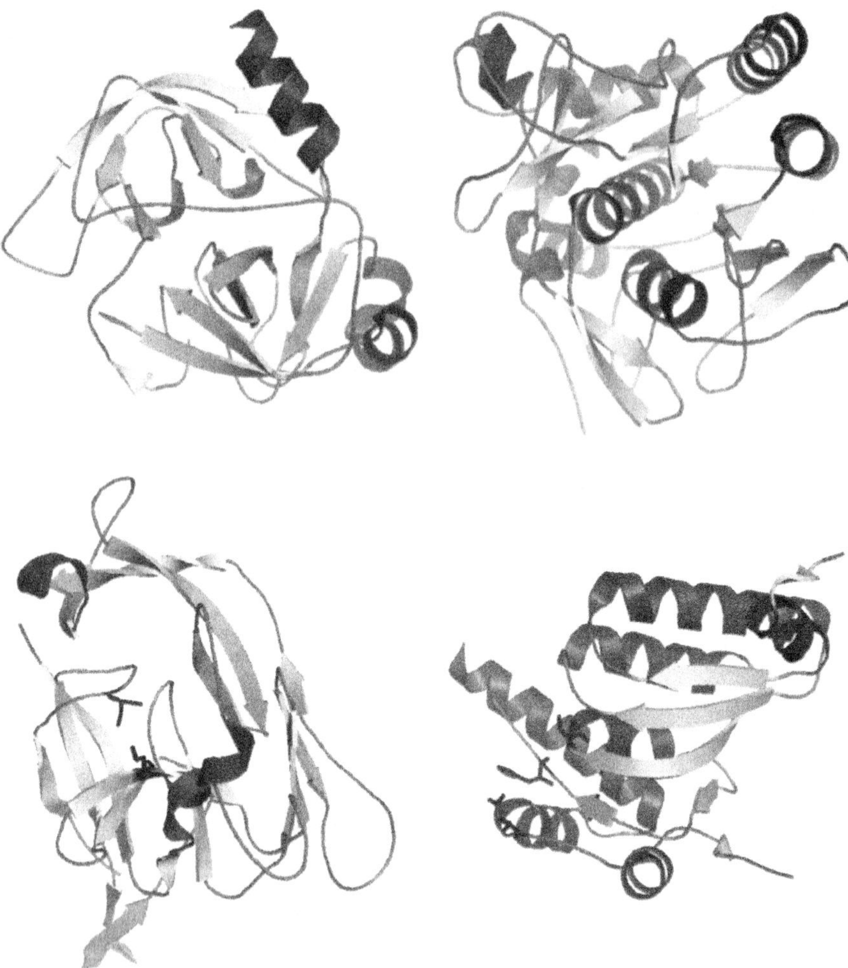

Figure 6.1. Mechanistic convergence in serine proteases. (**Top left**) Chymotrypsin (PDB 1GG6). (**Top right**) Subtilisin (PDB 1R0R). (**Bottom**) Side views of the same molecules, indicating the positions of catalytic residues (*black lines*). All images of protein structures in this book, except for Fig. 9.2, were produced using PyMOL (DeLano, 2002).

Another example of the same type from Doolittle's work is the molecular setup used for binding the phosphate group of ATP by two different superfamilies of kinases (Fig. 6.2). Most known small-molecule kinases have Rossmann fold. An important determinant of catalysis is the anion hole, which interacts with the gamma phosphate and in these kinases is made of a short, contiguous in sequence, known as glycine-rich loop, which contains a lysine residue (the GKS/T signature, also found in the ATP-binding P loops of helicases and other NTPases, some of which were mentioned in Chapter 4). On the other hand, kinases of the serine/threonine/tyrosine/lipid kinase superfamily have a different fold, called ATP-grasp, which also contains an anion hole made of glycine loop and lysine. In this case, however, glycine-rich loop and lysine are brought together from the different parts of the sequence—the loop from the N terminus and lysine from the middle (Fig. 6.2).

Analogous Structure

The high-level spatial structure of two proteins is the same, but there is no sequence similarity. This has to be elaborated further. On the one hand, it is commonly asserted that sequences

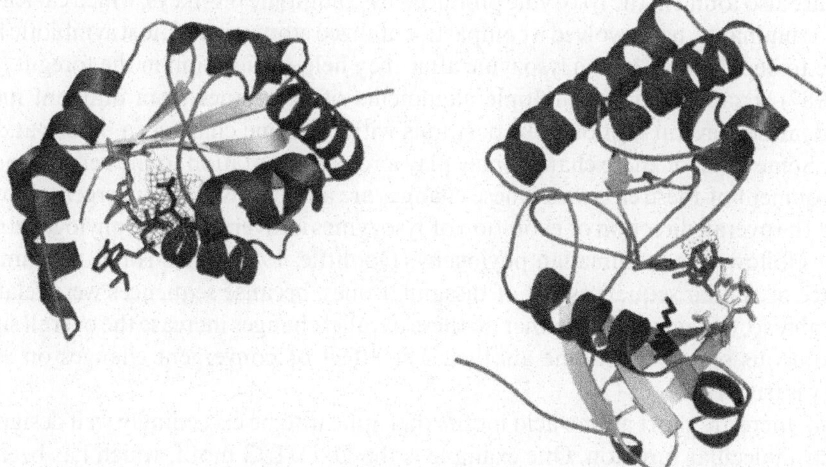

Figure 6.2. Mechanistic convergence in active centers of different kinases. (**Left**) Rossmann-fold kinase with Walker-type ATP-binding loop (gluconate kinase, PDB 1KOF). The main chain and the side chains of residues forming the loop are shown as *black lines*. In this case, catalytic lysine is part of the loop. (**Right**) ATP-grasp kinase (cAMP-dependent kinase catalytic subunit, PDB 1CDK). The ATP-interacting glycine-rich loop is shown as *black lines*. In this case, the lysine residue interacting with ATP is located in a different part of the sequence, brought close to the loop and to the ATP molecule by packing of strands in the beta sheet.

evolve faster than structures: Two sequences may diverge from the common ancestor to the point of random sequence similarity while retaining similar three-dimensional structures. On the other hand, it is widely accepted (and is probably true) that certain folds may be close to optimal, in the sense that they afford certain thermodynamically favored properties to proteins that adopt them. For example, dense packing of side chains, minimization of hydrophobicity on the molecule surface and its maximization in the interior of the molecule, and other factors minimize free energy of the polypeptide. If such folds are vastly more stable than the others (or, as sometimes is said, they are more designable), then perhaps they represent attractors in the space of structural evolution, and different sequences can converge into such optimal structures. It is clear that only the second case represents structurally analogous proteins, whereas in the first case the proteins are homologous, even though they are not similar.

Are there any known cases of structural convergence? To prove convergence, one needs to prove the absence of divergence. This is difficult, as discussed in more detail later and also in Chapters 9 and 10, in which we again discuss the interplay of sequence and structure evolution.

These cases of analogy at the molecular level presented by Doolittle have the following in common: They are examples of protein properties that, however similar, nevertheless do not indicate common evolutionary ancestry. It is remarkable that none of these types of analogy at the molecular level involves the analogous origin of long amino acid sequences. After several decades of sequence comparisons, there is not much evidence that two sufficiently long stretches of sequence can converge to statistically significant similarity.

The story is different when only a small number of residues are involved. Here, I know of at least two types of credible cases in which such local sequence convergence may be claimed.

First, there are cases of convergent changes inside homologous families. For example, Doolittle reviewed the classic work from Allan Wilson's lab at University of California-Berkeley showing that some of the amino acid changes observed in the lysozymes of ruminant animals are also found in the lysozyme produced by columbine monkeys, which eat leaves and, similar to ruminants, have evolved a compartmentalized stomach and host symbiotic bacteria that have to be killed off by the lysozyme after they helped digestion in the foregut (Stewart *et al.*, 1987). Examination of multiple alignments of lysozymes from different mammals indeed identifies several homologous positions with the same changes in bovids and langur monkey. Some or all of these changes may play a role in adaptation to the very low acidity in the environment of these enzymes. These changes are analogous and convergent. Note, however, that the overall direction of evolution of lysozymes is divergent: The phylogenetic tree of lysozymes follows the mammalian phylogeny (Doolittle, 1994). This is not the same as the emergence of similar sequences out of dissimilar ones, because sequences were related, and recognizably so, to begin with. Neither do these parallel changes increase the overall similarity of homologous sequences to one another. The effect of convergent changes on sequence similarity is trivial.

Second, there are short amino acid motifs that appear to be exceedingly well designed for a particular molecular function. One example is the DxDxDG motif, which has been found, with variations, in many proteins that tightly bind a calcium ion (Ridgen and Galperin, 2004). The best known structural context of this motif is a loop between two helices; an example of this structure is the EF-hand in such regulatory Ca^{2+} binding proteins as the best studied representative, calmodulin. Interestingly, in many Ca^{2+} binding proteins one or both of the helices are replaced either by a beta strand or by an unstructured region. In these cases, it is possible that the motif could have been inserted into different proteins by recombinational transfer of a short DNA fragment. At the same time, the motif is simple enough to have a nonnegligible chance of independent origin by mutation in many structural contexts. Similarly, a simple CxxS signature has a fairly specific role: It acts as a strong redox equivalent in almost all the proteins in which it is found, even though these proteins have quite different sequences and folds (Fomenko and Gladyshev, 2002). In this case, again, it is quite possible that the motif has evolved more than once, but recombinational transfer of this sequence between genes cannot be excluded either.

To conclude, analogous proteins may display parallelism or convergence in general or molecular function and in structure, either at a large scale or in a similar arrangement of a few crucial residues. In all these cases, however, the sequences are dissimilar overall, except for, perhaps, extremely short sequence motifs. In many of these cases, the three-dimensional structures of the analogous proteins are not similar. Moreover, if recombination plays a role in dissemination of short motifs, then at least DNA segments encoding such motifs are homologous by definition.

All this may sound trivial to those familiar with comparative anatomy and other areas of traditional biology curriculum. Similarities between morphological, anatomical, and physiological traits of different organisms have been known to scientists for a long time. Fins of fishes and finlike limbs of marine mammals; wings of various flying vertebrates, both extant and extinct; eyes of mammals and of cephalopod mollusks; and spikes covering stems of various higher plants—evolution of these and many other groups of organs that look and function similarly have been quite thoroughly studied. We cannot say that analogy has not been examined before.

The problem, however, is that many such examples of morphological convergence left open the question of fundamental mechanisms that are used to produce similar biological structures. Consider the popular example of tetrapod limbs. Wings of birds and bats perform similar functions and look similar in many ways, although, of course, there are also significant morphological and anatomical differences. Quite clearly, they have evolved convergently from the limbs of separate, nonflying ancestors. However, at the beginning, there was a limb of a primitive tetrapod, from which both birds and mammals have descended. Thus, as with lysozymes of cattle and langurs, there is an increase in similarity of characters that, in the first place, had evolved divergently from the common ancestor. Thus, the convergence in this case may not extend to the level of individual genes. Rather, the molecular mechanism of morphological covergence may be in the repeated activation of the same genetic program. If this is the case, winglike limbs may be morphologically analogous but genetically homologous. Possibilities such as this are frequently raised in the literature, and there is even an extreme opinion that most, if not all, morphological analogies between different organisms are the results of retention and reactivation of fundamentally homologous, once-produced genetic programs (Meyer, 1999).

Another morphological example, this time from botany, may argue that retention/reactivation of homologous genes is not sufficient to explain all cases of morphological analogy. Consider sharp, woody, needle-like spikes on the branches of taxonomically diverse plants. They include thorns, as in honey locust (*Fabaceae*); spines, as in barberry (*Berberidaceae*); and prickles, as in blackberry (*Rosaceae*). All these unpleasant spikes are morphologically very similar, and sometimes it is not easy to determine, by naked eye, to which type they belong. Although the genetic control of formation of spikes of any type is not well understood, they are all anatomically and developmentally different: Thorns are modified stems, spines are modified leaves, and prickles are the outgrowths of the stem's epidermis, bark, and some parenchyma tissues. All this becomes obvious at the histological and ultrastructural levels. It is difficult to believe that all genes controlling the development of thorns are exactly the same as those involved in formation of prickles: The latter are rather simple, whereas the former retain vascular tissue and can branch.

In recent years, comparison of genes involved in evolution of animal eyes has provided more indications of the interplay of divergence and convergence at the molecular level. Many fascinating details aside, there are essentially eight distinct optical solutions to seeing (Fernald, 2000, 2004; Arendt and Wittbrodt, 2001; Arendt, 2003). Several known components required for building all eye types are *Pax-6*, a transcription factor that serves as a major developmental switch starting the eye development pathway; opsins, the apoproteins of visual pigments; and crystallins, which are proteins packed in an orderly fashion with a refractory gradient in a specialized part of the eye (lens in vertebrates or various isofunctional organs in invertebrates). The evolutionary histories of these protein components of eye are all different. *Pax-6* belong to the superfamily of homeodomain genes, which appear to be eukaryote specific, having only distantly related homologs in bacteria and archaea (i.e., helix–turn–helix DNA-binding domains). *Pax-6* is orthologous in all animals. Its emergence in animal genomes predates eye formation: Nematodes and corals have no eyes but they have *Pax-6* orthologs, which may play a role in the formation of anterior end and sensory organs. Once recruited to control eye formation, however, *Pax-6* retained this function in all eyed invertebrates, as well as in vertebrates, albeit with modifications. Opsins also predate the origin of metazoa: Orthologs of opsins are found in most divisions of life, including Bacteria, Archaea, and Fungi (Zhai *et al.*, 2001; de la Torre *et al.*, 2003; Terakita, 2005). In different animals, there are different numbers of opsins, and their evolution included many gains and losses of paralogs. By and large, opsins of vertebrate eyes are paralogous to eye opsins of insects and nematodes (Terakita, 2005). Some of these genes have undergone mutations and selection and were tuned toward absorbing light of different wavelengths. The overall evolutionary trend here is divergent, with some parallel changes.

The situation with crystallins, however, is dramatically different. "Crystallins" are not a protein family. Major refractory proteins in vertebrates belong to the heat shock 20 family; in birds and crocodiles, the main crystallin is the enzyme lactate dehydrogenase; in cephalopods, crystallin is glutathione *S*-transferase; and in fruit fly, it is a unique protein, which consists mostly of long nonglobular regions and is not similar to any enzyme. All these classes of proteins are unrelated in sequence and structure, but nonetheless they are functionally analogous, playing the same role of refracting and focusing light in lens or in its structural analog. Also different are the signal transduction pathways between photopigment and the membrane, as well as the membrane polarization status. In rhabdomeric eyes of insects, G protein activates phospholipase C that produces inositol triphosphate, which ultimately causes a spike in membrane potential followed by depolarization. On the other hand, in ciliary eyes of vertebrates, G proteins signal via cyclic GMP phosphodiesterase, and the membrane potential initially declines and then is restored by hyperpolarization.

Thus, in the case of eye composition and development, the "null hypothesis," in which similar structures and functions are determined by homologous molecular components, is rejected. The alternative is the second fact of comparative genomics, which states that similarities in biological systems do not require homologous genes; instead, biological similarities are enabled by a combination of analogous and homologous genes.

The stage for the discovery of this interplay of analogy and homology had been set before the genomic era. Most microbiologists who were studying biosynthetic pathways in bacteria worked with one or two model organisms, such as *Escherichia coli* and *Bacillus subtilis*. Often, it was presumed that at least central biochemical pathways, such as biosynthesis of amino acids or coenzymes, are the same everywhere. But a handful of scientists were interested in comparing pathways in multiple species, and evidence of significant biochemical variation in properties of individual enzymes and in the layouts of the whole pathways has been accumulating. In 1976, Roy Jensen (then at the University of Buffalo, and currently at the University of Florida) presciently noted that "the unity theme for biochemical pathways may have been overemphasized in the literature" and "in molecular biology, the remarkable productivity of experimental systems such as *E. coli* and *B. subtilis* has tended to distort the generalized image of microbial characteristics." Also, Carl Woese and co-authors (Olsen *et al.*, 1994) stated with sarcasm, "To understand prokaryotes, we had only to determine how *Escherichia coli* differs from the eukaryotes. This was no invitation to creative thought, no unifying biological principle."

Only with the complete genome sequences could the interplay of unity and diversity in the genetic makeup of biochemical pathways be studied in quantitative detail. As far as I know, one of the first definitive analyses of this sort was performed by Eugene Koonin's group at the National Center for Biotechnology Information as part of reconstruction of minimal genome, which is discussed in more detail in Chapter 13. When the complete sets of genes in two species are known, one can define all pairs of orthologs in two genomes. What results from such an enumeration is evidence that although some functions are performed by orthologous genes in two species, this is not the case for every function: Sometimes, one and the same function is performed by nonhomologous genes. We called the lack of orthologous relationship in two isofunctional proteins "nonorthologous gene displacement" (Mushegian and Koonin, 1996a; Koonin *et al.*, 1996). I will now discuss several themes relevant to our understanding of the molecular level of this phenomenon, which I rename here displacement of orthologous genes (DOGs). One reason for this name change is that the new name is more conducive to puns; the other reason is that orthologous/xenologous gene displacements also exist and are of interest, and the new term would be applicable here without confusion.

Because most of the completely sequenced genomes belong to microorganisms, which are endowed with rich intermediate metabolism but have somewhat limited repertoire of other functions, in the rest of this chapter I mostly give examples that have to do with the enzymes that make up metabolic pathways and, to a lesser extent, with the proteins involved in genome replication and expression. The notion of analogy at the molecular level, however, remains valid for structural proteins, signal transduction circuits, and gene products involved in other classes of biological processes.

Mechanisms of DOGs

Suppose we have two species and two isofunctional but nonorthologous proteins, one in each species. How this could have come to be? There are only two possibilities: Either there existed an ancestor of two species in which both these proteins were present, or there has never been such an ancestor (Fig. 6.3). If such an ancestor did exist, then the most plausible mechanism of a DOG is differential gene loss. The common ancestor, of course, cannot be observed directly, but the support of the differential loss scenario comes from the present-day genomes: In many of them, especially in relatively large genomes of some free-living species, there coexist two dissimilar gene products able to perform the same function. The first example we consider is from the "bottom" or "triose" part of glycolysis (Fig. 6.4). This set of reactions is considered very ancient and is found in nearly every species. Most enzymes in this pathway are orthologous everywhere, but phosphoglycerate mutase, converting glycerol 2-phosphate and glycerol 3-phosphate, is an exception. There are at least two families of enzymes with phosphoglycerate mutase activity. One, 2,3-bisphosphoglycerate-independent phosphoglycerate mutase (the product of the *yibO* gene in *E. coli*), belongs to the alkaline phosphatase superfamily. The fold of this family consists of two three-layered cores with eight-stranded beta sheets (Fig. 6.5). The other, cofactor-dependent enzyme (GpmA and its paralog GpmB in *E. coli*), is a member of the acid phosphatase superfamily. This fold consists of a single core containing a six-stranded beta sheet of a different connectivity. *Escherichia coli* and other large genomes from the gamma subdivision of proteobacteria tend to have both types of these enzymes. However, diverse parasitic proteobacteria with small genomes, and also free-living proteobacteria with large genomes that belong to the alpha subdivision, contain just one of the two types. If we compare, for example, *Haemophilus influenzae* (gammaproteobacteria) and *Helicobacter pylori* (epsilonproteobacteria), we can see that phosphoglycerate mutase function in these two species is a DOG. If we assume that a nonparasitic proteobacteria is a more adequate model of the ancestral proteobacterial state than parasites *Haemophilus* and *Helicobacter*, and given that glycolysis is an evolutionary invention clearly preceding proteobacterial divergence, it is quite likely that this DOG was produced by differential gene loss. Moreover, in archaea, a distinct version of a cofactor-independent enzyme appears to have displaced other enzymes.

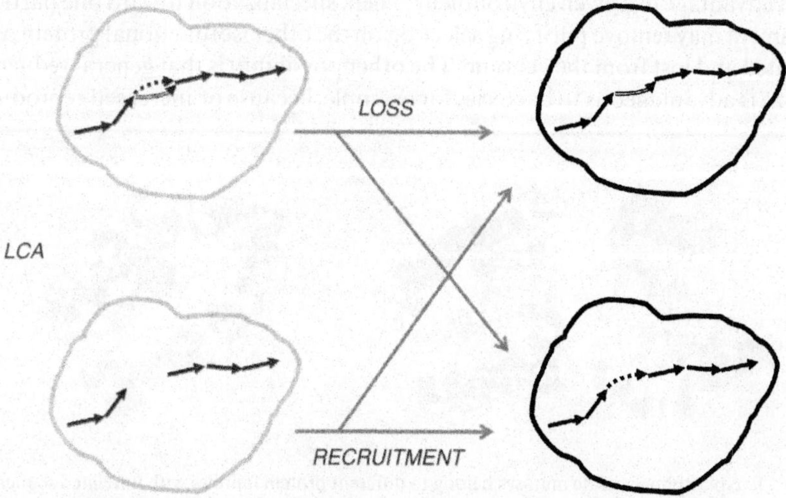

Figure 6.3. Two mechanisms of a DOG: differential loss and differential recruitment of isofunctional genes. LCA, last common ancestor.

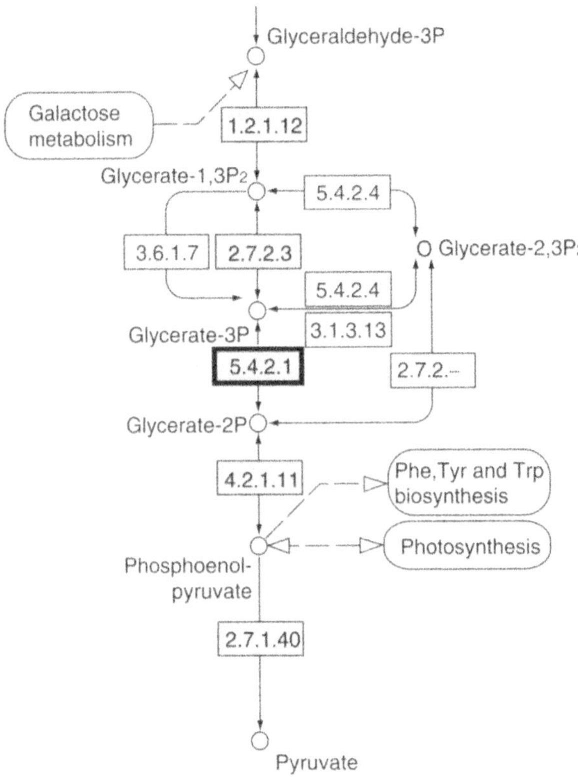

Figure 6.4. The "bottom" (triose) part of the glycolytic pathway. Phosphoglycerate mutase reaction that is prone to frequent gene displacement is indicated by the box outlined in boldface type. From *Kyoto Encyclopedia of Genes and Genomes* Web site (www.genome.jp/kegg; accessed May 15, 2006; Copyright 1995–2006, Kanehisa Laboratories).

There may be several forces facilitating differential gene loss. For example, two isofunctional proteins may be differently regulated, and it is possible that each mode of regulation offers a significant advantage in a given environment. Then, specialization toward one particular kind of environment may remove purifying selection on the other isofunctional protein, which may be inactivated and lost from the genome. The other possibility is that general reduction of the genome size is advantageous to bacteria, for example, because of increased reproduction rate

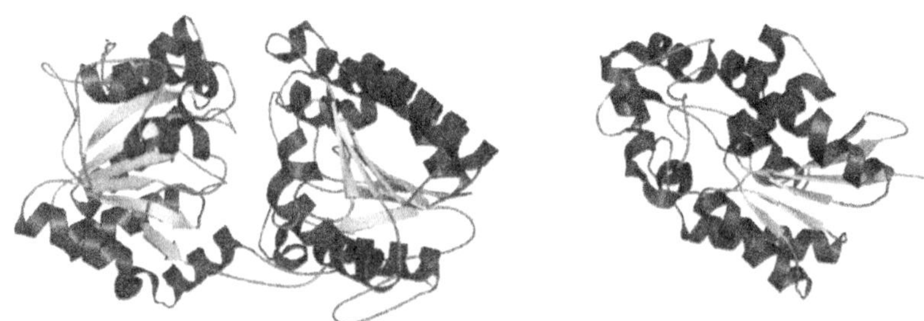

Figure 6.5. Two phosphoglycerate mutases belong to different protein families with unrelated sequences and distinct spatial folds. (**Left**) Cofactor-independent enzyme (PDB 1EQJ) from the alkaline phosphatase-like superfamily. (**Right**) Cofactor-dependent enzyme (PDB 1E58) from the phosphoglycerate mutase/acid phosphatase-like superfamily.

(Mira *et al.*, 2001) or because of decreased cost of regulation (Ranea *et al.*, 2005), and this deletion drive results in differential elimination of two phosphoglycerate mutates, which can be affordable if the organism spends at least part of its life cycle in rich medium (as is apparently the case in symbionts and parasites, in which typically most nutrients are abundant, although some, such as iron ions in human parasites, are limiting). Finally, a role in differential gene loss may be played by random DNA rearrangements. The relative contribution of these factors into gene displacement needs to be better understood.

The other mechanism of a DOG is independent recruitment of genes. This mechanism does not require coexistence of two isofunctional genes in any one genome at any time. Suppose that a metabolite in a cell may be converted into another metabolite in a single chemical reaction, but the appropriate activity was not available in the last common ancestor of the two species. After divergence of the two lineages, different genes are recruited to provide the "missing link" (see Fig. 6.3). Consider β-lactam antibiotics, resistance to which in many pathogenic bacteria became an important medical problem. Bacterial strains that are resistant to these antibiotics overexpress enzymes with β-lactamase activity that belong to at least two completely unrelated protein families, namely metal-dependent and serine-dependent enzymes. Both classes have multiple paralogs in most completely sequenced genomes, which possess a wide variety of hydrolase activities. It is likely that lactamases of each class have been separately recruited from different pools of hydrolases to fight the presence of a deadly antibiotic in the environment.

Independent recruitment also appears to mark certain crucial evolutionary events. One such event must have been the emergence of processive DNA polymerases. These main enzymes of genome replication appear to have been mobilized on two independent occasions, once in Bacteria and once in the Archaeal/Eukaryal lineage. The recruitment was from two different nucleotidyltransferase families: in Archaea/Eukaryota, the polymerase beta family with ferredoxin-like fold, and in bacteria, a unique class of enzymes with a large fold that contains pol-beta core but has no sequence similarity to archaeal counterpart (Leipe *et al.*, 1999; Lamers *et al.*, 2006).

Several cases of independent recruitment may be hypothesized for the tricarboxylic acid (TCA) cycle. It has been argued that the ancestral pathway was noncyclic and consisted of two independent pathways from pyruvate; "oxidative branch" may have lead to α-ketoglutarate, and the "reductive branch" to succinyl-CoA (Fig. 6.6). Two reactions are required to link these branches into a cycle, to produce the pathway as it has been described in textbooks. Both activities needed to complete the cycle exist in contemporary organisms as two nonhomologous versions, namely 2-ketoglutarate dehydrogenase/2-ketoglutarate oxidoreductase and succinyl-CoA synthase/succinyl-CoA-acetoacetate-CoA transferase. So it is possible that the two branches have been joined in the TCA cycle more than once in the evolution, using different enzymes for this purpose (Huynen *et al.*, 1999).

From where are the new genes recruited? One possibility is that a new gene is acquired from another organism by horizontal transfer. There are many molecular mechanisms of gene exchange between species, some requiring a specialized vector (virus, plasmid, or another extrachromosomal element) and others relying mostly on the cellular systems of export and uptake of macromolecules. These mechanisms have been reviewed, for example, in Bushman (2001) and are not discussed here (but see Chapter 11 for more on the role of horizontal transfer in evolution).

Access to genes from other organisms, however helpful for dissemination of existing genes, does not solve the question of how new genes come into being. It is likely that the major way of recruiting genes into new functions (and, simultaneously, to produce novel genes) is to use a copy of a duplicated gene from the same genome. Acquisition of new function in a duplicated gene most likely occurs by way of broadening specificity of a protein; results indicate that such

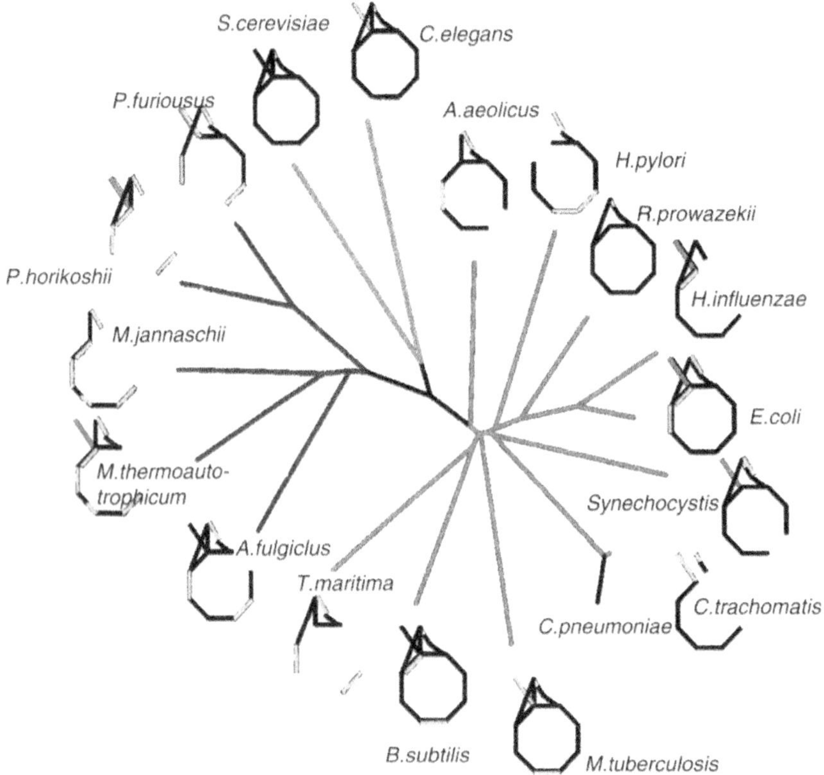

Figure 6.6. "Incomplete" or rearranged citric (tricarboxylic) acids cycle in different species. The graph includes connections with pyruvate and phosphoenolpyruvate as well as glyoxylate shunt. Nodes of the graph represent metabolites, and edges represent proteins that catalyze interconversions of these metabolites. When two nodes are connected by two edges of different shades in the same species, this means that there are two enzymes with such activity in this species; other species may have just one of the two, or none. The species' phylogenetic tree in the center is built using the distances derived from the fraction of genes shared by each pair of species. It is notable that the resulting phylogeny is very close to consensus phylogeny produced on the basis of many different characters (Wolf *et al.*, 2001a), suggesting that evolution of the TCA cycle and perhaps of other pathways proceeds by gradual gain and loss of genes. Reprinted from *Trends Microbiol.*, 7, Huynen, M., Dandekar, T., and Bork, P., Variation and evolution of the citric-acid cycle: A genomic perspective, pp. 281–291, Copyright (1999), with permission from Elsevier.

broadening may be easy to achieve by means of only a few point mutations (Aharoni *et al.*, 2005).

Thus, gene displacement can occur in many ways. Selection that chooses one gene with a given function among two or more, whereby a "displacer" gene eliminates a "displacee," is just one extreme case. More generally, displacement is a result of differential elimination or differential recruitment of two genes in two lineages, and in either case, the forces at play may be random (gene drift by gain, loss, and duplication) or nonrandom (selection of genes or gene variants for useful properties).

DOGs and the Types of Homology

When an ortholog of a gene in one species is displaced by an isofunctional gene in another species, what are the homology relationships of these two genes? In the case of phosphoglycerate mutase, the mutually displacing genes are evolutionarily unrelated at the sequence level, and, evidently, also at the level of three-dimensional structure of the protein products. Similar gene function in the absence of common ancestry is, by definition, analogy. In other cases, mutually displacing genes may share a common ancestral gene. Often, a DOG is between paralogous genes: Two paralogs are differentially recruited, or differentially eliminated, to end up doing the same function in two different lineages. A case in point is the helicase involved in replication of genomic DNA. In yeast, the main replicative helicase A (YKL017c) belongs to the helicase superfamily I, whereas in bacteria the main replicative helicase DnaB is thought to be specifically related to DNA-annealing ATPases from the RecA/Rad51 family. Helicases and RecA enzymes are distantly related, as demonstrated by sequence comparison and by the close similarity of the spatial folds, and the last universal common ancestor of present-day organisms is thought to have contained both a helicase and a RecA-like enzyme (Aravind *et al.*, 1999; see Chapter 13). Thus, helicases and RecA proteins are paralogs, and the relationship between replicative helicases in yeast and bacteria can be described as a paralogous DOG.

Yet another type of DOG is a displacement by an ortholog from a different species, after it has been introduced into the genome by horizontal gene transfer. This apparently has happened multiple times in the history of essential genes involved in translation. For example, several dozen DOGs can be discerned in the evolution of aminoacyl-tRNA synthetases, most dramatically in spirochetes, which appear to have at least six of their aminoacyl-tRNA synthetases displaced by orthologs from eukaryotes (in addition to other, more ancient displacements; Wolf *et al.*, 1999). Essential and highly conserved ribosomal protein S14 has also been transferred in this way (Brochier *et al.*, 2000). This is an orthologous/xenologous DOG.

Thus, based on the relationships of two isofunctional genes, a DOG can be analogous, paralogous, or orthologous/xenologous. If each of these relationships is combined with the two described mechanisms of a DOG (i.e., differential elimination and independent recruitment), there will be six distinct types of DOGs.

Interestingly, the ragworm *Nereis* (*Polychaeta*) has been shown to contain two types of photoreceptor cells, one with ciliate and the other with rhabdomeric-type receptors. Just like *E. coli* with its two types of phosphoglycerate mutases, ragworm with its two types of photoreceptors may be a model of the common ancestor of bilateral animals. One current hypothesis (Arendt *et al.*, 2002) is that each type was specialized—one involved in circadian clock and another in phototaxis—and in two lineages, different types of receptors were recruited to serve in the main visual organ. Thus, the majority of the pathway in insects and vertebrates may be a paralogous displacement, but several components, such as crystallins and the enzymes downstream of G protein-coupled receptors, are the result of additional, analogous DOGs.

Relationships of genes under a DOG are commutative but not transitive. If the same function in species *A* and *B* is performed by genes related by analogous displacement, and the same is true of species *B* and *C*, this gives no information about the relationship of functions in *A* and *C*: There may have been no DOG at all between these species.

DOGs and Horizontal Gene Transfers

Horizontally transferred genes can be recruited to perform a new function, or they can replace an old gene that had the same function. However, horizontal gene transfer (HGT) is not a sufficient condition of a DOG: A transfer may result in the coexistence of two isofunctional genes rather than elimination of one of them. As already mentioned, most genomes, prokaryotic and eukaryotic alike, contain only one of the two known types of phosphoglycerate mutase; some bacteria, such as *E. coli*, and rare metazoans, such as sea urchin, have both types. HGT is also not a necessary condition of a DOG. Differential recruitment of genes, resulting in a DOG, may occur without any intergenomic gene transfer—the recruitment is from within the same genome, after gene duplication/divergence or by way of gene polyfunctionality. For example, synthesis of mevalonate pyrophosphate, performed by phosphomevalonate kinase (PMK), is an essential step in the mevalonate pathway, which is used for isoprenoid biosynthesis by fungi, animals, plant chloroplasts, archaea, and a few bacteria. PMK typical of fungi, plants, and some bacteria, however, is not found in archaea. But side chains of archaeal lipids are made from isoprenoids, and many other enzymes in the mevalonate pathway are found in archaea; therefore, it has been suggested that PMK is displaced by another protein in archaea. Animals also have a mevalonate pathway but lack PMK of the fungal type. PMK in animals is displaced by an analogous enzyme, which does not have orthologs in Archaea (Smit and Mushegian, 2000). This, however, does not tell us which protein phosphorylates phosphomevalonate in archaea. The archaeal homolog of the preceding enzyme in the same pathway, mevalonate kinase, which is paralogous to PMK and is also found in fungi, plants, animals, and bacteria, has been shown to possess PMK activity *in vitro* (A. Osterman, personal communication). It is possible that, whereas in other organisms it is specialized for only mevalonate phosphorylation, in archaea it performs both phosphorylation reactions. Thus, the (still hypothetical) utilization of archaeal mevalonate kinase to phosphorylate phosphomevalonate is a paralogous DOG with regard to fungal-type PMK and an analogous DOG with regard to animal PMK. If this is the case, then the origin of this DOG in archaea must have been by recruitment, which did not involve either HGT or gene duplication.

One-for-One and One-for-Many Relationships in DOGs

The DOGs discussed thus far involved one gene in each of the two (or more) genomes. But the scope of analysis can be expanded to examine groups of genes. For example, a series of very ancient DOGs involves components of the DNA replication initiation complex.

In Bacteria and Archaea/Eukarya, several components of replicative complex are not orthologous. The aforementioned replicative helicase is a case of a simple, gene-for-gene DOG in this protein complex. Other parts of the replicative machinery are involved in more complex DOGs. In Bacteria, the opening of the origin of replication is achieved by the initiator ATPase DnaA, and RNA primer for DNA replication is synthesized by a single-subunit primase of the DnaG family. Both these proteins have multisubunit, nonorthologous counterparts in Archaea/Eukarya—respectively, an origin recognition complex made of six different proteins and a heterodimeric eukaryotic primase (Leipe *et al.*, 1999).

Two other examples concern the recently discovered anabolic pathways involving 1-deoxy-D-xylulose-5-phosphate (DXP). Isoprenoid biosynthesis, which proceeds by a five-step mevalonate pathway in many species, is displaced by the DXP pathway in most bacteria (there are at least seven committed steps to isopentenyl pyrophosphate). Similarly, there are two alternative pathways of *de novo* pyridoxal phosphate biosynthesis from glyceraldehyde 3-phosphate in bacteria, one requiring erythrose 4-phosphate (seven enzymatic reactions involving seven proteins) and another requiring ribulose 5-phosphate (not less than five enzymatic steps, which are apparently provided by only two proteins; Tanaka *et al.*, 2005). In some of these cases, the alternative pathways require different chemical precursors. One could argue that such pairs of pathways with the same end point but different starting points are neither "the same function," northe states of the same character. If, however, we allow a large set of precursors for anabolic reactions, the functions "biosynthesis of pyridoxal from ribulose 5-phosphate" and "biosynthesis of pyridoxal from D-erythrose 4-phosphate" can be represented more generally as "biosynthesis of pyridoxal from the available precursors," and the relationships between alternative pathways in different species can be seen as DOGs.

DOGs and Operons

Complete and annotated genome sequences can be examined for correlation between gene functions and their relative positions in the genome. Groups of proteins may belong to the same protein complex, or they may be components of the same metabolic or signaling pathway. In many prokaryotes, genes coding for such groups of proteins may be found close to each other on the chromosome. This trend was noticed many years ago in model species, *E. coli* and *B. subtilis*, which led to the idea of operons—that is, groups of genes that are involved in the same pathway, are located next to each other on a chromosome, and are expressed together as a single multigene transcript. In eukaryotes, polycistronic transcripts are relatively rare (viruses with genomic RNA being a special case; see Chapter 4), and where they exist, it is not likely that they represent operons in this established sense. Nonetheless, clustering of genes on chromosomes may reveal different types of functional and evolutionary signal in various species. We will examine the progress in this area in Chapter 8, but now we are concerned with DOGs and ask, What is the relationship between DOGs and operons?

It seems that the existence of operons puts constraints on DOGs. Suppose that operon-like arrangement of a group of genes increases fitness of the host, for example, by way of a more efficient coregulation of these genes. A DOG that consists of a loss of a gene within this operon, and a gain of an isofunctional gene elsewhere in the genome, may put the host at a disadvantage because the coregulation mechanism now has to be somehow reestablished. One can therefore expect that if clustering of a group of genes on a chromosome is important, then the DOGs will mostly occur *in situ*: The displacer gene will tend to occupy the same or almost the same position as the displacee. Evidence of frequent gene displacements *in situ* is indeed accumulating (Wolf *et al.*, 1999; Yanai *et al.*, 2002; Liu and Mushegian, 2004).

DOGs and Gene Competition

An intriguing aspect of DOGs is differential tolerance between the isofunctional pairs of genes or pathways. Some such pairs seem to be capable of peaceful coexistence in the same genome for a long time. For example, bacteria and eukaryotes have just one type of primase, whereas archaea have both bacterial-type and eukaryotic-type enzymes. As mentioned previously, nonparasitic gammaproteobacteria with relatively large genomes have two types of phosphoglyceromutases, not counting paralogs, whereas many other bacteria and even eukaryotes have just one. Often, however, isofunctional pairs of genes are found to the exclusion of one another. The reasons for such consistent intolerance are of special interest. Some such cases are likely to be fundamental lifestyle choices, related to the functioning of large multiprotein complexes that were "genetically annealed" early in the evolution of life and more recently have been largely banished from taking part in displacements (Woese, 2002). Such DOGs, exemplified by replication enzymes and also by significantly different sets of proteins involved in translation in Bacteria versus Archaea/Eukarya, are quite robust evolutionary markers (see Chapters 12 and 13).

Other gene pairs, however, exhibit the same tendency toward strong mutual exclusion without clear correlation to phylogeny. An interesting case of low mutual tolerance is the pair of thymidylate synthases—the better studied, folate-dependent ThyA, widespread in all kingdoms of life as well as in viruses—and the analogous enzyme, flavine-dependent thymidylate synthase ThyX, which is found in some bacteria, some archaea, some bacterial viruses, and also, oddly, in slime mold *Dictyostelium*. The lists of species in which each of the alternative thymidylate synthases are found are evolutionarily incoherent, especially for ThyX (see Fig. 5.7), which suggests multiple cases of horizontal transfer followed by rapid displacement.

A patchy distribution of a gene, when ones and zeroes in its phyletic pattern are not nested in the species' tree, may seem to indicate that the history of this gene contains nothing but horizontal transfer. But this is not necessarily the case: Generally, any phyletic pattern can be explained by vertical inheritance and gene loss, by horizontal transfer, or by some combination of these factors, and the difference is only in the scoring function, or penalty, that we impose in an attempt to reflect the probability of each type of such event. In Chapters 11–13, we will examine ways to reconstruct the most plausible combinations of these events using phyletic patterns in combination with evolutionary trees of genes and genomes. But regardless of the evolutionary history of ThyA/ThyX displacements, the result is dramatic: Two enzymes seem to avoid one another. The exception that supports the rule is the occurrence of both types of thymidylate synthase in mycobacteria, where one of the two enzymes is clearly contributed by an integrated prophage.

While on this topic, I would like to note that closer examination of the enzymatic mechanism of thymidylate synthesis indicates that ThyA/ThyX is not one-on-one displacement. Both enzymes transfer the hydroxymethyl group from a folate derivative to uridylate, with concomitant reduction of hydroxymethyl to methyl. In the case of ThyA, the redox equivalent, drawn from the same folate derivative, requires a coupled enzyme, dihydrofolate reductase. ThyX needs no partner protein because its redox equivalent is provided by a flavine cofactor (Myllykallio *et al.*, 2002). Phyletic patterns of ThyA and dihydrfolate reductase are very close to each other, so the DOG is actually between ThyA + DHFR, on the one hand, and ThyX on the other hand.

Competition between multiprotein pathways is even more dramatic. In isoprenoid biosynthesis, DXP and mevalonate pathways rarely coexist in the same organism. The only clade that contains a full complement of enzymes from both pathways is higher plants, but the two pathways are strictly compartmentalized there, with enzymes of the mevalonate pathway staying in the cytoplasm and DXP enzymes going to chloroplasts. All other species capable of isoprenoid biosynthesis have exactly one functional pathway to do it, although the remnants of the other pathway are detected in some genomes (Smit and Mushegian, 2000).

DOGs, RNA–Protein Displacements, and RNA World

Thus far, all examples of DOGs concerned protein coding genes. However, nothing forbids displacements between proteins and other sense-carrying units. One class of regulatory modules in bacteria, called riboswitches, provides examples. Riboswitch is an element on a polycistronic bacterial RNA that serves as a sensor of a specific low-molecular-weight compound inside the cell. For example, *S*-adenosylmethionine (AdoMet), an important intermediate in many biosynthetic and regulatory pathways (Kozbial and Mushegian, 2005), is synthesized *de novo* in most free-living bacteria. Production of the AdoMet precursor, amino acid methionine, is tightly controlled and can be turned on in response to a decrease in AdoMet concentration. Proteobacteria regulate methionine biosynthesis with the help of the protein, AdoMet-sensing transcriptional regulator (methionine repressor; Phillips and Stockley, 1996). In gram-positive bacteria, however, the level of AdoMet is sensed by a riboswitch—a highly conserved RNA sequence called S box that consists of approximately 150 extensively base-paired nucleotides located at the 5′ ends of polycistronic mRNAs—at least 11 of them in *B. subtilis*,—which control expression of at least 26 genes of sulfur metabolism and biosynthesis of methionine, cysteine, and AdoMet itself. At elevated concentrations of AdoMet, riboswitches tightly bind the ligand, which appears to cause premature termination of transcription (Winkler *et al.*, 2003). Thus, the same function is enabled by a protein coding gene in some species and by an RNA coding gene (or, at least, RNA fragment) in another species. This is a rather clear example of an RNA–protein DOG.

Most of the accepted scenarios of evolution of life on Earth include the RNA world—the stage at which both genetic material and catalytic machinery of the living creatures were represented by RNA molecules and massive RNA–protein displacements must have occurred at the time when ribozymes were being replaced by protein enzymes. A few ribozymes, however, appear to have never been displaced by protein enzymes and to persist in living organisms from the RNA world all the way to the existing species. These ribozymes include peptidyltransferase and decoding activities of ribosomal RNA (Steitz and Moore, 2003) and processing of tRNA 5' ends performed by RNAase P (Hartmann and Hartmann, 2003).

Detection of DOGs Using Phyletic Patterns and Phylogeny

Homology, analogy, and DOGs exist regardless of our ability to detect them. The DOGs we have discovered are most likely only a fraction of all DOGs that have occurred in evolution. How can we recognize a DOG, and how many of them remain undetected?

There are two prerequisites for finding a DOG. First, there has to be empirical evidence that similar molecular function is present in two lineages. Second, there has to be a way to robustly define orthologs and to assert their absence. The latter is only possible if both genomes are completely sequenced (see Chapter 3 for a discussion on how to define orthologs). But even if genome sequences are complete and protein lists in both genomes are accurate, false negatives and false positives in ortholog definition can occur, and they give rise to errors in DOG prediction.

One type of error occurs when orthologs have low sequence similarity, which is not found by a standard database search. Such was the case of "missing" RNAase H in the second completely sequenced genome, *Mycoplasma genitalium*. The authors of the original annotation (Fraser *et al.*, 1995) did not find RNAase H homolog in *M. genitalium* and correctly concluded that the proteome of this bacterium has to include the protein performing this essential activity (RNAase H specifically destroys RNA when it is hybridized with complementary DNA, and this is required for removing RNA primers left after initiation of replicative DNA synthesis). Eugene Koonin and myself did not do much better when we concluded that the homologs of RNAase H were indeed missing in *Mycoplasma* and proposed another putative nuclease as a DOG (Mushegian and Koonin, 1996a). When sequence similarity searches improved (in particular, when gapped BLAST and PSI-BLAST were introduced), it turned out that a proper ortholog, albeit with low similarity to RNAase H, had been present in *M. genitalium* genome all along (Bellgard and Gojobori, 1999). In Chapter 5, I noted that the recovery of proteins that have relatives in distantly related genomes (the PHIDO value) tends to improve with time, and with it, the number of correctly identified orthologs also rises. Therefore, the estimated number of DOGs will tend to decrease as the sensitivity of sequence comparison increases.

Other types of errors, on the contrary, result in the underestimation of DOGs. One case is orthologous/xenologous displacement. Suppose that a gene A_A in genome A has an ortholog A_B in genome B. It may be recognized as the ortholog using criteria given in Chapter 3 (basically, if we find no evidence of duplications). If, however, A_B has been transferred into genome B from the unobserved genome C, we may not find this out until we sequence the latter genome.

DOGs are also missed when paralogs are erroneously taken for orthologs. Consider differential loss of paralogs, when gene duplicate precedes speciation and one paralog is lost in each lineage. This may erase the evidence of duplication from the gene tree. Only if more genomes are examined, and a genome that still harbors both paralogs is found, may we know the truth. For example, elongation factor Tu, the GTPase that loads charged tRNAs onto ribosomes, most likely has been duplicated early in bacterial evolution, and one copy or the other was lost in most bacteria, except for gammaproteobacteria and *Deinococcus* (Lathe and Bork, 2001). Interestingly, recent work suggests that very similar duplication/losses have occurred in the evolution of eukaryotic homolog of EF-Tu, within the eukaryotic clade itself (Keeling and Inagaki, 2004). Differential loss of paralogs, when the remaining paralogs are able to play the same biological roles, is by definition a paralogous displacement. Thus, frequency of paralogous DOGs also tends to be underestimated.

In summary, even in the case of well-studied genomes, the precise set of DOGs is not easy to ascertain. Phylogenetic trees for genes of interest and for species as a whole are helpful for DOG finding: Any discordance between the two trees may suggest differential gene loss and differential recruitment. Phyletic patterns of isofunctional genes are also useful in this regard: They highlight "patchy" distribution of orthologs and alert to a possibility of the isofunctional, nonorthologous genes in those species in which orthologs are missing.

A nagging question, however, is how to prove the analogous origin of isofunctional proteins. It has been said that the ultimate proof is given by their three-dimensional structures. The hypothesis here is that similarity of three-dimensional structures supports the idea of common ancestry, and lack of such similarity argues for analogous origin of the common function. The problem with this two-part argument is that neither part is actually true. On one hand, suppose that we find two structures to be similar, even though the similarity of their sequences is at the random level. In this case, we are still left with the following dilemma: Do we observe preservation of function in homologous proteins that diverged beyond recognition, or is it rather convergence toward the common function of two sequences that never had common ancestor? On the other hand, suppose that structures are dissimilar. There is growing evidence that protein fold can change in the course of sequence evolution. If this is true, then how can we be sure whether we observe unrelated molecules or molecules that share a common ancestor but have diverged so much that they not only lost all sequence similarity above the random level but also experienced fold change in one or both molecules?

The problem is aggravated by lack of a good measure of structural similarity of proteins. When comparing primary structure of proteins, we have at our disposal the evolutionary model of sequence change and statistical theory of sequence similarity (see Chapter 2). In contrast, statistical theory of structural similarity is, for all practical purposes, unavailable. We will examine these problem is more detail in in Chapter 9. It is safe to say, however, that the proof of analogy and convergence in protein function has to rely on the compatibility of many lines of evidence. Consider again two types of thymidylate synthases, ThyA and ThyX. Their sequences are not alignable in any meaningful sense. None of the functionally important residues that are conserved in ThyA have a counterpart in ThyX, and vice versa. The monomer of each thymidylate synthase appears to fold into an alpha-beta structure with distinctive large beta sheet, but examination of functional forms of the enzymes shows that these beta sheets are positioned in a completely different way with regard to intersubunit contacts and to substrates (Fig. 6.7). Moreover, as discussed previously, the cofactor requirement of the two enzymes is quite different: ThyA needs a coupled enzyme, dihydrofolate reductase, from which the reducing equivalent is derived, whereas ThyX uses bound flavin nucleotide to the same end. Thus, ThyA and ThyX protein monomers may have superficially similar features of tertiary structure, but everything else—sequence, catalytic residues, patterns of oligomerization, functional coupling with other proteins, catalytic cofactors, and reaction mechanisms—is different. The case for a common ancestor of these two proteins is all but impossible to make. The hypothesis that no such ancestor existed seems to be the one best compatible with the evidence.

A dramatic proof of the existence of analogy at the molecular level is given by the RNA–protein displacements. Riboswitches discussed above are one such example. Reconstruction of minimal ribosome also suggests a possibility of RNA–protein displacements, when a function of a deleted segment of rRNA is taken over by a protein, or the other

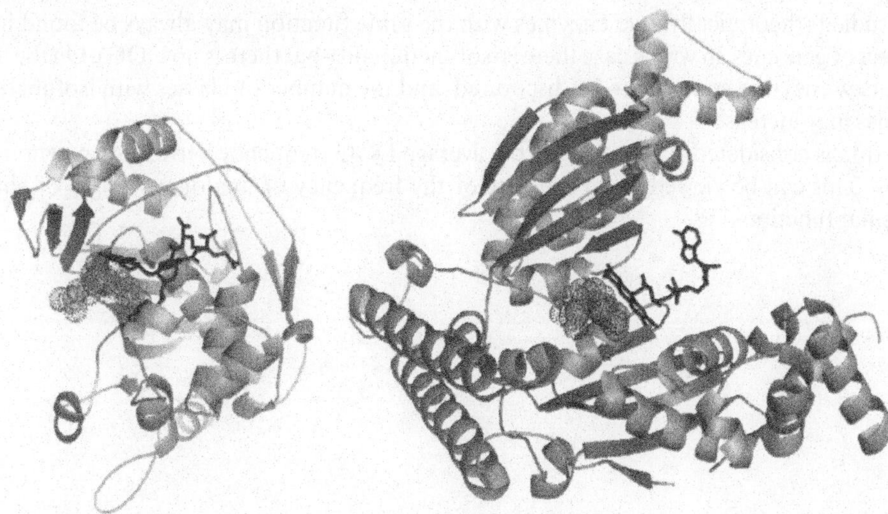

Figure 6.7. Two unrelated types of thymidylate synthases. (**Left**) Monomer of folate-dependent ThyA (PDB 1BJG). Cofactor methylenetetrafolate is rendered as the black stick model, and the substrate UMP is rendered as dotted spheres. (**Right**) Dimer (half of crystallographic tetramer) of flavin-dependent ThyX (PDB 1O29) with flavin mononucleotide cofactor (sticks) and substrate analog (spheres). Both protein subunits are needed to bind the substrate and the flavin mononucleotide cofactor.

way around (see Chapter 13). In these cases, isofunctional molecules are protein and RNA, which are definitely not homologous.

What may be the DOG frequency in nature? This can be studied from two different angles. First, we can consider a pair of completely sequenced genomes and ask how many isofunctional genes in two genomes have been displaced. Second, we can start with a group of isofunctional proteins, such as the same EC class, ask how many isofunctional proteins within a class are not homologous to each other, and examine their distribution across multiple species.

Neither of these approaches has been implemented on a large scale, but several observations are of interest. In an example of the first approach, when reconstructing a minimal gene set (Mushegian and Koonin, 1996a; see Chapter 13), we counted all putative nonorthologous displacements between the first two completely sequenced genomes, *H. influenzae* and *M. genitalium*; the two bacteria have approximately 1700 and 460 genes, respectively, of which 230 were shared (the first approximation of the minimal genome). Counting isofunctional but nonorthologous proteins, which appeared necessary to fill the gaps in the rudimentary metabolism of the minimal organism, the number of DOGs was estimated to be approximately 25, or close to 10% of the complete minimal genome (Mushegian and Koonin, 1996a). This number has been revised downwards, mostly because of improved recognition of distantly related orthologs (Mushegian, 1999; Koonin, 2001). Recently, I estimated the number of protein–protein DOGs between Bacterial and Archaeo/Eukaryal ribosomes at approximately 10; both minimal and ancestral ribosome are thought to contain 35–45 proteins, which gives the DOGs rate of 20–30% (Mushegian, 2005; see Chapter 13).

Using the second approach, Koonin and Galperin provided evidence that at least 10% of EC classes contain isofunctional but nonorthologous proteins (Galperin *et al.*, 1998; Galperin and Koonin, 1999). This is very close to the previously discussed estimates made by genome comparisons, although distribution of these isofunctional enzymes across the genomes is not well studied (theoretically, two enzymes with the same function may always be found in the same set of genomes, in which case there is isofunctionality but there is no DOG). On the other hand, new enzymes continue to be discovered, and the number of classes with isofunctional enzymes may increase.

All things considered, I suspect that the average DOG frequency between two genomes is 5–10%. This can be viewed as an estimate of the frequency of analogous origin of similar molecular functions.

7

Prediction of Function and Reconstruction of Metabolism from Genomic Data: Homology-Based Approaches

Currently, GenBank contains completely sequenced genomes of 385 biological species, counting strains of bacteria but not counting viruses and organelles, and also 339 draft assemblies, 462 genomes in progress, and several large "metagenomes" coming from sequencing of uncultivated environmental samples (data from www.ncbi.nlm.nih.gov/genomes/static/gpstat.html; accessed August 2, 2006). One goal of computational biology is to understand as much as possible about life of all these species from their genomic sequences.

"As much as possible" is a useful qualification here. Computational inferences can only be made at the background of biological knowledge, and they will be only as good as the models that we derive from that knowledge. This becomes very obvious when we annotate a novel gene based on similarity of its sequence to a better studied gene. In fact, establishment of homology is the foundation of any computational prediction of gene function. Computational methods are informed by previous knowledge of biology, and computational predictions are confirmed, or fail to be confirmed, by wet-lab experiments that involve gear such as petri dishes, Eppendorf tubes, mouse cages, sunblock and rubber boots, and so on. Computers may seem to be far removed from this physical world.

This is a good time to argue, however, that that computational, or dry-lab, experiments are in fact not so different from what is going on in the wet lab. Computer predictions, such as detection of homology or building a phylogenetic tree, are sometimes believed to be "theoretical" as opposed to "experimental" work, but I do not think this is right. "Theoretical" has to do with developing theories, and there may be elements of this in both wet-lab and dry-lab work. On the other hand, gel electrophoresis and BLAST search are quite similar in that both use certain genetically controlled, but chemically defined, properties of protein molecules in order to sort them by similarity. The difference is only in the nature of signals that the molecules display in each case.

All this is relevant to computational analysis of protein function because a common point of view is that the final word is always with wet-lab experiments, whereas computational analysis is "mere speculation." On the contrary, I believe that in the analysis of biological function, dry-lab and wet-lab methods should be used together for testing compatibility of different lines of evidence.

Therefore, how can we understand the biology of the newly sequenced genome from its sequence? In this book, I hardly mention such themes as sequencing strategy, genome assembly, and gene finding. Not that these problems are unimportant; on the contrary, it is clear that complete understanding of molecular function requires accurate identification of genes and their products. Fortunately, we have become quite good at finding a reasonably complete set of genes encoded by a genome sequence (which is not the same as successful prediction of the complete sets of RNA and protein variants encoded by each gene; for discussion of the approaches to these important problems, and some initial insights into evolution and function of alternative transcripts and protein variants, see Kriventseva *et al.*, 2003; Kim *et al.*, 2004; McCullough *et al.*, 2005; Roth *et al.*, 2005; Nakao *et al.*, 2005; Kimura *et al.*, 2006; Yamashita *et al.*, 2006). Thus, we will assume that we have a list of genes predicted in a newly sequenced genome, together with the location of each gene in this genome. We also have such lists for all genomes that have been sequenced before. Often, the newly sequenced genome belongs to a species that has not been well studied. We may know something about its living conditions and interactions with the environment, and, in the case of cultivated microorganisms, we sometimes know the set of nutrients that are required to grow this species in the laboratory, but functions of individual genes and their products are not known to us.

The genomes sequenced earlier are all different, and some of them, like our newly sequenced species of interest, have not been extensively studied. Others, however, may have been examined in great detail, and the knowledge about their genes, proteins, cells, bodily functions, and interactions with the environment is available in scientific books and journals and also, increasingly, in specialized on-line databases. (I am omitting the discussion of these databases; as with all Internet resources, the best way to learn about them is to perform a search or to consult the indispensable yearly database issue of *Nucleic Acid Research* for the types of databases that are of the most immediate interest to us; for a meta-review, see Galperin, 2006). Our goal is to make functional inferences about the molecular setup of a poorly studied species, based on the information about other, better studied species.

This became known as the task of computational reconstruction or molecular pathways, or metabolic reconstruction—the term first used, as far as I can judge, by Selkov *et al.* (1997), who, however, defined it in a slightly narrower way as "an attempt to formulate a model reconciling the sequence data with known biochemistry."

As with some other new developments in comparative genomics, different people understand metabolic reconstruction differently. There are automated and manual reconstructions; reconstructions covering multiple genomes and reconstructions focused on just one genome or one pathway across many genomes; flat files of tentative assignments and sophisticated relational databases of predicted protein functions; static charts of known and missing functions; as well as attempts to model kinetics of metabolic fluxes. Most of these efforts are interesting, and some are more popular than others. What is not well studied is the accuracy of these approaches.

Arguments about what constitutes good or not so good metabolic reconstruction occasionally get quite passionate, sometimes approaching "behavior that would not be condoned in other, more mature fields of science" [Joseph Felsenstein's (2003) recollection of the early period of another area of computational genomics]. I have been at the receiving end of some such behavior, all the while being accused of the same (see Kyrpides and Ouzounis, 1999; Mushegian, 2000), but examination of those episodes is no longer of interest. Rather, my goal here is to examine what currently appears to be achievable in the art and science of metabolic reconstruction, which scientific ideas enable these achievements, and which problems remain unsolved. In discussing all this, I use "new genome," "new sequence," "new protein," etc. to indicate the species with completely sequenced genome for which a metabolic reconstruction is sought. The sequences that are deposited in the database can be "annotated," "better studied," etc., or they can be "uncharacterized." All new sequences are uncharacterized by definition.

Databases of Biochemical Information

Good databases of empirical knowledge about biochemical pathways are essential for making inferences about functions of new genes. And perhaps one of the most important components of a good database is the indication of the primary evidence, on the basis of which the annotation of the representative family member was done. In the past, the evidence supporting a statement about the gene may have been scattered across journals and could not be easily accessed from a sequence database such as GenBank. Later, an improvement was afforded by partial incorporation of SWISSPROT information into GenBank entries). However, to evaluate the credibility of even this better curated information, one sometimes has to undertake detective work. For example, in 1994–1996, while annotating bacterial genomes, Eugene Koonin and I had to deal with the case of the protein then known as HemK. In *Escherichia coli*, the *hemK* gene is found downstream of another gene of porphyrine biosynthesis. Biochemical studies suggested that, among other activities required for the synthesis of mature heme, there should be an enzyme protoporphyrinogen oxidase. This enzyme was not known, but some genetic evidence attached this function to the aforementioned HemK. The SWISSPROT database incorporated this evidence and correctly indicated that it was flimsy, at least as concerned the *E. coli* protein; the homologs from other species, however, were sometimes annotated as simply "protoporphyrinogen oxidase," without "predicted" or "putative," even in SWISSPROT.

The problem is that HemK is not an oxidase. The inference based on the position of the *hemK* gene was wrong. This is not to say that analysis of gene neighborhoods is always futile; on the contrary, we will examine the productive uses of such an approach in Chapter 8. Alas, it did not work in that particular case. On the other hand, sequence similarity analysis indicated that the HemK family shares a set of conserved motifs with diverse methyltransferases, which belong to the class I of methyltransferases, characterized by canonical Rossmann fold (Schubert *et al.*, 2003). It became clear that HemK, renamed into PrmC, is conserved in almost every genome and has a very particular substrate specificity—namely, it methylates a specific glutamine residue in peptide chain release factors (proteins that are required for termination of translation) and this modified glutamine is essential for protein function (Nakahigashi *et al.*, 2002). Incidentally, peptide chain release factors in Bacteria and Archaea/Eukarya are a nonorthologous DOG (see Chapter 6), but in an apparent case of local sequence convergence, both types of release factors contain the GGQ tripeptide, where Q is the target of methylation by HemK.

In 1995, the true biological function of HemK was not known, but it was quite clear that it is a methyltransferase. This was mentioned in the study of *E. coli* proteins (Koonin *et al.*, 1995), which we examined in Chapter 5. However, a newly sequenced ortholog of HemK would most commonly be annotated by transferring a more widely available annotation—that is, "protoporphyrinogen oxidase" (with or without "putative"). Finally, several years ago, the correct annotation caught up with the databases and is no longer difficult to find.

Another problem with the database annotations of gene and protein functions is that, for much of the 20th century, such functions were studied in a relatively small number of species—either common model organisms or species that are of direct relevance to our existence. The biology of mammals, yeasts, and laboratory strains of *E. coli* and *Bacillus subtilis* continues to influence not only our view of metabolism in other species but also our definition of metabolic pathways and our understanding of protein function. For example, a wall map of biochemical pathways contains the tricarboxylic acid (TCA) cycle, giving the impression that it is a biochemical mainstay of every cell. But, as already discussed in Chapter 6, this set of reactions takes place mostly in proteobacteria, including gammaproteobacteium *E. coli* and alphaproteobacteria, as well as in eukaryotic mitochondria, which are the descendants of an ancient alphaproteobacterium. In many living species, however, the orthologs of the TCA cycle enzymes are not arranged into a cycle. Out of the eight enzymatic reactions (not counting shunts and variations at the entrance point), most genomes have six genes or less, with interruptions all over the cycle, and mycoplasmae as well as spirochetae have no TCA cycle enzymes at all (see Fig. 6.6; Huynen *et al.*, 1999).

One can argue, on the other hand, that eukaryotes and proteobacteria are species that matter most to us, so a version of any pathway found in these organisms is a sensible null hypothesis after all. In addition, some order in the metabolic map is better than no order at all. However, when we want to determine what was going on in a phylogenetically distant species, or in ancestral living forms (see Chapter 13), it is important to remember that the "metabolic map" as we know it is heavily influenced by studies in a small number of species, not necessarily representative of every species we may want to examine.

Homology and Orthology: How to Use Them for Prediction of Protein Function, and What Can Go Wrong

As discussed in Chapter 3, two orthologs, one in each of two species, are more likely to have the same molecular function, and the same biological function, than two paralogs. Paralogs are more likely to evolve new, if related, functions. This has been called the principle of phylogenomic by Zmasek and Eddy (2002). As an aside, *phylogenomics* is a term coined by Jonathan Eisen (1998) to summarize the idea that the evolutionary tree can help us understand how biological functions are distributed across the members of a protein family. However, other authors use this term to refer to other things—most often to large-scale phylogenetic analysis, when gene/protein phylogenies of many genes in a genome are examined together. It will be interesting to see which meaning will be used in the future.

The justification of the principle of phylogenomics is common sense. Orthologs, especially single-copy genes in two species of similar complexity, are maintained in evolution, most likely because they are required for the same function, which, ostensibly, their common ancestor also had. Gene duplications, with divergence among paralogs, are thought to be the main mechanism for evolving new functions (Ohno, 1970; Lynch and Conery, 2000; Kondrashov *et al.*, 2002).

The phrase "more likely," which occurs twice in the principle of phylogenomics, is important. The link between homology (of any type) and function is loose—functions of any homologs, including orthologs, can change in evolution. In general, however, it is less risky to predict function by transferring annotation between orthologs than between paralogs. If biological function of an ortholog in any species is known, it can often be transferred directly. When a new protein has no orthologs with known function but has paralogs with known function, direct transfer of annotation is more problematic, but there is a good chance to define molecular function—for example, the class of chemical reactions that the predicted enzyme might catalyze.

Another aspect of the problem is that there is no rigorous definition of biological function—in general and in the case of each particular protein. Often, despite all biological evidence, it is impossible to say whether the functions of two proteins are "the same" or only "similar." As with characters, which exist at many levels and have complex relationships, biological function has many facets.

Consider, for example, the case of presenilin-1, the protein product of a gene that is specifically mutated in a large percentage of Alzheimer's disease patients. Presenilin-1 gene has homologs in all multicellular eukaryotes and even in unicellular organisms, including archaea (Ponting *et al.*, 2002). If the "biological function" of human presenilin-1 is to protect the owner from Alzheimer's disease, then the function of plant and prokaryotic homologs of presenilin-1 is most certainly not the same (or even similar) as in the human protein because plants do not get Alzheimer's disease. Even if we learn molecular details, such as that presenilin-1 is an intramembranous protease processing amyloid-beta propeptide (and several other substrates, including developmental protein Notch, which was the first presenilin target, discovered in a genetic screen in *Drosophila*), this does not cover "biological function" of plant homologs—*Arabidopsis* has presenilin-like genes but does not have Notch or amyloid-beta. Note that the last fact about plants is known because we have a completely sequenced plant genome; thus, the definition of a protein function is influenced by genes that are present in the same genome—and by genes that are absent from it.

We can redefine the function of presenilin in a way that will also cover functions of its orthologs and paralogs in different species. For example, "intramembranous proteolysis of various secreted or membrane-bound proteins" might be a sufficiently general description of the molecular function of all presenilin-like proteins. However, imagine that species *A* contains two paralogs of such intramembranous protease, species *B* lost one of them after the split between *A* and *B* but retained the other, and species *C* also lost one but evolved four in-paralogs (and, moreover, as *C* is a multicellular organism, these in-paralogs work in different types of cells). Even after sorting all evolutionary relationships out, it is not possible to predict, on the basis of this information, which homologs have which exact function.

All these are good problems to have, compared to the difficult cases of conserved protein families about which nothing is known. In Chapter 8, we will discuss what to do with them. But in fact, the percentage of protein families that are completely uncharacterized is not huge. Given that the PHIDO and PHISO values (see Chapter 5) are each more than 50% in every completely sequenced genome, annotation by homology provides functional clues to a large percentage of proteins in any genome. Therefore, every practical way of computational reconstruction of metabolism of a poorly studied species relies on finding all sequence homologies to better studied proteins and partitioning the homologs into orthologs and paralogs.

Let us consider a predicted protein A_A in a newly sequenced genome *A*, which has a homolog A_B in another genome *B*. Two possibilities are of interest for functional annotation of A_A by homology. First, we may apply tests of orthology, discussed in Chapter 3, and find out that A_B is the ortholog of A_A, and the function of A_B is known. Second, we may find out that A_B is the ortholog of A_A, the function of A_B is not known, but in species *B* there is also a paralog A'_B with known function. (Yet another possibility, that there is no ortholog of A_A in species *B* but there is a paralog A'_B with known function, has exactly the same utility for functional annotation of A_A as the previous one).

Jonathan Eisen, then of Stanford University and currently of the University of California at Davis, pointed out that if the terminal branches of an evolutionary tree of a gene/protein family are labeled with the available functional information, the distribution of labels can be used to infer the functions of the unknown homologs in that tree (Eisen, 1998). Eisen, however, did not propose a practical algorithm implementing his ideas; this was first done by Zmasek and Eddy (2002). The basic idea is to examine a position of an uncharacterized gene with regard to a group of those homologs that have the same function. If it can be shown that the uncharacterized gene is nested among isofunctional genes, then the new gene may be functionally annotated by "borrowing" the annotation from this group of proteins. If, however, the new gene falls outside of all clusters that have representatives with known functions, then the specific function of such a gene cannot be inferred with confidence, although a more general description may be possible (Fig. 7.1; the details of the algorithm and the statistical criteria of annotation reliability proposed by Zmasek and Eddy are not examined here).

The selection between specific and more general functional assignments is aided by the catalogs of biological functions. As many other traits of biological systems, functions of individual genes and gene products can be classified in a hierarchical fashion. One such hierarchical system is the IUPAC Enzyme Classification (EC) mentioned in Chapter 4. A similar approach has been extended to all gene products, with and without enzymatic activity, to produce Gene Onthology (GO; Schulze-Kremer, 1997; Ashburner *et al.*, 2000). Two enzymes that share all four numbers in EC have the same enzymatic activity, and two proteins that share all numbers in GO hierarchies have the same molecular function (in fact, there are three different hierar-

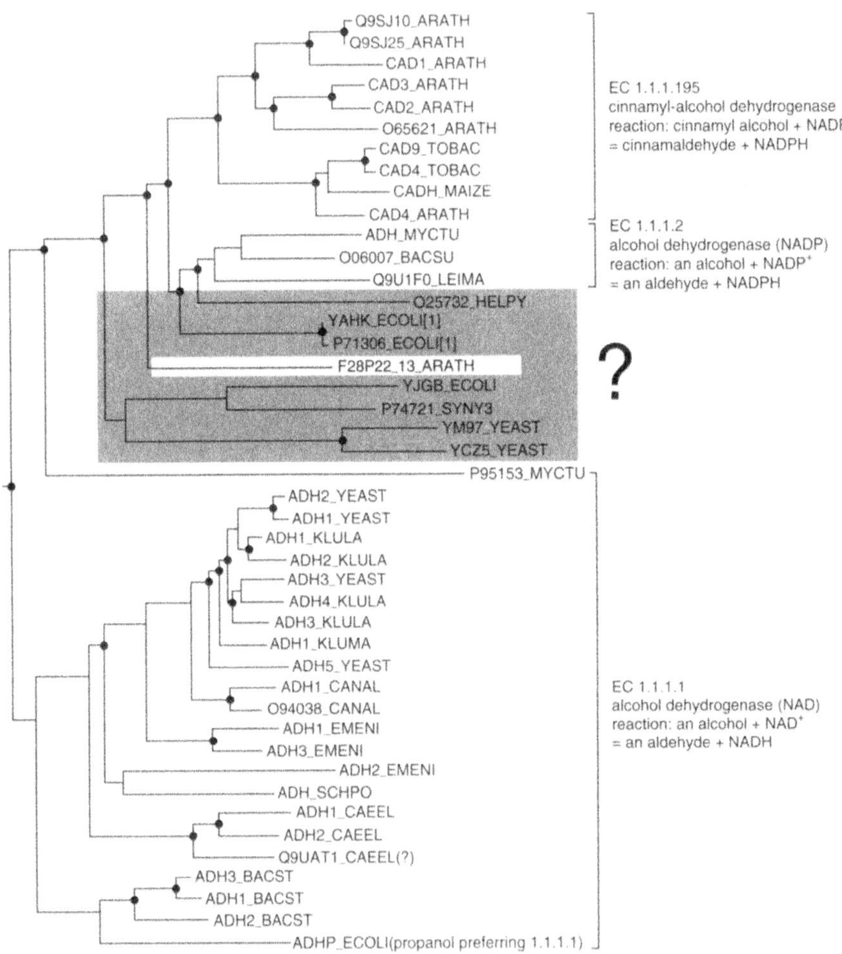

Figure 7.1. Mapping of known functions on the phylogenetic tree of a protein family. The query protein (F28P22.13 of *Arabidopsis thaliana*; *white box*) belongs to the NAD/NADP-dependent alcohol dehydrogenase superfamily, but its neighbors in the tree are all uncharacterized proteins (*gray box*), and there is no specific relationship between the query and any homolog with the precisely defined substrate specificity. Reprinted from Zmasek, C. M., and Eddy, S. R. (2002). A simple algorithm to infer gene duplication and speciation events on a phylogenetic tree. *Bioinformatics* **17,** 821–828, by permission of Oxford University Press.

chies in GO—one for molecular function, another for biological function, and the third for cellular localization; any gene product may be annotated within all three hierarchies). Inasmuch as orthologs tend to preserve the same function, they also tend to share all four EC or GO numbers. Paralogs, however, tend to evolve different functions and often different, albeit related, activities; therefore, they share only the first two or three EC numbers and may have different GO numbers, at least in some of the GO hierarchies. Note that shared EC or GO identifiers do not imply much about sequence similarity and common evolutionary origins: Two enzymes with perfectly matching EC numbers may be nonorthologous to each other. This is because of functional convergence, as discussed in Chapter 6.

Ultimately, however, the success of the whole enterprise of labeling the trees by orthologs and paralogs, and inferring functions in that way, depends on the proportion of homologs that have been experimentally studied and on the distribution of these well-studied homologs over the tree. Recently proposed probabilistic approaches suggest that even the moderate proportion of annotated genes/proteins in the tree (approximately 50%) may be sufficient to annotate new function (Sjolander, 2004; Engelhardt *et al.*, 2005)

Annotations by Homology and Errors Inherent in the Process

Bork and Koonin (1998), Devos and Valencia (2003), Park *et al.* (2005), Green and Karp (2005), and others studied several types of mistakes that are common when annotation is transferred from a database entry to an uncharacterized homolog. In effect, there are three groups of errors. First, the homology may be inferred incorrectly—either true homologs are not found or unrelated proteins are taken for homologs. Second, the homologs may be found correctly, but the orthologs and paralogs may be assigned incorrectly. Third, an error can occur even when the orthologs are properly assigned. Let us examine each group of errors in more detail.

One possibility is that a homolog of the new protein is present in the database, but it is not found. This often happens when an arbitrary threshold is used as a filter in sequence similarity searches. For example, suppose that we ignore BLAST matches with E-value higher than 10^{-10} and conclude that there are no "significantly similar" sequences in the database. As discussed throughout this book, different sequences evolve at different rates, and some pairs of orthologs from two species are less similar to each other than other pairs of orthologs from the same two species. For example, we and others failed to notice the ortholog of RNAase H in completely *Mycoplasma genitalium* and tried to explain it by way of a DOG (see Chapter 6). The truth was more simple: RNAase H is the enzyme that exhibits much less sequence similarity than many other essential enzymes. The result of such error is undetected similarity and lack of prediction—a clear false negative.

A milder version of the same mistake may be committed when some, but not all, homologs of a new protein are detected. For example, if I pay attention only to the top 10 BLAST matches, or do no more than three PSI-BLAST iterations (both cutoffs are, of course, completely arbitrary), then I may end up with only uncharacterized homologs but will miss more remote, yet better annotated, homologs. The result of this error is a failure to recover a homolog suitable for functional annotation; in the context of function prediction and metabolic reconstruction, this is also a false negative. The set of uncharacterized homologs, however, may provide a lead: A more sensitive model can be built using this information, and different comparison regimens may be employed to find more homologs, some of which may be better studied.

Another way to miss relevant homologs has to do with the multidomain composition of proteins. Multidomain proteins are found in all divisions of cellular life (reviewed in Ponting *et al.*, 1999; Copley *et al.*, 1999; Ponting and Dickens, 2001; Copley *et al.*, 2002). If a database search program is tuned to detect pairwise local similarity (as, for example, is the case with PSI-BLAST), and if the best conserved domain has many homologs in the database, the matches to this domain will dominate the results. However, sometimes the highest scoring domain does not provide enough clues to protein function. For example, one family of DNA helicases, widely distributed in many species, is named after the *E. coli* homolog, RecQ, involved in recombination and repair. There are many RecQ homologs in multicellular eukaryotes. All eukaryotic RecQ-like helicases, however, are multidomain proteins. To the N terminus of the longest helicase domain, there is an additional N-terminal domain that is not found in *E. coli* RecQ. There are two main classes of such N-terminal domains. One is mostly an alpha-helical, coiled-coil domain, presumably involved in protein–protein interactions; this domain is found in human RecQ-like protein mutated in Bloom's syndrome patients and in its yeast ortholog Sgs1. The other, three-layer alpha-beta-alpha domain is a 3′–5′ exonuclease, related to the famous Klenow nuclease, RNAase D, and many other nucleases. This domain is found in human Werner disease protein as well as in the RECQL4 protein, mutated in Rathman–Thompson and RAPADILINO syndromes. Nonetheless, similarities in the helicase region tend to overwhelm the outputs of BLAST searches; the nuclease domain in Werner syndrome protein, WRN, therefore was not discovered until several years after cloning of the *WRN* gene and its identification as a RecQ helicase [computational discovery was published by Mushegian *et al.* (1997) and Moser *et al.* (1997); the wet-lab verification was given by Kamath-Loeb *et al.* (1998)], whereas the nuclease domain of RECQL4 seems to have not been reported before.

In all these cases, focusing all attention on one, perhaps best-conserved domain of a protein results in underprediction of its molecular function; moreover, if the best conserved domain was functionally uncharacterized, the outcome is a false negative.

The opposite occurs when the new protein has no homologs in the database but we mistakenly think that such homologs do exist. Or we can correctly identify a family of homologs, all of which are functionally uncharacterized, and then, instead of quitting, erroneously conclude that this family is homologous to another, better studied family of proteins, when in fact there is no evidence of such a homology (or, even more dramatically, there is convincing evidence of homology with a completely different family). In one example of such error, a candidate archaeal protein was sought that could perform conjugation of cysteine and cognate tRNA; methanogenic archaea are notorious for lacking cysteinyl-tRNA synthetase homolog, and it is of interest to understand the DOG that apparently has occurred there. Using an original, but apparently not well-validated, computational method, the authors of one recent study detected their candidate, MJ1544, and assigned the putative CysRS function to it (Fabrega *et al.*, 2001).

More careful computational experiments, using standard techniques of sequence comparison and well-established statistical methods, do not support this part of archaeal metabolic reconstruction at all. Direct PSI-BLAST search immediately finds a moderately sized family of MJ1544 homologs, scattered across a handful of archaeal and bacterial genomes, most of which have a separate and easily recognizable CysRS. Moreover, when this family of uncharacterized homologs is converted into a hidden Markov model (HMM) and compared to other HMMs using HHsearch (Soding, 2005), it shows high, statistically significant similarity to many protein families and conserved domains with polysaccharide hydrolase activity. The functional relevance of these sequence similarities is underscored by the fact that they cover most of the length of MJ1544 and correspond to the known catalytic domains of these glycosylhydrolases. On the contrary, none of these families is known to have affinity to amino acids, tRNA, or ATP; thus, it is highly unlikely that these substrates are recognized by MJ1544 at all.

On the strength of this evidence, functional assignment of MJ1544 as the novel type of CysRS appears to be a false positive, resulting from error in homology inference, i.e., suggesting homology when there is none. (The "mystery of experimental verification of false predictions" is further examined in Iyer *et al.*, 2001).

Another way to misidentify a molecular function is by erroneous domain assignment. Previously, we saw that examining just one of many domains may lead to false negatives or at least to underprediction. But false positives are also easy to come by: For example, several groups suggested that the protein MG262 of *Mycoplasma genitalium* is a DNA polymerase (Bult *et al.*, 1996; Ouzounis *et al.*, 1996), even though it was already known that mycoplasmas have not one but two DNA polymerases encoded by other genes. In reality, MG262 does not contain a polymerase domain; it consists of a single domain with predicted nuclease activity, which is, indeed, highly similar to nuclease domains found in family A of DNA polymerases. Thus, a protein is named after a wrong domain, which may be present in its homologs but is missing from the protein itself. Analogously, in the course of genome annotation of archaeon *Archaeoglobus fulgidus*, several proteins have been annotated as inositol monophosphate dehydrogenases (Klenk *et al.*, 1997). In reality, all these *Archaeoglobus* proteins consist of non-catalytic CBS domains, which are also present in inositol monophosphate dehydrogenases and in many other proteins. What the *Archaeoglobus* proteins do not contain are the inositol monophosphate dehydrogenase catalytic domains. This is a clear case of mistaken annotation, originating from erroneous domain assignment and resulting in a false positive.

In another variation of these themes, false positives and false negatives can be caused by the error of sequence filtering. The so-called "simple" or low-complexity regions are segments with unusual amino acid or nucleotide composition. Homopolymer regions—that is, strings made of just one residue, such as polyglutamine expansions implicated in many hereditary diseases (Ding *et al.*, 2002)—are the extreme cases of a low-complexity region, but there are also more subtle patterns, such as enrichment in one or a few residues (Wootton, 1994; Wan and Wootton, 2000) or in perfect or degenerate repeats (Letunic *et al.*, 2004). Compositionally biased, non-globular regions are found in at least 20% of bacterial proteins (and in higher percentages in parasites than in free-living organisms; Koonin *et al.*, 1997). Many such regions have important functions. For example, they serve as flexible connectors between globular domains, as interfaces of protein–protein interactions, and they may also contain signals for protein sorting or degradation. Statistics of similarity search, however, breaks down in such regions because of their skewed sequence composition. As a result, matches to low-complexity regions are usually not informative, even as they seem statistically significant. The methods for obtaining correct statistics in these regions have been proposed only recently (Yu *et al.*, 2003).

In the 1990s, when the genome projects gained momentum, the issue of propagation of erroneous annotations caused concern to many of us. Annotations of newly sequenced genomes were increasingly done by computers, whereas finding bugs in the annotation pipeline, and cleaning up the results, requires slowly working human beings. There was a worry that errors will propagate to the point at which the databases will become completely unreliable—a meltdown compared by some researchers to mutational catastrophe in mathematical genetics (Eigen, 1971; Tannenbaum and Shakhnovich, 2004). There has even been a mathematical model of error spread in the databases that used formalism from the percolation theory and seemed to indicate that a large fraction of errors will spread fast (Gilks *et al.*, 2002).

Fortunately, as of today, the disaster did not quite come to be. The research community, by and large, is aware of the problem and is up to the job of improving and correcting annotations and functional predictions of proteins discovered in genomic projects—certainly not with perfect efficiency but still promptly enough to avert chaos. A particularly important source of reliable sequence annotations is a set of databases of conserved sequence domains. Behind each such database, there is a team of curators, to whose attention the community can bring the errors. Even more important, when wet-lab experiments provide serious evidence that contradicts dry-lab predictions, eventually this is reflected in such databases. To wit, the database annotation of HemK started to improve even before its biological role became clear; first it was recorded that there was no evidence for protoporphyrinogen oxidase activity, next the similarity to methyltransferases was noted, and most recently, information about the substrate specificity appeared. Of course, once a correct annotation is present in the databases, it will also be propagated automatically.

However, even if we infer homologs correctly, our prediction of protein function may still be incorrect. One set of reasons has to do with errors in sorting homologs into orthologs and paralogs. In Chapter 3, as well as earlier in this chapter, we discussed the principle of phylogenomics, which states that orthologs are more likely to share the same molecular, and even biological, function, whereas paralogs are more likely to have diverse, even if mechanistically related, molecular functions (their biological functions may still remain unknown : although most glutathione *S*-transferases are involved in maintaining redox balance in the cytoplasm and detoxification of xenobiotics, recall that one paralog in cephalopods is the main lens protein, utilized for its refractory properties rather than enzymatic activity).

Suppose that I have an uncharacterized paralog of a well-studied protein. If I erroneously think that the two proteins are orthologous, I am going to be overly specific when transferring functional annotation. This is overprediction, which can also occur as a result of inadvertent mix-up between molecular and biological function, For example, upon re-annotating the first completely sequenced genome of an archaeon *M. janaschii*, we came across a family of proteins that seemed to have expanded in this species and was related to a group of diverse, metal-dependent hydrolases found in many different species and possessing different biological functions (Koonin *et al.*, 1996). One of the members of this family was known in yeast and mammals as a component of the protein complex involved in processing of the 3′terminus of mRNAs, which is one step in the mRNA polyadenylation pathway. We chose to annotate the members of this family in *M. janaschii* as "putative metal-dependent hydrolases," inferring metal dependence from the pattern of conserved histidines, which, in at least one structurally characterized homolog, were involved in chelation of a zinc ion. However, in the absence of robust orthology assignment in 1996, we called these proteins "putative metal-dependent hydrolases, most likely nucleases." The alternative proposal (Kyrpides and Ouzounis, 1999) was that ours was an underprediction, and that proper annotation for at least some members of the family would be "homolog of cleavage and polyadenylation specificity factor subunit." On the contrary, I think that their suggestion, although literally true, is in fact an overprediction. Indeed, it draws attention to specific biological functions of this enzyme in eukaryotes, none of which have been demonstrated in archaea. At the same time, it does not state what is really known about this protein, namely, that it is firmly predicted to have a metal-dependent hydrolase activity.

One way to avoid these types of mistakes is to know your orthologs and paralogs and to employ a well-controlled vocabulary. Underprediction, though it may be a problem in some cases, can also be used as the annotation strategy: In a sense, it is safer to underpredict than to overpredict a function. When I call a protein "putative mevalonate kinase," most people would think that this is the predicted function of this protein. If I choose to call the same protein "mevalonate kinase homolog," there is an ambiguity: The statement is formally correct, inasmuch as it is stating that related, better studied proteins include mevalonate kinase, but it is not obvious from that description whether or not I really intend to say that the novel protein indeed has mevalonate kinase activity. However, if I decide to forego the last digits in the EC number and change annotation to a more general one, such as "predicted kinase of GHMP family," my meaning becomes less specific but in a sense more clear: I know to which class the enzyme belongs, and I am making it explicit that I am not sure about its specificity. It is known that the GHMP kinase family includes sugar kinases, mevalonate and phosphomevalonate kinases, homoserine kinases, and other small-molecule kinases, as well as some proteins that may not even be formally described as kinases, such as diphosphomevalonate decarboxylase (Bork *et al.*, 1993; Smit and Mushegian, 2000). My "generic" description of the protein function leaves all these possibilities open, gauging the level of our ignorance.

Thus, partitioning into orthologs and paralogs, if done correctly, may help to determine the appropriate level of functional prediction and annotation. In practice, however, there is always some overprediction and underprediction. One reaction to all these difficulties is to use annotation schemes that avoid sorting into orthologs and paralogs altogether (Ouzounis, 1999). I do not think that this is the best strategy: If we do not define orthologs, we simply assume "reasonable doubt" about transferring annotation from a better studied homolog to an uncharacterized protein across the genome. This mostly increases the underprediction rate.

It has also been proposed to define in a more precise way the meaning of qualifiers such as "putative," "possible," and "predicted." Many of such terminological suggestions are not very intuitive, and they have not quite caught on.

It is nevertheless useful to incorporate the evidence for each prediction into the sequence database, and it is perhaps better to do it in plain English than by using code words. For example, the annotation may contain a summary of the experiments that led to the prediction, the mutant phenotypes that have been observed, and so forth. This is quite practical in the case of well-studied model organisms and indeed has been implemented for some of them. For example, the following is the definition line of one yeast gene product, taken from GenBank:

> *Part of actin cytoskeleton-regulatory complex Pan1p-Sla1p-End3p, associates with actin patches on the cell cortex; promotes protein–protein interactions essential for endocytosis; previously thought to be a subunit of poly(A) ribonuclease; Pan1p [Saccharomyces cerevisiae].*

There are several desirable properties in this annotation. It contains information about the biological role of Pan1p; complexes that it forms; and earlier, probably incorrect, guesses of its function. What is missing is information about sequence similarities and the known homologs of this protein, but in the on-line databases, these relationships are only a few mouse clicks away (in this case, Pan1p is a large protein containing two relatively short regions of homology to other proteins, namely two EH domains, consisting of paired EF hands, in an arrangement typical of several endocytosis factors).

As we have previously seen, even when an uncharacterized protein has a correctly recognized ortholog in the database, and even if this ortholog is encoded by a genome of well-studied species, the annotation of the ortholog may be outright wrong. One notorious case is a eukaryotic and archaeal pseudouridylate synthase, the enzyme involved in base modification of rRNA, which for many years existed in the sequence databases under the name of "centromere/microtubule-binding protein." That annotation came from *in vitro* experiments prompted by the spurious observation of a short C-terminal KKD signature, which was thought to be a specific microtubule-binding signal (Jiang *et al.*, 1993). The orthologs of this protein are conserved in eukaryotes, archaea, and some bacteria, as is commonly the case with the proteins involved in RNA metabolism and translation (Anantharaman *et al.*, 2002). Some of these archaeal proteins acquired incorrect annotation from their eukaryotic homologs. Obviously, proper centromeres and microtubules are found only in eukaryotic cells (even though cell division GTPase FtsZ, found in most bacteria and archaea, is the ortholog of eukaryotic tubulin, the main protein component of microtubules), and "centromere-binding protein" in a prokaryotic cell does not make much sense. In the context of classification of protein function, such as EC or GO, this type of error is a serious misclassification.

Another large class of errors can also occur after orthologs and paralogs has already been defined. The function of many proteins can only be understood in context by examining other genes encoded by the same genome. Such was the case of GlyA, serine–glycine hydroxymethyltransferase, in *M. genitalium*. Homolog of GlyA is found in *E. coli*, and so the *Mycoplasma* GlyA protein was assigned to a group of proteins involved in biosynthesis of amino acids—undoubtedly prompted in part by a quick look at the name of the enzyme. This is not problematic in the context of the metabolic capacities of *E. coli*, which, indeed, possesses many enzymes for biosynthesis, degradation, and salvage of various amino acids—some of which are incorporated into proteins, and others serve different roles. In*M. genitalium*, however, there are no other enzymes for *de novo* biosynthesis of any amino acid (if we do not count a three-subunit complex that in Gram-positive bacteria and their relatives is required for amidation of a particular species of Glu-tRNA to form Gln-tRNA). Serine hydroxymethyltransferase does not seem to have any role in amino acid biosynthesis in *M. genitalium* because mycoplasmas receive their amino acids from their environment.

Better understanding of the role of GlyA in *M. genitalium* comes from the knowledge that conversion of serine and glycine by this enzyme is one step in the pathway of C_1 turnover by folic acid, which is highly conserved in bacteria and eukaryotes and has several essential functions, most notably a role in the thymidylate synthase reaction. In Chapter 6, we discussed a DOG involving two types of thymidylate synthases; each requires a hydroxymethyl group, donated by folate, in order to complete synthesis of thymidylate from uridylate. Without thymidylate, there is no DNA; without folate cycle, there is no thymidylate; and without GlyA, there is no folate cycle. Thus, the molecular function of GlyA (i.e., transfer of the hydroxymethyl group) can be properly placed in more than one pathway or functional category: In addition to amino acid biosynthesis, it has roles in the metabolism of folate (the category "coenzyme biosynthesis"), thymidylate ("nucleoside biosynthesis"), and, additionally in bacteria, special initiating amino acid formylmethionine ("translation"). And although molecular function of GlyA was recognized correctly all along, the correct interpretation of its biological role in mycoplasmas and other parasitic bacteria depends on knowing which other genes are also found there and which ones are missing.

Placing Predicted Functions onto Metabolic Maps and Filling the Gaps

After functions of gene products have been predicted on the basis of their homology to proteins in better studied species, these functions can be "placed on the metabolic map." The idea is simple: Buy a wall chart of metabolic pathways or use an on-line metabolic pathway database, start at the top of the list of gene predictions for the newly sequenced genome, and match every gene function to the same name in the database or on the chart. Continue gene placement until you run out of genes on your gene list, and you will have a sketch of metabolism in the genome of interest.

This is where the effort to use a controlled vocabulary for functional annotation, and to employ hierarchical function classification schemes such as EC and GO, pays off. Matching names and numbers can be automated, and the results of this matching can be displayed. However, there are at least two complications.

One obvious problem is that, whatever we do, there will be vacant spots in the metabolic database (i.e., known functions for which there are no good candidates in the genome that we are annotating). At the same time, there will be many genes in the gene list that cannot be placed on the map with any confidence. What needs to be investigated, then, is whether any of the latter, "orphan" genes should be assigned to some of the former, "vacant" functions, and how to do that with reasonable degree of accuracy. (Of course, wet-lab experimentation should in principle allow us to determine biological and molecular functions of all orphan genes, but our quest at the moment is to determine what can be done by computational genomics approaches). Here again, protein functions may transpire from the examination of genome context.

Before we discuss examples of such approaches, another problem has to be mentioned. As discussed at the beginning of this chapter, what metabolic maps really show is a consensus set of reactions, generalized from the information obtained by studying a relatively small number of organisms. A novel genome may belong to a species that performs chemical reactions still unknown to humankind and not represented in the biochemical databases. For example, the molecular function responsible for the ability of some microbes to incorporate a fluoride ion into organic compounds, forming a rare C–F bond, was identified only recently (Dong *et al.*, 2004), and it was not until 2005 that the first-ever cadmium-dependent enzyme was discovered (Lane *et al.*, 2005). The fluorination activity requires a single protein molecule that had been known for some time in several bacterial species under the name of COG1912. The fluorination of organic compounds, however, was not in the biochemical databases. Thus, the complete metabolic map is a work in progress, and taking differences between species into account is also an ongoing task. Our goal, then, is to start with the existing, if imperfect, metabolic map, on which many predicted functions of genes in the newly sequenced genome have already been placed, and to determine how far can we extend metabolic reconstruction.

For many gene products that cannot be confidently placed at a specific location in a map, some functional annotation is in fact available. For example, the genome of alphaproteobacterium *Agrobacterium tumefaciens* encodes approximately 40 predicted *S*-adenosylmethionine-dependent methyltransferases (some of which are misannotated, including the HemK ortholog, which is still called "protoporphyrinogen oxidase"). The general, or molecular, functions of these enzymes are known: They are transferring methyl (or carboxypropyl) groups from *S*-adenosylmethionine to some substrate. We know this because these proteins have significant sequence similarity to methyltransferases. However, in approximately half of all cases, the orthologs of these proteins in other species have not been functionally uncharacterized. The identity of their substrates, therefore, has not been established, and the exact biological function, or position on the metabolic map, remains unknown for all these enzymes.

From this simple example, we see the essence of the "candidate list" approach: Genes with general functional predictions can be compared to the lists of functions that are "missing from the map" in search of suitable matches. If, in the course of analysis of one pathway or another in *Agrobacterium*, biochemists conclude that there must be a methyltransferase performing a specific reaction, the list of orphan methyltransferases may be examined and two dozen candidates may be expressed and tested for a specific biological role or molecular activity. Although this may seem a less satisfactory way of prediction, compared to pinpointing of an ortholog with the known function, it is in fact a quite good way of planning confirmatory experiments.

There are several limitations inherent in the candidate list annotation, the most important of which is that we are selecting candidates only from the proteins that are already known to perform similar chemistries, so we will miss an isofunctional protein if it belongs to a completely uncharacterized COG or family. The other problem is that the function we are searching for may be truly and completely missing in the newly sequenced genome.

Notwithstanding these difficulties, functional inferences from homology information can be made for a large fraction of all conserved proteins in any genome. In most cases, one can expect to infer functions of at least 50% of the proteome with some confidence. Of course, the exact number depends on many factors, the most important of which is the choice of an appropriately sensitive method of finding homologous sequences.

This concludes our examination of methods that allow us to predict functions of uncharacterized proteins directly from their homology (better yet, orthology and paralogy) to other proteins. Some such functions are predicted unequivocally; others are tentative, perhaps in the form of candidate genes imprecisely matched to lists of candidate "missing" functions.

Our next task is to shorten these candidate lists and to predict functions of those proteins that either have no homologs in the databases or have only homologs that are not functionally characterized in any way. In Chapter 8, I discuss what is sometimes called the "nonhomology" approach to function prediction and show that homology analysis is implicit in this class of methods, too.

8

Prediction of Function and Reconstruction of Metabolism: Post-Homology Approaches

In addition to sequence homology, proteins have other recognizable features that are useful for functional annotation, including signal sequences, which serve to facilitate localization or retention of proteins in specific cellular compartments; hydrophobic and hydrophilic regions, which can be used to find transmembrane, extracellular, and intracellular sequences; and regular (e.g., periodic) patterns of amino acid distribution. Sometimes, these traits of proteins are called "intrinsic features" to indicate that they are computed directly from the protein sequence, without database searches. This is slightly misleading because the most accurate approaches to intrinsic feature prediction usually rely on machine learning (i.e., training of the algorithm on a set of sequences that have a desired property). Thus, there is often an implicit special-purpose database of reference sequences in this case, too. In any case, this information may help in pinpointing the candidate gene for a "missing" function, for example, by selecting those that are more likely to be targeted into a compartment of interest.

Other methods predict protein function by simultaneously examining the behavior of many genes in multiple completely sequenced genomes. We examine two approaches that require no other information than the genome sequences with correctly recognized open reading frames. One of these methods is based on the analysis of conservation of gene order and gene clusters across multiple species, and the other is based on analysis of the presence and absence of homologous genes, also across multiple genomes.

We start with gene clustering and gene order conservation. This approach can be outlined as follows: For gene A_A in newly sequenced genome *A*, define its neighbor genes; also define the orthologous gene(s) A_B in genome *B*, A_C in genome *C*, and so on, and find neighbors of A_B, A_C, etc. Then compare the lists of neighbors.

There are actually two separate ideas here. One is that genes with related functions may have a tendency to be clustered in the genome. This can be tested within a single genome and, indeed, the basic concept of operons—groups of genes that are involved in the same process or pathway and are (transcriptionally) regulated as one unit—was proposed without resorting to interspecies comparison. The other idea is that if such clustering of genes with related functions on the chromosome is selectively advantageous, then the same arrangement of the orthologous genes may be observed in many different species: If A_A and B_A in genome *A* are neighbors, then their orthologs A_B and B_B in genome *B* also stand a good chance of being neighbors. The important question is how to define and measure the "neighborliness" of genes.

Clustering of genes on the chromosome may occur in three different forms. The most extreme one is translational fusion, when two proteins are combined into one multidomain protein (of course, one or both of the fusion partners may already be multidomain proteins). In fact, one of the definitions of a multidomain protein is that its components can exist as separate proteins, or can have different fusion partners, in another species. Translational fusion of genes is a strong indication that they are involved in the same biochemical pathway or a functional system.

Another level of gene clustering is transcriptional fusion. Here, one RNA encodes several open reading frames, which can direct synthesis of several distinct proteins, using either bacterial strategy of internal initiation of translation or, as discussed in Chapter 4, various virus-like mechanisms of translating many proteins from one transcript. In this case too, the products of these gene clusters (operons in bacteria) tend to work together, although exceptions, again, are known. The definition of "working together," however, is broader for transcriptionally fused proteins than for those proteins that are fused at the polypeptide level. Operons often code proteins that act in the same metabolic or signaling pathway, but this does not always mean that these proteins are physically linked or are produced in equimolar amounts in the cell. In contrast, both properties are observed in translational fusions.

The third type of gene clustering does not involve polycistronic transcripts or polyproteins. It is simply a tendency of certain genes to be positioned close to each other on a chromosome. Some of these clusters may be revealed by analysis of a single genome, for example, by searching for regions in which intervals between adjoining genes are shorter than the average for this genome, or by finding zones with overrepresentation of particular types of pairwise gene arrangement (head-to-head, tail-to-tail, or head-to-tail). Other clusters are discovered by comparing lists of gene neighbors across many genomes and identifying genes that are found in clusters in several genomes. Some of these clustered sets of conserved genes ("neighborhoods") may be functionally coregulated, although often the connection between them is more remote, for example, these groups may display similar levels of expression (Rogozin *et al.*, 2002; Boutanaev *et al.*, 2002; Kalmykova *et al.*, 2005) or similar tissue specificity (Li *et al.*, 2005).

One example of successful functional inference by gene proximity is desiphering of biosynthesis of terpenoids in Lyme disease spirochete, *Borrelia burgdorferi.* There are at least two pathways of terpenoid biosynthesis in living organisms. The mevalonate pathway, which was discovered first, operates, with variations, in fungi, animals, in plant cytoplasm, and in some bacteria, including *Borrelia.* The trunk pathway comprises six enzymes, five of which have orthologs in *B. burgdorferi.* Yeast isopenthenyl pyrophosphate isomerase (IPPI) has orthologs in plants and animals, and it belongs to a vast Nudix family, which includes mostly enzymes with pyrophosphatase activity (IPPI, however, is isomerase, not pyrophosphatase, and it is thought to retain pyrophosphate-binding ability, which is handy for interacting with the substrate). There are no orthologs of IPPI in *B. burgdorferii*, and all known enzymes from the Nudix group in *Borrelia* are "taken" (i.e., unrelated functions can be assigned to them). However, five genes of the mevalonate pathway in *Borrelia* are arranged in a row, close together in the same DNA strand. This looks very much like an operon, and in the middle of the same operon there is the sixth gene. A few years ago, we predicted that this protein is a founding member of the new class of isopenthenyl pyrophosphate isomerases (Smit and Mushegian, 1999), and this prediction turned out to be correct (Kaneda *et al.*, 2001).

The already discussed case of HemK methyltransferase, however, shows the limitations of inference-by-proximity. The *hemK* gene in *Escherichia coli* is one gene away from *hemA*, another gene with the role in protoporphyrinogen biosynthesis, but HemK is involved in modifying translation termination factor. Thus, gene clustering on the chromosome does not guarantee functional linkage.

The accuracy of inference improves if we compare many genomes and using better definitions and more quantitative approaches. But what is a neighbor? Is "closeness" or "neighborliness" a binary (all-or-none) trait, or could there be degrees of it? Any two genes are separated by some distance on a chromosome, and this distance can be measured, for example, in base pairs or in the number of intervening genes, but what is the maximal distance between two proteins at which they are still considered neighbors? How should we measure clustering of several genes in one chromosome and how should we compare the extent of clustering in several genomes? How should we handle insertions, deletions, and chromosome rearrangments? What kinds of functional linkage exist, and what kind of linkage is possible to infer from gene closeness?

The interplay of evolutionary, structural, and functional information is important here. Gene clusters may be functionally significant, telling us about the involvement of genes in the same pathway; this is what mostly concerns us in this chapter. However, clusters of genes in two closely related species may reflect only the gene order in the common ancestor of the two species. Finally, some clusters may be the by-product of the chromosome rearrangements. We would like to distinguish the latter two signals from the functional connections between clustered proteins.

Before the era of complete genomes, most evidence on chromosome colinearity came from comparative karyotyping of eukaryotes and the study of chromosome translocations (Wolfe *et al.*, 1991; Sankoff *et al.*, 1992; Hannenhalli *et al.*, 1995). In the 1980s and 1990s, many organellular genomes were sequenced, and the comparative studies of chloroplast genomes indicated that gene order in the chromosome is affected by rearrangements (often inversions) of large DNA segments. This is the area in which many interesting mathematical results and new computer algorithms have been produced, as explained in much detail by Sankoff and Blanchette (1998, 1999), Pevzner (2000), and Eichler and Sankoff (2003). These algorithms for counting and ordering chromosome rearrangements, however, reveal mostly information on the evolution of the chromosome as a physical entity and do not tell us much about protein function. Since we are examining metabolic reconstruction in this chapter, we will not discuss these studies in detail. Here, our attention will be mostly on the genomes of prokaryotes (bacteria and archaea), in which gene order carries significant functional signal.

Anecdotal evidence of isofunctional operons in distantly related bacteria began to accumulate as early as the 1960s. Ribosomal genes were organized in similar operons in many distantly related bacteria, and partial sequencing of *Salmonella* species, a gammaproteobacterium diverged from *E. coli* approximately 150 million years ago, revealed long regions of near perfectly conserved gene order in two species (Neidhardt *et al.*, 1996).

The first complete genome sequence of *Haemophilus influenzae* was published in 1995, when a significant portion of *E. coli* genome was also available. *Haemophilus influenzae* is a gammaproteobacteirum that is more distant from *E. coli* than *Salmonella. H. influenzae* also has a much smaller genome than *E. coli* (less than 1800 protein coding genes in *Haemophilus* vs. approximately 4200 in *E. coli*). There is essentially no long-range colinearity in gene order between *Escherichia* and *Haemophilus* (Tatusov *et al.*, 1996). At higher resolution, however, there were 226 "gene strings" (i.e., the sets of adjacent orthologs running in the same order in both genomes, allowing for one or two gene indels per string). In total, these strings contained 825 genes. The majority of strings contained 2–4 genes, and a few strings were relatively long (e.g., 28 genes in one of the operons that consisted mostly of ribosomal proteins). These genes accounted for 78% of all orthologs shared by *H. influenzae* and *E. coli* genome. Notably, only half of these strings, and 40% of all genes that were included in strings, belonged to operons in

E. coli (Tatusov *et al.*, 1996), and the other half of the strings comprised genes without clear functional connection to one another. Many, perhaps most, of these strings were probably conserved because the ancestral gene order has not been completely disrupted by ongoing genome rearrangements, not because of selective forces preserving functional links between adjacent genes. In conclusion, at the evolutionary distance of several hundred million years, approximately half of gene colinearity seemed to be explained by functional connections between genes, and the other half appeared to be due to the evolutionary signal.

At the other extreme of divergent evolution in bacteria, we have compared the almost complete genome of *E. coli* with a deeply branching blue-green *Synechocystis* sp. and the first completely sequenced archaeal genome *Methanococcus janaschii* (Koonin *et al.*, 1997). Although there were approximately 400 orthologs shared by all three species, only one-tenth of these orthologs were arranged in conserved strings in all three genomes. Thus, at very large evolutionary distances, the conservation of gene order is only barely seen. This is probably because the selective advantages of gene clustering on the chromosome, such as the ability of species to coexpress functionally linked genes as a single polycistronic transcript or a polyprotein, are not infinitely high: Given enough time, any two adjoining genes will be set apart by DNA insertions, deletions, and recombination, and these processes will not be offset by selection for gene clustering on the chromosome.

Examination of the most conserved gene strings showed that each such string comprised orthologs related in one and the same, quite specific, way: They were not merely involved in the same biological function or metabolic pathway but were parts of stoichiometric multiprotein complexes. Examples of such strings included ribosomal proteins, two of the largest subunits of RNA polymerase, and a few others (Mushegian and Koonin, 1996b). Our observations have been confirmed and extended by Peer Bork's group at the European Molecular Biology Laboratory in a paper with a self-explanatory title, "Conservation of Gene Order: A Fingerprint of Proteins That Physically Interact" (Dandekar *et al.*, 1998). Stoichiometric amounts of proteins would be easier to obtain when all proteins are translated from the same transcript. Furthermore, a polycistronic transcript may perhaps be viewed as a cellular microcompartment, where locally high concentrations of emerging proteins facilitate their recognition and interaction. However, experimental evidence for these hypotheses is still missing.

Analysis of "gene strings" relied on a narrow definition of conserved gene order: There had to be two or more orthologs in the same order in two or more chromosomes, with no more than two gene indels per species. This approach is intended to recover the most conserved, collinear sets of genes. In fact, some of the conserved strings conforming to these constraints may be quite "patchy" (Fig. 8.1). It seemed quite clear that there may be additional conservation of local gene order, which we would not see if we did not consider longer indels and local permutations of genes.

In 1999, Overbeek and co-workers, then of Integrated Genomics, Inc., proposed a broader definition of a gene pair that is "close on the chromosome." Two genes are a "close pair" if they both belong to a group of genes encoded in the same DNA strand and none of these genes is separated from its immediate neighbors by more than 600 base pairs. The number 600 is not altogether arbitrary: It may be seen as an estimate of intergenic distance in the known operons (perhaps on the higher side). Thus, some genes may be considered "close" in this sense even if they are separated by many genes, and a pair remains close if gene order is shuffled as long as all genes stay in the same coding strand (Fig. 8.2). The "conserved pair of close genes" in genomes *A* and *B*, then, is really not one pair but two pairs of orthologs A_A, A_B and B_A, B_B, such that A_A and B_A are close in the genome *A*, and A_B and B_B are close in the genome *B*. The authors noted that there are approximately 1000 conserved close pairs of genes in 4 genomes and

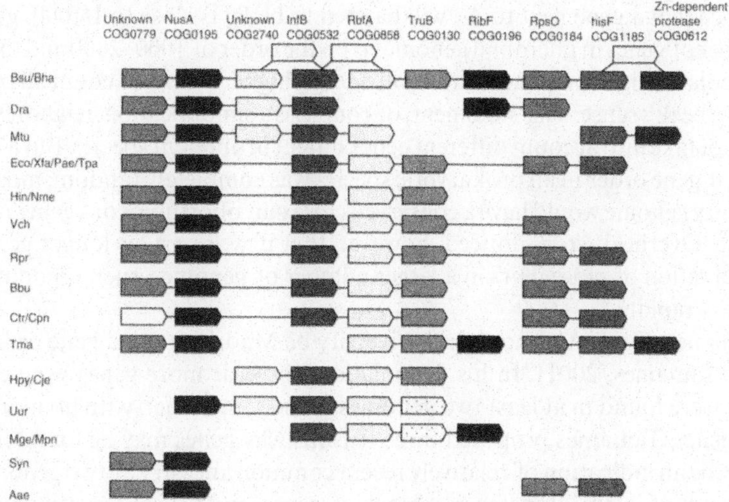

Figure 8.1. Gene strings with partially conserved gene order. Orthologs are indicated by similar shading, and inserted genes in *Bacilli* without homologs in other species are shown at the top. Reproduced from Rogozin, I. B., Makarova, K. S., Murvai, J., *et al.* (2002). Connected gene neighborhoods in prokaryotic genomes. *Nucleic Acids Res.* **30,** 2212–2223, by permission of Oxford University Press.

nearly 58,000 such pairs in 31 genomes. One-third of all genes included in conserved close pairs were functionally connected to each other.

This study is notable for the extended definition of conserved close pairs, but important questions remained unanswered. Some of the "conserved close" genes are functionally linked, but some are not: Why is this so, and what can be said about each of these subsets of genes? Do all close genes that are functionally linked belong to operons, or are there other forms of clustering? What are the forces that preserve gene clustering on the chromosome, if these genes are not functionally linked?

Overbeek and coauthors also asked what would be the expected number of randomly generated close conserved pairs. It is easy to see that the random component is not negligible here: If A_A and B_A are adjacent in genome *A*, and both genes have orthologs in genome *B*, the

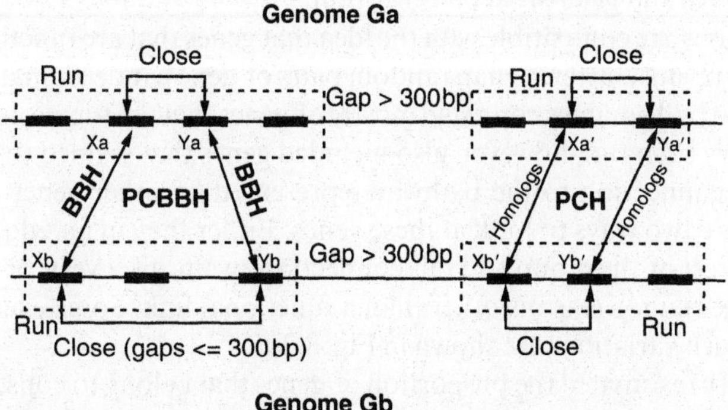

Figure 8.2. Close conserved pairs of genes. Reproduced from Overbeek, R., Fonstein, M., D'Souza, M., Pusch, G. D., and Maltsev, N. (1999). The use of gene clusters to infer functional coupling. *Proc. Natl. Acad. Sci. USA* **96,** 2896–2901. Copyright (1999) National Academy of Sciences, U.S.A.

random chance that a gene next to A_B will happen to be B_B is close to 1/1000, given that the average number of genes in microbial genomes is on the order of 1000–2000 and that each gene has two immediate neighbors. The chance will be even higher if we also count all genes that are "close" in Overbeek's sense. This statement, of course, is a simplification; it ignores gene polarity and does not take into account different gene content in different species. But all things considered, even if gene order in all prokaryotic species was completely random, most close pairs of genes in every genome would have a conserved close pair of orthologous genes in at least one other genome. Overbeek *et al.* noticed, however, that if we consider longer gene strings or require conservation of close pairs in a larger number of genomes, such random probability diminishes very rapidly.

Javier Tamames of the Autonomous University of Madrid studied gene order conservation further (Tamames, 2001). In his formulation, three or more genes were considered a "run" if they were found in at least two genomes in the same order, with no more than three intervening genes. Tamames proposed three reasons why genes may be found in such runs: (1) A run may be an indication of relatively recent common ancestry of two genomes,(2) a run may be the result of lateral transfer of a block of genes, and (3) conservation of gene order may be advantageous for the organism because of functional optimization that is provided by gene adjacency. It may seem that only the latter case represents functional signal, whereas the other two are purely evolutionary (i.e., they reflect only gene order in the common ancestor). But, of course, some of the gene clusters in that common ancestor may have included functionally linked genes, too, and therefore the first and last explanations are not mutually exclusive. Regarding lateral gene transfer, there is a hypothesis that transfers of whole operons may be more advantageous to the recipient than transfers of single genes or of groups of functionally unrelated genes (Lawrence and Roth, 1996). This is because operon may be more likely to code for a complete pathway and thus may be more likely to endow the recipient with a complete new function. Clearly, this explanation of gene clustering also contains a functional component.

In the same work, Tamames studied the shape of the curve that relates the number of genes in runs and the evolutionary distance between species (Tamames, 2001). This curve is sigma-like, indicating that in many groups of closely related species and genera, the gene order remains very much the same, whereas in extremely diverged species the gene order is almost randomized, with the exception of the small number of extremely well-conserved operons. Most of these operons are dominated by ribosomal proteins, as had been previously noticed. However, the middle, close-to-linear, part of the curve was considerable, suggesting that at a broad range of medium evolutionary distances, there is an almost linear relationship between evolutionary distance and gene order conservation.

These observations are compatible with the idea that genes that are functionally linked may be more prone to staying clustered than random pairs of genes. Indeed, many of the best conserved runs noticed by Tamames consisted mostly of genes that are functionally linked to each other. Almost every such run, however, also included genes that seemed to be "out of place" (i.e., without any connection to the pathway represented by other genes in the same run). Obviously, there are two ways to look at these genes: Either they in fact do have a functional connection to the rest of the group and this connection awaits discovery, or they are clustered with the rest of the run for reasons other than a functional link. Some examples of the conserved runs with such variations are shown in Fig. 8.3.

Wolf *et al.* (2001b) estimated the proportion of genes that belong to conserved sets, defined along the same lines (at least two genes per a run and no more than two indels per a run). If we ignore species that are closely related (i.e., species of the same genus and some closely related genera, such as two enterobacteria *Escherichia* and *Yersinia*), the proportion of genes falling into strings shared by at least two genomes is in the range of 5–25%, and require that the

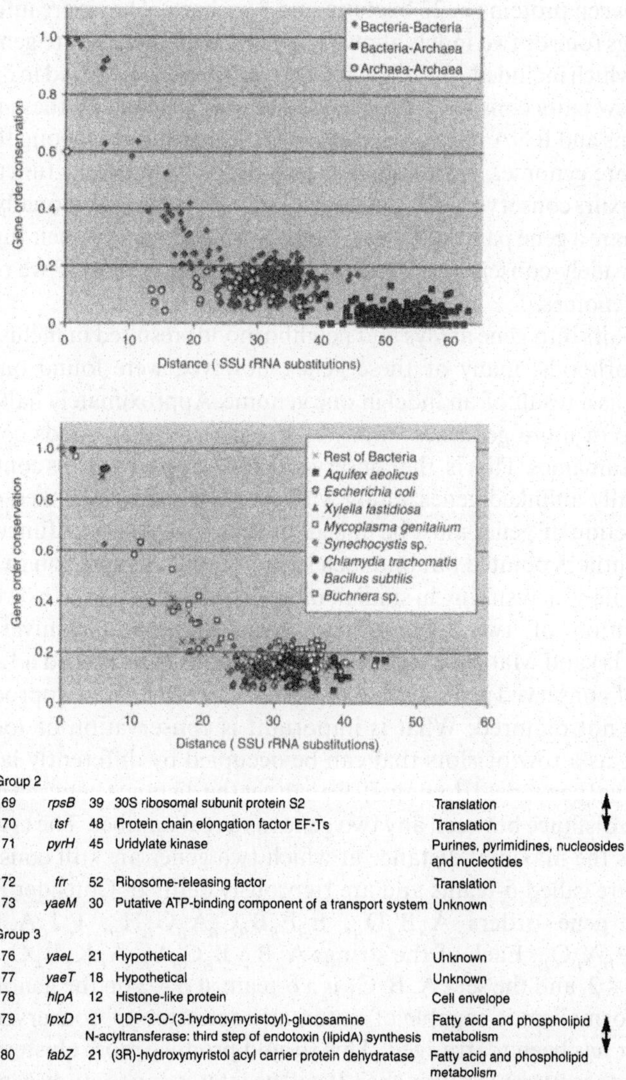

Figure 8.3. Partially conserved gene order in prokaryotes. (**Top**) The function that relates conservation of gene order and evolutionary distance between species has a sigmoid shape, indicating rapid decay in colinearity at the intermediate evolutionary distances. (**Bottom**) Conserved order of genes is not always easily explained by their involvement in the same metabolic pathway. Reprinted from Tamames (2001) under BioMed Central Open Access license agreement

string is shared by at least three genomes, the proportion of genes that belong to strings declines sharply. This indicates that the chance for an uncharacterized gene to become functionally annotated on the sole basis of its closeness to genes with known functions is relatively low.

Rogozin *et al.* (2002) analyzed extended regions with conserved gene content in some detail. They defined "neighborhoods" as regions more complex than strings or runs. For example, strings $A_A B_A C_A$, $A_B C_B$, $B_C C_C D_C$, and $C_D E_D$ in genomes *A*, *B*, *C*, and *D* give rise to a neighborhood consisting of genes A, B, C, D, and E. These and more complex configurations of neighboring genes can be detected by a well-defined algorithmic approach. As in most other studies, we start with conserved gene pairs. Rogozin and co-authors used COGs to define orthologous relationships between proteins in 23 bacteria and 8 archaea. They were interested in pairs of adjacent orthologs (one or two indels allowed), shared by three or more genomes. There were 1505 such pairs, which included 1337 COGs. Most pairs were conserved in 3–13 genomes, and there were very few pairs conserved in all genomes (as expected, all these pairs consisted of ribosomal proteins and RNA polymerase subunits). Interestingly, among the gene pairs conserved in 6 or more genomes, approximately two-thirds were clearly functionally linked; in contrast, among pairs conserved in 3–5 genomes, only 36% were functionally linked. Thus, the more genomes share a gene pair, the more reliable is the inference of their functional link. The number of such widely-conserved pairs, however, decreases rapidly if we require them to be present in more genomes.

Extension of pairs into gene arrays and neighborhoods resulted in inclusion of 6611 genes into 118 neighborhoods; many of these genes, however, were found only in one or two genomes, mostly as a result of an indel in one genome. Approximately half of all genes were inherited in three or more genomes. Analysis of these neighborhoods confirmed previous observations by Tamames. That is, the conserved sets of clustered genes contain a fair proportion of functionally unlinked genes, along with some functionally linked ones. In this case, again, a large fraction of genes must be linked for reasons other than functional interaction. Rogosin and coauthors pointed out one such likely reason—adaptation to a similar mode of transcription regulation, resulting in similar, most likely high, levels of expression.

The collaboration of two laboratories, Anne Bergeron's (University of Quebec, Montreal, Canada) and Matthieu Raffinot's (CNRS-Evry) produced a further generalization of the idea of conserved gene clusters (Luc *et al.*, 2003). In their approach, conservation of gene order is not required: What is important is conservation of local gene content. Treating genome as a row of slots that can be occupied by differently labeled genes (e.g., COGs or individual conserved domains), we can set the distance between adjoining genes at 1 and express the distance between any two genes as a real number. The crucial parameter in the analysis, δ, is the maximal distance at which two genes are still considered close. The neighborhoods are called δ-teams and are rigorously defined. Consider four chromosome fragments with gene orders $A_AB_AD_A$, $E_BF_BB_BC_BA_BG_BH_B$, $I_CJ_CA_CK_CB_CC_CL_C$, and $M_DN_DO_DC_DA_DP_DA_DQ_D$. Each of the strings A_AB_A, $B_BC_BA_B$, $A_CK_CB_CC_C$, and $C_DA_DP_DA_D$ are δ-chains at $\delta = 2$, and the set {A, B, C} is a δ-team at $\delta = 2$ on the genomes $\{A, B, C, D\}$. The gene team formalism is capable of automatic detection of conserved gene clusters in which gene order has been rearranged. This should produce more clusters, and more genes per cluster than other current approaches. It will be interesting to study gene teams extracted from different sets of genomes at various values of the δ parameter. I expect, however, more of the same theme with variations: Along with novel functional links, there will be clusters of genes that evade functional connection.

In this chapter, I view multidomain proteins and gene neighborhoods on the chromosome as different, but ultimately related, modes of gene clustering because there is no clear-cut evolutionary boundary between different types of gene fusion. Shuffling of discrete genes and of protein domains occurs through fundamentally the same process of DNA recombination. Teams include more genes than neighborhoods, neighborhoods include more genes than operons; operons are fusions of genes at the transcriptional level, but some of the genes can be fused even further, at the translational level, into multidomain proteins. Gene–neighborhood and gene–team approaches can be naturally extended to deal with individual protein domains (Pasek *et al.*, 2005).

We now examine another approach to inference of protein function—the one that makes inferences from the presence and absence of orthologs in different genomes. In Chapter 5, I introduced phyletic patterns. They were invented by Tatusov, Koonin, and Lipman (1997) as a way to represent the distribution of any COG across different phylogenetic lineages. The authors called them "phylogenetic patterns" and demonstrated how they can be used to track gains and losses of genes in evolution. Matteo Pellegrini and co-workers, in David Eisenberg's laboratory at the University of California at Los Angeles (UCLA), suggested that the same constructs, which they called "phylogenetic profiles," can also be used to infer functional links between genes and proteins (Pellegrini *et al.*, 1999). In another terminological quibble, I prefer "phyletic" to "phylogenetic," because a pattern explicitly tells us about gene presence or absence in each phylum, whereas the phylogenetic history of this gene remains implicit. (A separate and interesting problem of using phyletic patterns to learn more about evolution is discussed in Chapters 12 and 13).

Phyletic pattern is a record of presences and absences of a gene (or else domain or COG) in several completely sequenced genomes. Phyletic pattern most naturally takes the form of a binary vector—that is, a string of numbers, where each number (binary coordinate, corresponding to one genome) is set at either one, if a gene is present in a species, or zero, if a species has no such gene. Let us therefore call the whole construct "phyletic vector." The set of coordinates of a vector (i.e., the list of species that are examined) can go in any order; one useful way to order the coordinates is by traversing the tips of the phylogenetic tree (see Fig. 5.7). Any gene found in only one species has phyletic vector with only a single coordinate set to one. If a vector corresponds to a COG, there will be, by definition, at least three nonzero coordinates in it (see Chapter 5). In the rest of this chapter, we examine only vectors that correspond to COGs.

With two possible values for each coordinate, there can be $\sim 2^{120} \approx 1.3292 \times 10^{36}$ COGs in the ~120 species included in the current release of the NCBI COG database. In fact, there are only 14,669 COGs; obviously, only a tiny fraction of all possible phyletic vectors is encountered in nature. Moreover, the number of phyletic vectors is smaller than the number of COGs because there are groups of COGs that share the same vector (Fig. 8.4). More than 90% of all COGs are found in less than one-third of all species, and only 70 COGs are found in every species.

Thus, the observed set of phyletic vectors is far from random: Some vectors are overrepresented, and most vectors are not found at all. One reason for missing vectors is trivial: There are only approximately 10^7 genes in the genomes that are included in the COG database. However, distributions of each gene across genomes are not independent: For many pairs of genes, their presences and absences across many genomes correlate. One type of such nonrandom distribution is when two or more genes are simultaneously present in the same set of genomes and absent from the other genomes. The idea of Pellegrini and coauthors was that a major reason for such coinheritance is functional connection between the coinherited proteins. Their proposal was to find, for each gene, all genes with the same or similar phyletic pattern, where "similar" was defined as any vector that has three or less coordinates different from the vector of interest (in mathematics, this measure is called Hamming distance). Genes that are functionally linked to the gene of interest were expected to be well represented among those "phyletic neighbors" (i.e., genes with similar phyletic patterns).

This leaves many open questions. First, functional interaction may not be the only reason why genes are coinherited: As with clustering on the chromosome, we may expect that the copresence of some groups of genes in the same genomes reflects mostly evolutionary closeness of these genomes. Second, the allowance of Hamming distance up to three between the vectors of coinherited genes is arbitrary—why not four or five? Finally, Pellegrini *et al.* produced no actual discoveries using this approach in their original paper, limiting validation of the approach only to demonstrating that genes that are known to be functionally linked indeed tend to be coinherited. Thus, the proof of concept for this provocative idea deserved better. Within several years, however, several groups used this principle to find candidates for "missing" functions, and several such discoveries were confirmed in the wet lab. In one example, phyletic vectors aided in finding novel components of terpenoid biosynthesis pathway in bacteria. Most bacteria are unlike the aforementioned *Borrelia*: They do not produce terpenoids

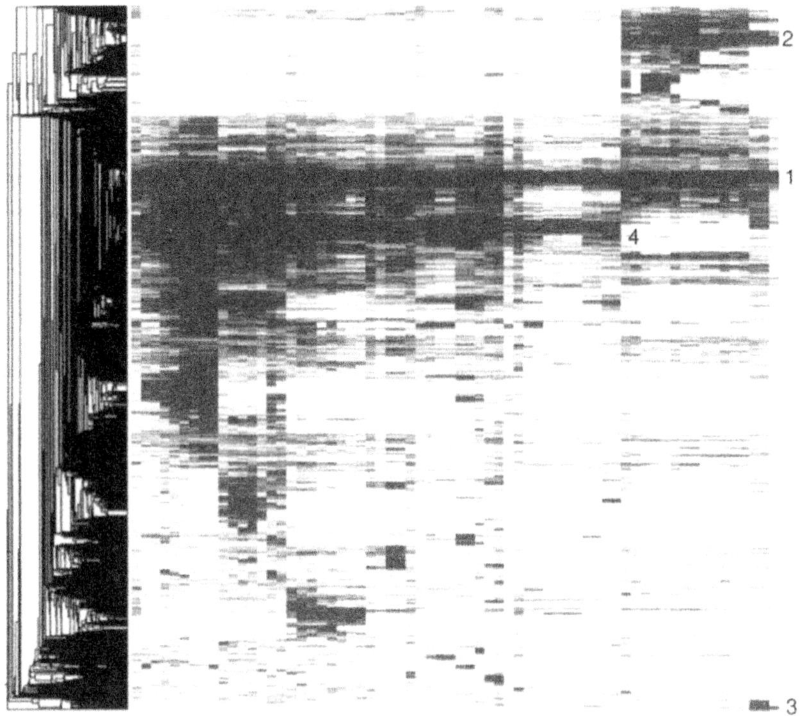

Figure 8.4. Hierarchical clustering of the space of phyletic vectors. Rows are COGs, and columns are completely sequenced genomes. 1, Cluster of COGs present in all living species; 2, cluster of Archaeo-Eukaryal ("unbacterial") COGs; 3, small cluster of eukarya-specific COGs (all three eukaryotic species in the data set are fungi or related microsporidia); 4, pan-bacterial COGs missing in Archaea (note that some of them are present in Eukarya as a result of mitochondrial origin by symbiogenesis). Reprinted from Glazko and Mushegian (2004) under BioMed Central Open Access license agreement.

by the mevalonate pathway. Instead, they make terpenoids starting from 1-deoxy D-xylulose, and in the late 1990s the first three genes in the corresponding pathway (*dxs*, *dxr*, and *ygbP*) were cloned. Then Luttgen *et al.* (2000) searched for genes with the same phyletic pattern as these genes and found *ychB*, a gene whose product indeed turned out to catalyze the remaining reaction in the pathway.

Carlson *et al.* (2004) studied biosynthesis of selenocysteine, the 21st amino acid that is encoded by certain UGA codons in a subset of archaea and eukaryotes. They wanted to find a kinase that phosphorylates minor seryl-tRNA, the intermediate in selenocysteinyl-tRNA biosynthesis. Only 2 of the 14 archaeal genomes completely sequenced at the time insert selenocysteine into proteins, and search of the COG database revealed 27 COGs with exactly that phyletic pattern ("one" in both selenocysteine-positive genomes, and "zero" in the rest). Two of these 14 proteins were paralogs of known kinases. One of them, annotated as "predicted sugar kinase," had no orthologs in eukaryotes, and the other, annotated as "predicted nucleotide kinase," had orthologs only in humans, fruit flies, and nematodes (i.e., the eukaryotes that also are able to insert selenocysteine into proteins) but no orthologs in higher plants and fungi, which lack the selenocysteine insertion system. The protein product indeed turned out to have strong and specific phosphoserine-tRNA kinase activity.

In both these cases, some of the genes involved in the pathways were known, and their phyletic vectors were used to query the database for all other phyletic vectors. But an interesting pattern query can also be made up. For example, Patrick Forterre of Universite Paris-Sud, currently at Pasteur Institute, searched for genes associated with a particular phenotype, namely the ability to thrive at very high temperatures (Forterre, 2002). Among 26 prokaryotes covered at the time by the COG database, there were 6 hyperthermophiles (4 archaea and 2 bacteria with growth optimum from 80 to 106°C), 1 thermophile with growth optimum much lower than 80°C, and 19 mesophiles. Phyletic vector with six coordinates set at one and 20 coordinates set at zero was compared to the COG database, and only two COGs were associated with such a vector. One COG was eliminated when a few more genomes of hyperthermophiles were sequenced and turned out to lack this COG. The only remaining COG was reverse gyrase, the enzyme that can positively supercoil covalently closed DNA, a property useful for preventing excess denaturation of a double-stranded molecule at high temperature.

Another way to compare phyletic vectors is to search for a "complementary pattern" (i.e., for a vector in which all the ones and zeroes are reversed. The thought to flip the coordinates occurred to me under slightly amusing circumstances. In 1998, I came across the paper by Aurora and Rose of Johns Hopkins University in which they used their own method to predict structure from sequence (Aurora and Rose, 1998). They applied that method (details of which were never published) to search for thymidylate synthase in archaea. I had analyzed proteome of *M. janaschii* a few months earlier as part of a large project of comparative analysis of several completely sequenced genomes (Koonin *et al.*, 1997), and in my opinion, there was a perfectly good candidate for thymidylate synthase function in that species—MJ0757. Aurora and Rose saw it but dismissed it. Their reason for doing so and their suggestion of the alternative candidate are not of much interest, but this prompted me to examine the phyletic distribution of thymidylate synthase. To my amazement, several completely sequenced genomes indeed lacked any recognizable homolog of thymidylate synthase. This is significant because the only type of thymidylate synthase known at the time, specified by *E. coli* ThyA protein, is found in a wide variety of eukaryotes, prokaryotes, and viruses. ThyA family members are exceptionally well conserved in all these diverse species; in fact, this is one of the best conserved protein sequences among those that cover such a large evolutionary span. Yet, *Synechocystis*, *Helicobacter*, and several other bacteria and archaea have no ThyA homolog at all. Then I examined the COG database and found that there was exactly one phyletic vector that had every coordinate reversed compared to the vector for ThyA. This COG was already annotated as "thymidylate synthase-complementing protein Thy1" based on the activity of a homolog from the slime mold *Dictyostelium*, which was picked from the expression library for its ability to rescue thymidylate auxotroph on thymidylate-lacking medium (Dynes and Firtel, 1989).

I discussed the possibility that Thy1 is another thymidylate synthase, a displacement of "missing" ThyA homolog in a subset of bacteria and archaea, with Eugene Koonin and Michael Galperin, who extended the observation of "complementary" or "mirror" phyletic pattern of ThyA and Thy1 to a larger number of species. This prediction was confirmed by showing the activity of Thy1 homolog, renamed ThyX, from *Thermotoga maritima* (Myllykallio *et al.*, 2001).

The main problem with matching phyletic vectors, whether in "direct" or "complementary" manner, is that the process of gain and loss is never perfectly synchronous in different genes, even if these genes and their products interact with each other. Gene losses, and related pathway remodeling, perturb the pattern of vector coinheritance. The optimal way to find groups of related phyletic vectors, and to understand what signal is represented in these groups, still remains to be discovered. In the past few years, however, several advances toward this goal have been made.

The key question in any comparison of sequences, numeric vectors, or anything else, is the choice of distance or similarity measure. Phyletic patterns are binary vectors, and there are many ways to compare them. Galina Glazko and I have noted that there is a crucial requirement that needs to be satisfied by any distance measure between phyletic vectors (Glazko and

Mushegian, 2004). Consider four proteins (or genes/domains/COGs) A, B, C, and D, with patterns across seven genomes, A = (1011110), B = (0111110), C = (1000000), D = (0000001). We are interested in whether there is a functional link between A and B, and between C and D. It can be said that A and B proteins tend to be inherited together, and it is quite obvious that C and D are not coinherited. Distances that are derived from correlation or mutual information (in fact, these two distances are equivalent in the case of binary vectors; Li, 1990), where zero coordinates do not contribute to the distance or similarity, are helpful for resolving this problem, whereas Hamming distance, Euclidean distance, and other so-called L_p-norm distances are inadequate (Glazko and Mushegian, 2004). Apart from these preliminary observations, not much is known about the properties of distance spaces between phyletic vectors, and more work in this area is needed (Glazko *et al.*, 2005).

Chromosome proximity and coinheritance are two types of genomewide information that have been hailed as the "nonhomology methods of function prediction." This name is somewhat misleading: Both methods rely, in a major way, on establishing the sets and positions of homologous genes across many genomes. Only after all pairs of homologs (ideally, orthologs) have been accurately defined can we hope to use their clustering or phyletic patterns for functional inference. Perhaps "post-homology" may be a more appropriate moniker for these methods.

So, what has been contributed, at the genome scale, by the post-homology methods of metabolic reconstruction? The answer is surprisingly difficult to come by. Even David Eisenberg's group at UCLA, which did much to raise the profile of these approaches, never estimated the efficiency and accuracy of these approaches in any detail. There are several on-line databases in which a number of functional predictions are made for proteins in many completely sequenced genomes, but these predictions are typically scored on the basis of complex, not well-explained systems of joint inference from many types of evidence, starting with sequence similarity, followed by post-homology methods, and further supplemented by analysis of genome-scale profiling of gene expression and of protein–protein interactions.

One of the few studies in which the power of post-homology methods was evaluated directly is the work by Huynen *et al.* (2000). They studied the performance of chromosome clustering (including domain fusion) and phyletic pattern analysis in predicting functions of proteins in *M. genitalium*. Important aspects of performance that they measured were coverage (percentage of all proteins for which a given method produces an inference), the type of function or interaction that each method is able to predict, and the overlap with homology-based function assignment. Conservation of gene order covered 37% of all genes, gene fusion covered 6%, the co-occurrence of genes in operons without conservation of gene order covered 8% of genes, and the coinheritance across genomes covered 11% of genes. The total fraction of genes annotated by at least one of these post-homology methods was 50% because some genes were annotated by more than one method. All this, however, resulted in only a modest increase in functional understanding because, in the end, new functional features, not already evident from sequence similarity alone, were predicted for only 10% of *M. genitalium* genes. Clearly, functional inferences on the basis of sequence homology dominate every metabolic reconstruction. On the other hand, for most genomes the gain of 10% translates into new functional information for many hundreds of genes.

Interestingly, the same post-homology methods can be used to examine genetic elements that do not encode proteins. One quite obvious application of the chromosome clustering approach is to search the upstream regions of coregulated genes for conserved nucleotide motifs, which may then be evaluated as the candidate regulatory elements. Recently, this has been combined with phyletic vector analysis: Rodionov and Gelfand (2005) discovered a ribonucleotide reductase control element (NrdR) upstream of many known ribonucleotide reductase genes and then found all occurrences of this element in bacterial genomes (which identified additional genes coregulated with ribonucleotide reductases). Next, they matched phyletic vectors of nucleotide element NrdR and vectors of protein coding COGs. In this way, the predicted DNA-binding protein with Zn-ribbon and ATP-cone motifs (COG1327) was identified as the cognate transcription factor binding to NrdR.

The post-homology methods of functional annotation of genes and proteins are still in their infancy and lack robust statistical framework. If there is a general conclusion from everything that has been learned thus far, it is the evidence that gene content in genomes is extremely variable. Although protein sequences even in distantly related genomes follow the first fact of comparative genomics, and conserved regions account for the majority of the proteomes, at the same evolutionary distance there is much less conservation of gene order or of the repertoire of shared genes.

9

Structural Genomics: What Does It Tell Us about Life?

Computational analysis of protein structure is a vast part of science, impossible to cover in one chapter. In my dreams, I see a physicist, a crystallographer, a molecular biologist, an evolutionist, a statistician, and a computer scientist joining forces and covering, in one book, all computational aspects of structural biology, from deciphering the patterns of x-ray diffraction to classification of protein shapes and functions. This book is not that book, but if we want to be true to the theme of evolutionary, structural, and functional signals—in genes, gene products, and genomes—we need to understand what role is played by structural biology in all this. We also want to know what the era of complete genome sequencing and the large-scale structural genomics initiatives, funded on the heels of genome sequencing projects, are telling us about protein structure.

Before examining the evidence, let me make it clear that the problem of predicting three-dimensional structure from sequence, which seems to immediately enter any conversation on computer analysis of protein structure, is not really the focus of this chapter. The attempts to solve this difficult problem in one way or the other account for a substantial part of all literature in the field, and it has even been called the "Holy Grail" of computational structural biology (or, for some people, of biology as a whole). But I have my doubts about that—the chalice, in my opinion, is located elsewhere. In fairness, however, before searching for this "elsewhere," a few words about prediction of structure from sequence are presented.

As with any predictive modeling, it helps to agree on the measures of success and failure of structure prediction. Suppose we have built a model of a protein molecule. Do we want the set of atomic coordinates of this model to have minimal distance from the real set of coordinates, whenever the latter becomes known? Or should we look to minimize some difference at another level of structural organization of proteins—perhaps the difference in the number of equivalent secondary structure elements connected in the same order or the difference in the number of amino acid residues falling into these elements? These and many other parameters may be relevant to comparison of protein structure, yet each of these properties can be measured in a variety of ways, and there is no principled argument as to what should be optimized there. For example, the atomic coordinates are often compared by root mean square deviation (RMSD) of the Euclidean distance between some superimposed atoms. But even the most basic aspects of this distance measure have never been justified: We do not know which atoms and how many of them should be superimposed, we do not know what the best way to superimpose the atoms is, and we also lack theory that would relate the RMSD measure to any aspect of protein evolution or function.

Notwithstanding these difficulties, it is indeed possible to produce good, useful models of unknown protein structures. But blunders, in the form of completely wrong models, also happen. Examples of both success and failure are given by recent nearly simultaneous prediction, by two groups, of the structure of ataxin-3, the product of a gene mutated in Machado–Joseph disease. One group, in a somewhat convoluted narrative replete with expressions such as "high homology" (see Chapter 3 to recall why this is not helpful), suggested that predicted secondary structure of ataxin-3 and its database relatives consists mostly of alpha-helices and reported BLAST match with E-value of 2×10^{-7} between the N-terminal regions of ataxin-3 and of adaptin-like domains involved in protein–protein interactions (mostly with clathrin, in the context of endocytosis and membrane trafficking). The purported similarity was thought to be further supported by genome context—that is, by the observation that some ataxin-3-like proteins, also called josephins, form domain fusions with ubiquitin-interacting motifs, and so do some of the adaptins (note the same spirit as in the approaches discussed in Chapter 8). The title of the paper was "Structural Modeling of Ataxin-3 Reveals Distant Homology to Adaptins" (Albrecht *et al.*, 2003), but said revelation was wrong. Indeed, almost simultaneously, another group found remote sequence similarity between the ataxin-3 family and one of the two classes of deubiquitinating enzymes (DUBs). Analysis of sequence similarity and sequence-guided prediction of secondary structure indicated that ataxin-3 is likely to adopt the same serine protease-like alpha-beta fold as this class of DUBs, very different from all-alpha adaptin-like fold (Scheel *et al.*, 2003). The conserved residues that are required for catalytic activity of deubiquitinating proteases were also preserved in josephins. The hypothesis, then, was that ataxin-3 and other josephins were DUB-like proteases involved in ubiquitin-mediated degradation of some protein target. This helps to explain the role of ubiquitin-interacting motifs in ataxin-3 and indicates that the genomic context argument proposed by Albrecht *et al.* was misleading (as contextual arguments sometimes can be; see Chapters 7 and 8). The deubiquitination activity of ataxin-3 has since been demonstrated experimentally, and structural study indicates serine protease-like, and not adaptin-like, fold in josephin domains (Burnett *et al.*, 2003; Mao *et al.*, 2005).

The josephin case is not an exception. For years, the Critical Assessment of Techniques for Protein Structure Prediction (CASP) series of meetings has been giving researchers, as well as fully automated servers, a chance to predict structures from sequences. After all predictions are in, the real structures of the target proteins are made public. This is the format specifically designed to assess the state of the art, and the results clearly demonstrate that there are better predictors and worse predictors among both humans and robots, and that best predictors submit models of much better quality than the average ones, in every measurable way (Moult, 2005; Ginalski *et al.*, 2005).

Several factors appear to play a major role in the success of structure prediction. First, predictions that rely on the comparison and consensus of several different methods are usually better than those that use just one method. Second, consensus of fully automated methods, which can be derived automatically using the "metaserver" Web portals, works very well, often comparable to the best human experts. Third, the most successful approaches essentially converge at a consistent application of increasingly sensitive methods of searching sequence databases for distantly related homologs with known structure. Obviously, these approaches work only when such a homolog exists in the database, and only for those investigators who know sequence analysis well enough to detect it. Thus, structure prediction works mostly because computational matching of sequences has become very sensitive and specific (see Chapter 2), and once a template and its alignment to the query are known, the question of the three-dimensional structure is, in a sense, settled (Ginalski *et al.*, 2005).

Reading about the CASP meetings, and participating in three of the last four of them, I could not help noticing where most of the progress is made from one competition to the next.

In CASP3, which took place in 1998, the most successful prediction approaches were those that used the PSI-BLAST program, which was still a novelty at the time. In CASP4 (2000), everyone caught up on the importance of a probabilistic iterative search of large sequence databases (and also on the importance of searching the NR database, and not only the space of sequences with known structures, which is more sparsely populated). The overwhelming advantage that year, however, was with the Rosetta method, which had been developed by David Baker's group at the University of Washington. This program starts with a sensitive probabilistic search to find homologous domains and switches to the ingenious approach of assembly from small template fragments if there are no homologs (the details of how it is done are scattered across many publications throughout the years, so the reader interested in the Rosetta algorithm has to assemble it from small fragments, too; see Simonst *et al.*, 1999; Bonneau *et al.*, 2001a,b, 2002; Bradley *et al.*, 2003, 2005; Chivian *et al.*, 2003).

In CASP5 (2002), all of the best predictors employed metaservers—that is, Web sites that interrogate a worldwide network of automated methods, collect structure predictions from each server, and then use some rules to determine which known structure is most consistently selected as a template for a given sequence. It is interesting to see which primary method has the most impact of the metaserver outcome: In my experience of using metaservers, I usually find that the structure with the best overall score is the same structure that is predicted by probabilistic sequence matching algorithms, usually PSI-BLAST or profile-to-profile alignment programs. Before CASP6, the most successful program of the latter sort was FFAS03 (Jaroszewski *et al.*, 2005; this also used to be the default search engine in Rosetta), but recently the upper hand seems to be had by the HHsearch algorithm, which compares a hidden Markov model (HMM) made from a query and its relatives to the library of known HMMs (Soding, 2005).

A different view, quite common in the literature, is that we have at our disposal a variety of significantly different methods, many of which do not ("merely") match sequences but, rather, directly compare sequence to structure using some physical or geometrical features of the structure template, such as the relative spatial position of every pair of amino acids. Some of these methods are called "threading," although I no longer understand what this really means. These "hybrid" methods may start with probabilistic sequence matching but then supplement it with the physical terms that improve the discovery of the correct template. It is true that some such capacity is part of many threading algorithms; however, as far as I know, it has never been definitely proven that physical terms in the scoring functions employed by threading algorithms give any specific advantage over probabilistic sequence comparison. On the contrary, one recent assessment suggests that, in fact, most if not all relevant information about physical rules or protein folding is already captured in the function that is used for scoring sequence similarity. A review written by some of the best predictors and the assessor of CASP competitions (Ginalski *et al.*, 2005) states this observation thus:

> *Since the first communitywide benchmarking of servers ... in 1998, such "sequence-only" algorithms have proven to be competitive in structure prediction tests. Until today, the advantage of using the structural information available for one partner in comparing two protein families has not been clearly demonstrated in benchmarks.*

In the same review,

> *Profiles generated with sequence alignment methods, such as PSI-BLAST, already include the mutation preferences imposed by the native conformation. Threading algorithms would then be required only if insufficient information exists about the sequences of proteins homologous to the template protein. However, the majority of protein families with known structure do have sufficient homologs to calculate local substitution preferences from multiple alignments. This observation gave rise to hybrid methods, which were designed to utilize sequence information from multiple sequence*

alignments if available, but also added terms such as residue-based secondary structure preferences or preferences to be buried in the core of the protein.... Such methods have been successfully applied in genomewide structure prediction experiments and claimed a higher fold assignment rate than that obtained with PSI-BLAST, which is routinely used as performance reference. However, direct comparison with profile–profile alignment methods turned out to be surprisingly favorable for the latter ones, which became serious competitors in protein structure prediction. Presently, the advantage of including the structural information in the fitness function cannot be clearly proven in benchmarks.

The prevalence of probabilistic sequence matching in successful structure prediction may not be surprising for those of us who spent years studying sequence conservation and evolution of sequence families. But even as recently as 10 years ago, sequence analysis was not necessarily expected to be the main engine of structural computational biology. For decades, the inspiration in structure prediction came from the Anfinsen postulate, which states that protein sequence contains all information sufficient for protein folding into native spatial structure, characterized by the minimum of free energy of the protein molecule. This postulate still holds (if we leave aside the quantitative aspects of folding efficiency and control, which may require additional components *in vivo*, such as chaperone proteins), and it motivated many laboratories to seek ways to compute the native conformation by finding that minimum.

The task of computing this minimum on the basis of protein physics turned out to be exceptionally difficult. Nonetheless, the struggles of biophysicists and computational structural biologists have provided many useful heuristics that result in relatively quick and relatively accurate, if approximate, estimation of many parameters involved in the determination of protein structure. In the future, new ideas and more powerful computers may allow us to solve the problem in principle. But the fact of the matter is that the physics-inspired approaches still have a relatively minor impact on structure prediction, especially as it is currently practiced (i.e., on a large, genome scale). Sequence comparison continues to be the main approach to structure prediction.

But the problem with even the best structural models of proteins is that their accuracy is usually wanting. It became more or less accepted that a model built on the basis of the structure of a homologous sequence will be much closer to that structure than to its own, when the latter eventually becomes known. Again, the model can be improved by iterative multiple alignment of related sequences, but physics-inspired approaches continue to be of only limited use for such models.

Whatever the past and future promises of structure prediction might be, no one doubts that, in a sense, it is a mature field: One can get proper education and acquire substantial practical skills in it. Even a casual user can utilize a metaserver, receive the best prediction, and convert a resulting alignment into a set of coordinates of a structural model of his or her favorite sequence. Moreover, the combination of algorithmic innovation, better software engineering, and, perhaps most important, focused production of X-ray and nuclear magnetic resonance structures of many proteins, selected for diversity and coverage of the sequence space, gives us the ability to predict folds for many proteins encoded in complete genomes.

The fraction of proteins with predictable folds is highest in bacteria and lowest in eukaryotes, reaching 50% in many species in the past few years (Elofsson and Sonnhammer, 1999; Liu and Rost, 2001; Orengo and Thornton, 2005). This percentage is likely to increase with time: Newly sequenced species will undoubtedly produce proteins with novel folds, but this is likely to be handily offset by the available structural templates and improved methods of structure prediction. Moreover, the number of folds in nature is thought to be limited (see Chapter 10).

Yet, in the midst of all this good news, we may ask: Why do we want to predict protein structure from sequence in the first place? One answer to this question is idealistic: We want to do it because this is a difficult problem and a worthy challenge to the human mind. However, in more practical terms, for those of us who are doing science here and now, what are we hoping to achieve by building a model?

For decades since the advent of the suitable technology, direct determination of protein structure was lengthy and costly. Also, the main expectation for structure prediction was that a model will be used to infer something about protein with yet unknown structure and function. At the beginning of the 21st century, however, experimental crystallography has become much more efficient. This is closing the gap between the sequence and the unknown structure, perhaps faster and certainly in a more definitive way than modeling is currently able to do. For example, finding a highly specific ligand by docking it on a predicted structure remains quite imprecise, unless the model and the template are very similar at the sequence level, and unless we are willing to ignore the flexibility of both interacting partners. If, however, both protein and ligand change shapes in the process of interaction, the modeling becomes so complicated that determination of the real structure with bound ligand may rapidly approach cost-effectiveness.

Perhaps the true goal of structure prediction would be to computationally convert a sequence into its three-dimensional structure—with accuracy higher than what is achievable by x-ray diffraction. This challenge was posed by Milton Saier from the University of California at San Diego at a bioinformatics meeting in Atlanta in 2001. Since then, indeed, the National Institutes of Health has invited researchers to submit proposals in the area of such very high-accuracy modeling. I will be glad to see any takers on this challenge and will be thrilled to see them succeed.

However, imagine that one morning we wake up and learn that the structures of all proteins have been experimentally determined. What would it do to fold prediction? Perhaps low-accuracy prediction will no longer be needed (although, of course, protein design tasks, such as targeted changes of local conformation of protein chain, will still require computational experiments). What will become the Holy Grail of computational molecular biology? And why not start the quest for that other chalice today?

I believe that the real goal of structural biology is to decipher evolutionary and functional signals that emerge from the known, as well as predicted, biological structures. And to begin this work in earnest, we need to examine the relationship between similarity of sequence and similarity of structure.

The issue is really not too controversial, but its discussion in the literature is often confusing. For example, time and again we read that "protein structure is better conserved in evolution than protein sequence." Let us call this "the first claim of evolutionary structural biology" and examine it for what it is worth.

First, conservation of which structures and which sequences are we comparing? Obviously, we are not talking about sequences and structures of two randomly selected proteins. Their sequence similarity is likely to be random, and so may be their structural similarity: It is not of much interest which of the two is higher or lower. The real meaning of "the first claim" is, more or less, that protein structure is better conserved than protein sequence in the evolution of homologous proteins. But even in this case, how we can start to compare the conservation of sequence to conservation of structure, when sequences are linear strings of symbols and structures are sets of points in the three-dimensional space defined by their coordinates? The two types of similarity are measured in completely different ways. Someday, perhaps, we will invent a transformation that relates these different types of measurements to some sort of unified scale, but it is currently not available. Instead of this task, give me comparison of apples and oranges any day: At least they are all juicy and round, and some of them are even similar in size.

One way out of this confusion is to say that "better conserved" really means that two homologs may have no similarity at the sequence level, but their structure can still be recognizably similar. This statement is, of course, true. The proposal here is to pay attention not to the degrees of similarity but to its discrete states. So let us take up this proposition and examine the ways in which common ancestry and commonalities in sequence, structure, and function may interact.

First, two proteins are either homologous or they are not (see Chapter 3). Second, sequences of any two proteins may be either similar or not. Unlike homology, sequence similarity is not really an all-or-none trait, but we just agreed to discretize it and to threat it as a binary character. Thus, there are four combinations of these properties. Unlike many other schemes that enumerate different combinations of key components (recall, for example, the discussion of virus expression strategies in Chapter 4 and of phyletic vectors in Chapter 8), in this case each combination of traits is actually encountered in nature.

Sequences that are both similar and homologous are one of the main objects of study in computational molecular biology. Most of this book is in one way or another devoted to discussing them. Homologous but not similar sequences are also of great interest; these sequences have diverged beyond recognition, but as the methods for analysis of sequence similarity improve, so too will our ability to match them and move at least some of them into the first category.

A trivial, and perhaps the most common, case is when two sequences are neither similar nor homologous. A randomly selected pair of sequences will most likely belong to this class. Lastly, it is also possible that two sequences are similar but not homologous. In Chapter 6, we examined various ways in which sequences can converge and concluded that the nontrivial level of global similarity cannot be achieved by convergence. But convergent origin of local simple patterns is possible, and there are also some other special cases, such as simple periodic structures convergently forming similar (often fibrillar) regions in proteins. The challenging question is whether more complex, aperiodic folds can also evolve by convergence.

In principle, this classification covers all possibilities without overlap—each pair of proteins belongs to exactly one class. In practice, however, when presented with a pair of proteins, we may not always be able to place this pair into a correct class. This limitation is technical, not substantial.

If the state of structural similarity is also a binary trait, then adding it to the mix gives eight classes. In this case, too, all classes are occupied (Table 9.1 and Fig. 9-1). Let us examine each class, starting from the very last category. Similarities in Class VIII comprise all pairs of proteins such that two proteins in the pair are unrelated to one another in every way -they are not homologous and not similar, at either structural or sequence level.

At the other end of the spectrum, there are similarities that belong to class I. This class includes pairs of homologous proteins with similar sequences and similar structures. In the Structural Classification of Proteins (SCOP) database, the authoritative classification of all protein structures (Andreeva *et al.*, 2004), such pairs of proteins are related at the family or

Table 9.1. Eight Classes of Sequence and Structure Similarity between Homologous and Nonhomologous Proteins

Common evolutionary origin (homology)	Sequence similarity	Structural similarity	Similarity class
Yes	Yes	Yes	I
	Yes	No	II
	No	Yes	III
	No	No	IV`
No	Yes	Yes	V
	Yes	No	VI
	No	Yes	VII
	No	No	VIII

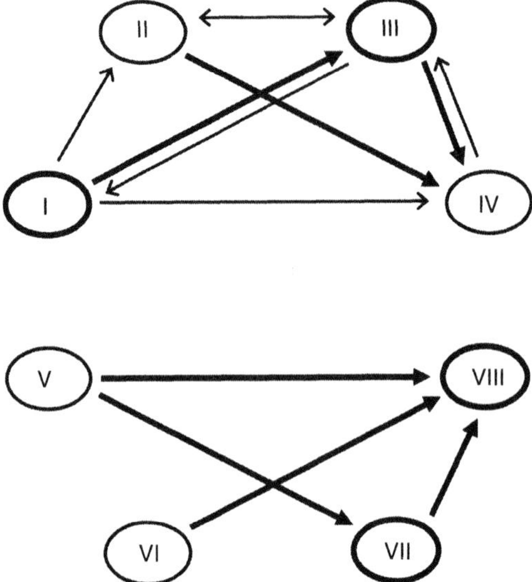

Figure 9.1. Evolution of the eight classes of similarity shown in Table 9.1. Classes that are more common in nature are indicated by boldface ovals, and transitions that are more likely to occur are indicated by boldface arrows.

superfamily level. The degree of similarity may vary for different pairs of proteins, but a frequently cited threshold number is 30% of identity in a global alignment, which extends to the whole protein domain (Orengo *et al.*, 1994). At this level of sequence similarity, recognition of structural similarity is not problematic, and finding sequence homologs in the context of a protein database search is also trivial. Some of the protein pairs, however, may be assigned to this class on the basis of much lower pairwise sequence similarity. As elaborated in Chapter 2, this is because of high sensitivity and specificity of protein sequence analysis: In an extreme case, multiple sequence alignment and construction of probabilistic model allows one to prove sequence similarity and predict similar structure for groups of sequences such that many of their pairwise similarities are at the level indistinguishable from the random background.

Similarities of classes II, III, or IV also deal with homologous proteins. Class II includes such pairs that have sufficiently close sequences to establish their homology, and yet they adopt different structures that belong to distinct folds. This type of similarity is of the utmost interest: It represents the phenomenon of fold change in the evolution of proteins. Convincing examples of this phenomenon are accumulating, and there is even some understanding of the elementary evolutionary acts, or "moves," that enable such fold changes, such as addition and deletion of helices or strands, circular permutation of the entire fold, strand invasion or withdrawal, beta-hairpin flip, and swap of subdomains (Grishin, 2001a,b; Krishna and Grishin, 2005; Vesterstrom and Taylor, 2006). These evolutionary reconstructions are seconded by the observations of fold changes that may happen in the lifetime of one protein, as part of a natural function of that protein [see Carrel and Huntington (2003) for discussion of interprotein strand invasion in the process of interaction between proteases and their inhibitors serpins].

Deletions and rearrangements of the small number of elements, even if they produce a new fold, may not affect the majority of the old fold. After just one elementary move, the old and new folds would contain a comparable number of helices and strands, most of which are connected to each other in a similar order. But give it enough evolutionary time, and after several small moves the difference between the starting fold and the final product may become profound. To illustrate this, Nick Grishin at the Southwestern Medical Center at Dallas proposed a hypothetical scenario of a series of evolutionary steps that could lead to an extreme change, such as between an all-beta and an all-alpha protein (Fig. 9.2). All the structures in Fig. 9.2 are those of the real proteins, and at least four of the seven moves are plausible from the evolutionary standpoint, as a and b are homologs, and e, f, g, and h are most likely also homologous to each other (although, of course, there is no claim that any of these proteins are ancestral to one another). The existence of class II similarity between proteins is in direct contradiction to the first claim of evolutionary structural biology.

Class III covers homologous pairs that do not have sequence similarity but retain structural similarity. It is for members of this class that "the first claim" holds true. Presumably, class III pairs evolve from class I pairs as sequences continue to diverge (unless, of course, the structure changed abruptly at the point when sequence similarity is still recognizeable, as in class II). In structure databases, this class of similarity is thought to be represented by pairs of sequences that belong to different SCOP superfamilies within the same SCOP fold.

The number and diversity of folds is widely believed to be constrained, although the nature of these constraints is only partially understood. It has been proposed that many, although probably not all, existing folds are in some sense optimal. It has also been hypothesized that at least some of these optimal folds may be "attractors" in the evolutionary space (i.e., that unrelated sequences might converge toward certain structures) (Ptitsyn and Finkelstein, 1980; Finkelstein and Ptitsyn, 1987; Finkelstein *et al.*, 1993; Babajide *et al.*, 1997; Xia and Levitt, 2004; Wagner, 2005; Zeldovich *et al.*, 2006).

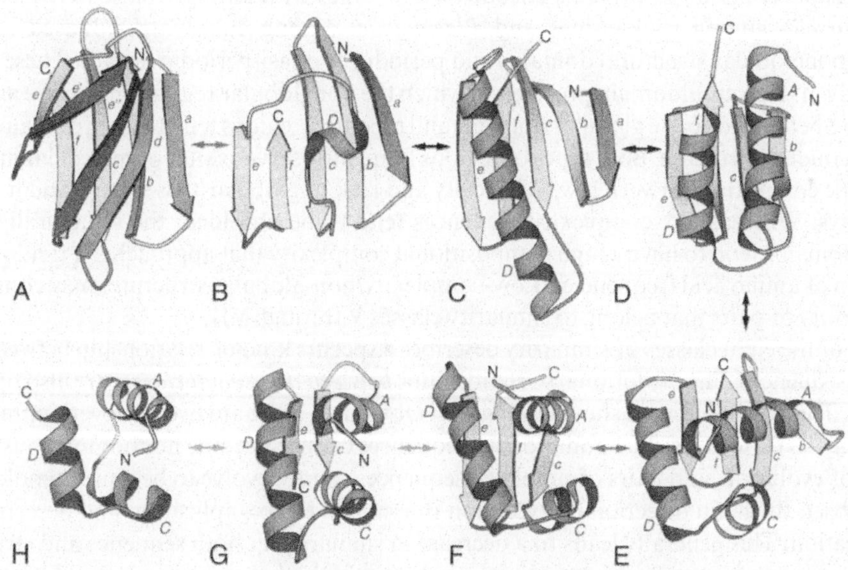

Figure 9.2. The conceptual evolutionary path from an all-beta to an all-alpha protein (**A**) C-terminal domain of alpha-amylase (PDB ID 1BPL); (**B**) C-terminal domain of G4-amylase (2AMG); (**C**) N-terminal domain of the gamma subunit of glycogen phosphorylase kinase (1PHK); (**D**) N-terminal signaling domain of sonic hedgehog (1VHH); (**E**) C-terminal domain of catabolite gene activator protein (1CGP); (**F**) N-terminal domain of biotin repressor (1BIA); (**G**) C-terminal domain of ribosomal protein L11 (1FOW); (**H**) DNA-binding domain of HIN recombinase (1HCR). There are evolutionary connections between a and b, possibly between these proteins and c, and between proteins e–h. Reprinted from *J. Struct. Biol.*, 134, Grishin, N. V., Fold change in evolution of protein structures, pp. 167–185, copyright (2001), with permission from Elsevier.

Any pair of proteins that shares structural similarity but has only random-level sequence similarity is potentially a class III candidate. To decide whether it indeed belongs to class III, we need to know if two proteins are homologous (then it is a class III pair) or not (in which case this pair belongs to class VII). In the practice of sequence and structure analysis, this is one of the most difficult determinations to make. Devising the ways to distinguish between class III and class VII pairs is a major challenge of computational structural biology.

The situation with class IV is even more dramatic. There is no reason to believe that it is empty. Evolutionary processes that produce class II and class III pairs may act simultaneously or consecutively, and this can produce homologs that no longer share either similar sequences or similar folds. Inference of homology for these pairs of proteins is perhaps the most difficult.

The remaining three classes (V–VII) describe three kinds of similarity that are possible among nonhomologous proteins. Class V comprises pairs of evolutionarily unrelated proteins that converged at both sequence and structure levels, and class VI includes pairs of unrelated proteins that have similar sequences but different structures. These classes are sparsely occupied: In Chapter 6, it was shown that in most examples of sequence convergence, only small parts of sequence can really converge. However, there are special cases of sequence evolution, which result in convergence that we are able to recognize as such. These are sequences with unusual, or statistically biased, amino acid composition. Regions that differ in frequency distribution of the constituent monomers—nucleotides or amino acids—are frequent in biopolymers (Salamon and Konopka, 1992; Wootton, 1994; Wootton and Federhen, 1996). At least one-fourth of all residues in protein databases are in compositionally biased regions, and more than one-half of proteins have at least one such region (Wootton, 1994). These numbers are somewhat lower for prokaryotes, but higher for eukaryotes. Regions with skewed composition are especially common in proteins encoded by genomes with strongly biased A+T content (e.g., *Borrelia* among prokaryotes and *Plasmodium* among eukaryotes). Biased regions include nonglobular structural domains and periodic or quasi-periodic repeats. These occur especially in large multidomain proteins, in which the nonglobular regions may serve as flexible hinges between discrete globular domains and repetitive regions tend to form domains with supersecondary structure. Both types of regions frequently serve as sites of molecular interactions. The connection between low complexity and lack of globularity is strong, and it works both ways: Whereas low-complexity sequences tend to be unfolded, the structurally well-folded domains tend to have a high compositional complexity that approaches a random distribution of amino acid frequencies. Low-complexity/non-globular structures may constitute the majority of pairs that belong to similarity classes V through VII.

Each of the eight classes of similarity describes a specific kind of relationship between two protein sequences. The evolutionary relationships *between these eight classes* are also of great interest. The possibilities are shown in Fig. 9.1. Obviously, there are two nonoverlapping subsets (quartets) of classes: The homologous sequences cannot become nonhomologous in the course of evolution, and pairs of unrelated sequences cannot evolve to become homologs. In each subset, the main direction of evolution is assumed to be duplication and descent with modification. This generally leads to a decrease in similarity, at both sequence and structure levels. An increase in similarity can also occur, but everything that we know thus far suggests that, at least at the sequence level, this is a countervailing trend.

If sequences and structures of all proteins were known (or if structures of proteins could be accurately inferred from their sequences), and if all evolutionary relationships between proteins were also known, each pair of proteins in the universe would be assigned to one of the eight classes of similarity shown in Table 9.1. Of course, scientists have not reached this stage yet. But learning as much as possible about all eight types of sequence and structure similarity between proteins is a goal that, in my opinion, forms the basis of a worthwhile research program in computational structural biology. More specifically, we want to know

how proteins are distributed between the eight classes;
what are the practical ways of distinguishing between these classes;
what are the pathways of evolution of different types of similarity; and
how to use this information to infer protein function and understand principles of protein organization.

This program is already under way, by the efforts of many colleagues throughout the world, and it will not become obsolete even after all protein structures are obtained or after the problem of computational protein folding is solved.

Not everyone is convinced of the worthiness of such an approach. For example, when discussing the protein folding problem, Blundell and Johnson (1993) stated the following:

> *Attention was also distracted by an often fruitless argument on evolution. It seems likely that many protein structures have converged by the evolution of stable, common folds. Equally many proteins have evolved by swapping exons corresponding to structural, and sometimes functional, modules to give rise to complex multidomain structures. But it is difficult to be confident of divergent evolution, and in any case the knowledge is not very useful. Karl Popper reminds us that a hypothesis is of little value unless an experiment can be devised that might falsify it; this is certainly difficult for hypothesis about divergent evolution of protein folds.*

The general idea of that quotation seems to be that the evolution of proteins should not be studied because it is a difficult problem irrelevant to protein folding anyway. But in this book, I argue exactly the opposite, namely that the distraction takes place and confusion sets in not when researchers care about evolution but rather when they stop paying attention to it. This applies as much to protein structures (including protein fold prediction) as to genome sequences. Speaking of Karl Popper, I do not think that *difficulty* of obtaining a falsification is the point of his theory. Rather, *falsifiability in principle* is what distinguishes a scientific theory from nonscience. I also do not recall Popper suggesting that scientists should stay away from difficult problems. More important, a trouble with invoking Popper's approach is that it has little to say about the workings of historical method (although, notably, Popper did not exclude observational sciences that try to infer past events from the realm of true science). One way to provide falsification (and verification) in historical reconstruction is by probabilistic inference, in which Popper did not seem to be very interested [further discussion of the problems with Popperian justification of evolutionary studies, with emphasis on inference of phylogenetic trees, can be found in Chapter 10 of Felsenstein (2003)].

Let us now return to the research program that is proposed here. It includes several exciting, open questions, but perhaps the most interesting among them is how to decide between classes III and VII (see Table 9.1 and Fig. 9.1).

As already noted, there is no statistical theory that would allow us to distinguish homologous from analogous proteins based on similarity of structures alone. Many structurally similar proteins are known to be homologous, but they are so known because sequence similarity of these proteins is high enough to infer homology. The real problem is when two proteins have similar structure and yet sequence similarity between them is at the random level. It is in these circumstances that we need to make a nontrivial choice among two hypotheses: one about common ancestor and the other about independent origin of these two proteins.

Several types of approaches may be helpful in answering this question. All of them try to find support for the first hypothesis (i.e., the common ancestry of two proteins). There does not appear to be a truly independent way to prove the lack of common ancestry in protein structures.

The first approach to substantiate the hypothesis of the common origin of two similar structures is by improving the way to detect statistically significant sequence similarities between proteins. During the past 20 years, there have been several scientific breakthroughs in algorithms and scoring functions for protein (see Chapter 2). With each of these achievements, the protein universe has become more linked, the families of protein sequences acquired more members, and members of each family became more diverse. Along the way, some pairs of structurally similar proteins turned out to be also similar at the sequence level. To use the language of the SCOP database, improvement of probabilistic methods of sequence comparison results in reassigning some fold-level similarities to a lower level of hierarchy (i.e., superfamily level). And to use the definitions introduced previously in this chapter some similarities that could belong either to class III or to class VII were reassigned to class I when methods of sequence comparison became more sensitive. This process may be expected to continue, as long as we do not run out of new ideas in sequence analysis. What remains unknown, however, is whether each fold will end up being just one superfamily—that is, whether a common ancestor will be established for every fold. If we could show that this is the case, it would follow that no structural convergence ever occurred, at least up to the fold level—a nontrivial result indeed.

The second approach to infer common ancestry of similar protein structures relies on introduction of novel characters. A 20-letter alphabet is most likely not the only one that is relevant for evolutionary comparison of protein sequences. As discussed in Chapter 2, each amino acid residue possesses not just one property, such as hydrophobicity or hydrophilicity, but rather a whole array of properties. For example, serine is a hydrophilic amino acid with the hydroxyl group on the small-sized side chain. Because it is small, it can be found in the turns of the main chain, as can glycine and alanine; however, unlike these latter two, serine also commonly serves as a nucleophile in catalysis (most often in hydrolases), which makes it similar to aspartic and glutamic acids, as well as to cysteine. Although nucleophile serine in the conserved catalytic center of a protein is physically the same as a small serine in a nonconserved loop, they are in a sense different amino acids, with distinct functional roles and distinct evolutionary trajectories. Serine, moreover, is a target for posttranslational modifications, most prominently by phosphorylation (in this respect, it is unlike G, A, D, E, or C but is quite obviously more similar to threonine and tyrosine—two other amino acids with hydroxyl groups in the side chains, which are also subject to phosphorylation). In this case, even the physical identity of amino acid side chains is different (phosphoserine vs. serine). New analytical techniques, such as mass spectrometry, allow us to characterize many, eventually most, covalent modifications that happen to proteins. The databases of the known amino acid derivatives and their locations in proteins will continue to grow. More characters in the amino acid alphabet would result in a better signal-to-noise ratio in protein sequence comparison. The work on such enlarged alphabets has barely started.

An even more radical alphabet change would be to consider other characters besides the identities and modifications of amino acid side chains. The protein sequence can be viewed as a string on the alphabet that is much longer than what is provided by the amino acid chemistry. Not much work has been done on this issue, but a few existing proposals include using dipeptides or other short words (Gonnet *et al.*, 1994); matching elements of secondary structure directly (Russell *et al.*, 1996); and searching for certain "peculiar structural hallmarks," such as kinks and bulges in the protein main chain (Richardson *et al.*, 1978; Hemmingsen *et al.*, 1994). In fact, the latter class of characters appears to play a role in Alexei Murzin's (1998) assignments of fold and superfamily level in the SCOP database. The statistical foundations of all these approaches remain to be worked out.

The third approach is to use the post-homology evidence (see Chapter 8). Suppose that two proteins with similar structures but random sequence similarity are found in similar genome context across many different species: For example, they are surrounded by the same set of homologous genes on the chromosome, or they are fused with the same set of homologous domains. The idea is to treat such additional information as an indication of the possible common ancestry of such genes. Note that this is an argument different from using genome context to infer similar functions of proteins. As can be recalled from Chapters 6 and 8, gene function can indeed be inferred from similar genome context, but this does say much about their common origin. In the context of the current task, however, this is not a problem. Indeed, precisely because structure and function of genes in similar genome context does not have to be similar, it is all the more telling when the *structurally similar* proteins are encoded by genes that share the same genome context. In special cases, such as when two domains of the same protein have very similar structure, the hypothesis of protein origin by intergenic duplication seems all but proven. Note, however, that, as discussed in Chapter 8, more general quantitative theory of genome context and its evolution remains to be developed, and even in the most straightforward cases, such as those of duplicated domains, nothing has yet been put into a quantitative framework that allows us to make a probabilistic inference.

All these, and probably other, new approaches will be worked out in the near future. The fourth, perhaps most promising, new approach may be to study the structurally similar fragments of unrelated folds, where "unrelated" means that the cost of conversion of one fold into the other is evolutionarily or physically prohibitive. This would lead to the collection of truly convergent pairs of protein fragments with similar structure (Nick Grishin, personal communication).

However, there is a major obstacle to these programs of study—the unsolved problem of measuring structural similarity between proteins. The classification outlined in this chapter is only possible if we can determine which pairs of structures are similar and which are not. Moreover, although we agreed to treat structure similarity as a binary (all-or-none) character, in reality similarity has degrees, and some pairs of similar sequences are more similar than the others. What is the best way to measure structural similarity?

Often, there are multiple ways of measuring the distance between the same objects, and they are not all equal in their ability to recover the signal in which we are interested. So how to choose the right measure—in this case, the best measure of structure similarity?

The choice of the distance measure is more straightforward when there is a hypothetical model, or several competing hypotheses, of the process that generates data. In that case, the likelihood of the models given the data can be compared, and the best model can be selected using, for example, likelihood ratio test. This approach is applied, for example, when studying evolutionary relationships: One compares several models of character evolution, determines the likelihood of each model given the phylogenetic tree, and selects the best model, in this way obtaining the most plausible distance measure between characters. For protein structures, however, there is no causative process model.

Hierarchical clustering is a useful way to organize the data on sequence and structure similarity. Phylogenetic trees are perhaps the most familiar example of hierarchical clusters that are studied by biologists; in this case, hierarchical structure of the cluster represents its evolutionary history, as long as the distances between the sequences containing evolutionary information. There is only one, but major, problem in applying the clustering algorithms to structures: The differences between distances are informative only when the distances are relatively small (distances are good for quantifying similarity but not as good in comparing the degrees of dissimilarity). In other words, if two structures are very similar (as is the case for most orthologs in two evolutionarily close species), they can be superimposed, which means that a large number of homologous residues in both structures can be forced to occupy the same space. But as evolutionary distance between sequences grows, structures undergo several simultaneous types of change.

First, there are amino acid substitutions. When we score sequence similarities, the function that describes the cost of each type of substitution is straightforwardly derived from the database of aligned protein families. But what is the effect of substitutions on the three-dimensional shape of proteins? Although, by definition, only the side chain atoms are replaced in this case, the effect goes beyond the differences in side chains. Substitutions between small and bulky amino acids may affect the configuration of the main chain, changed patterns of hydrogen bonding between the replaced side chains affect the interactions between elements of secondary structure, and the length of these elements may also change.

Second, there are short insertions and deletions of a relatively small number of amino acids, which also affect the length of strands, helices, and loops, as well as their interactions. Third, some combinations of substitutions and indels cause even more extensive remodeling, such as the aforementioned changes in the number and orientation of helices and strands.

Thus, as dissimilarity between sequences grows, so too does the dissimilarity between protein structures. While sequence similarity is relatively high (by that we mean more than just pairwise similarity; we assume that probabilistic searches in the databases of sequence and sequence domains are also possible), there is no problem: We can use statistical theory of sequence similarity to compute the distances between sequences and use them to indirectly compare distances between different proteins with the known structures. This is useful in many regards, but for the goals of direct structure comparison it is neither here nor there: It does not provide any way to state which two shapes are the same, which are different, and by how much.

The same problem remains if we decide to forego sequence similarity completely and to develop some geometry- or topology-based measure to compare sequences. Many such measures have been proposed. They typically compare positions of certain atoms in the Euclidean space or use some vector-based representation of protein structure. Most of these methods have been optimized to recover, as closely as possible, the information on the relationships between protein structures as they are captured in the existing databases, for example, in the manually curated SCOP database, and they perform reasonably well (in the range of 80–90% accuracy) on this data set. But the problem is that the method's accuracy is an average. When newly determined structures are submitted for similarity search, many stand an excellent chance of being correctly linked to their closest relatives, whereas some will find no relatives at all or will produce only matches of uncertain relevance. Typically, the structures in the first category, which find relatives in unequivocal manner, match their structural relatives also at the level of sequence similarity. On the other hand, the structures that do not match anything may be either truly novel folds or extensively modified versions of already known folds (indeed, what is reported as a novel fold is not infrequently assigned to already existing folds by the SCOP database curators after more thorough visual examination). Thus, the automated methods of structure comparison work best for structures that can also be compared at the sequence level, and they are less useful when they are needed most.

Thus, structural similarity is a concept that remains poorly defined, and the existing scoring functions for structural similarity work well only at a relatively narrow range of high similarity. Moreover, most of the existing methods of structure comparison rely on some sort of mapping of equivalent structures and thus are not well suited for comparing sequences with different folds. The state of the art can be summarized by another quotation from Ginalski *et al.* (2005). The authors discuss one aspect of structure comparison, namely, the way to compare quality of different CASP models of the same protein to its native structure. Obviously, this should be a relatively easy task because a good homology model is by definition quite similar to the native structure. However,

> *the structure prediction community has failed so far to define a standard model assessment algorithm. The main reason for this is the lack of an exact definition of similarity between the native structures of two proteins. Different structural classifications of proteins, such as SCOP, CATH, or FSSP, disagree in many cases when assessing a weak similarity between two native structures, which is at the*

> *level of similarity between models for difficult to predict targets and the correct structure.... As a result of the ambiguity of structural classification, the annotation of a model as "correct fold" or "incorrect fold" remains rather arbitrary.*

Herein lies the need of real understanding of structural diversity of proteins and the process that generates them. Much in that realm, from fold recognition to structure classification, can be accomplished without even looking at the structures (i.e., at the sequence comparison level). However, it is hoped that in the future purely structure-based approaches will prove useful to the supertask of deciphering functional, evolutionary, and structural signals in proteins. (The quest for structural signals in protein structures may seem tautological, but in fact the ability to discern the elements of structure that contain a signal, as opposed to noise, is not trivial.)

The task of building a robust classification of structures without using sequence similarity remains formidable. One approach to this problem is to study physical restrictions that are imposed on the ways in which proteins fold. This work started long ago, before nucleotide sequencing became practical, and very general rules of what is allowable in protein folding were discerned by examination of a small number of then-available protein structures (Levitt and Chothia, 1976; Ptitsyn and Finkelstein, 1980; Richardson, 1981). Although the first impression from high-resolution protein structures was that they are much less organized than regularly ordered nucleic acids, this bewilderment was quickly replaced by the understanding that proteins are not random conglomerates of alpha helices and beta strands, but they "obey the restrictions of symmetry, simplicity, and good design." Based on the analysis of propensities of protein chains to engage in various types of physical interactions, Ptitsyn and Finkelstein (1980) proposed the following as the main principle of protein structures: "The dominating majority of polar groups either form intramolecular hydrogen bonds or are exposed on water while the maximum number of nonpolar groups are shielded from water."

This simple idea has powerful implications. Most important, in order to practically achieve these exposures and shieldings, protein molecules rely on the finite set of chemical groups and on the limited repertoire of bond lengths and angles that are compatible with protein chemistry. This imposes a set of powerful restrictions on the architecture of a protein fold. The following are some well-known rules. First, turns between beta strands tend to be right-handed (Fig. 9.3A). Second, knots (i.e., crossings of any two elements of secondary structure) are rare (although examples of crossing loops are known; see Fig. 9.3B). Third, as a corollary of the second restriction, two adjacent elements almost never cross, and two adjacent beta strands are therefore usually antiparallel, whereas two beta strands separated by a helix are more commonly parallel to each other, with each strand being antiparallel to the helix. Fourth, beta sheets have a twist (Fig. 9.3B), a feature for which different physical rationales have been offered but that is not fully explained. Finally, most protein folds are arranged in two, three, or four planar layers.

In the 1970s, these rules were used by several investigators, most notably Jane Richardson of Duke University as well as Oleg Ptitsyn and Alexei Finkelstein of the Institute of Protein Research, USSR, to enumerate various arrangements of the elements and layers in proteins with known folds. One extensively studied example is a commonly encountered arrangement of alpha helices and beta strands known as parallel alpha/beta topology due to the fact that the basic unit of this type of fold is a beta strand that may be followed by either a loop or an alpha helix, and when several such units are concatenated, the strands are parallel in the sense that their C-termini point in the same direction (Figs. 9.3 and 9.4). The simplest and largest class of structures with this topology is the four-layered alpha/beta barrel, which we will also discuss later in this chapter. The other, also very large, class is the so-called doubly wound parallel beta sheet—usually a three-layer structure that includes, in addition to the sheet, two layers of alpha helices on both sides of the sheet. Jane Richardson (1981) says

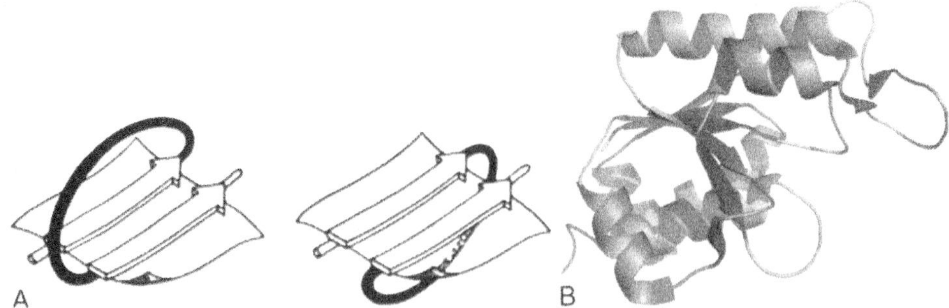

Figure 9.3. (A) Left-handed connection between adjoining beta strands. Reprinted from *Adv Protein Chem.*, 34, Richardson, J. S., The anatomy and taxonomy of protein structure, pp. 167–339, Copyright 1981, with permission from Elsevier. (**B**) A three-layered parallel alpha/beta structure of a tRNA-N1-guanyl methyltransferase TrmD of the SPOUT family (PDB ID 1P9P) showing two of the features described in the text: the twist of the beta sheet, which is very common, and the crossing of two loops (at the foreground), which is very rare.

"with right-handed crossovers the simplest way of protecting both sides of the sheet is to start near the middle and wind toward one edge, then return to the middle and wind to the other edge." Richardson has collected all variations of this arrangement, starting from the six-stranded sheets with closest adherence to this organizing principle and then expanding to folds that deviate from the standard by addition and deletion of stands and helices, as well as changes in the "wiring order" of the elements (Fig. 9.4). She noted that the scheme grouped "these domains into five gradually loosening levels of topological similarity, without attempting to make any definite decision as to where the dividing line lies between divergent and convergent examples."

Another class of beta sheets, known as Greek-key topology, was simultaneously studied along similar lines. The key element in all folds with this topology can be visualized as a beta hairpin with two strands in each half, the upper half of which is bent to the left. Many proteins, with unrelated sequences and different molecular functions, have folds that can be seen as variations or elaborations of this theme. The most important variation is the addition of one or more beta strands, which may be added at the N-terminal side of the main element, the C-terminal side, or both sides. Their position and connectivity determine the extent of the overall twist of the beta sheet, which may curl into a barrel, a half barrel, or end up being flattened into two layers. Some variation is also introduced by occasional addition of alpha helices. In 1980, Ptitsyn and Finkelstein enumerated 13 distinct all-beta Greek-key domains. By "domain," they meant a group of protein structures that had recognizable sequence similarity and could be treated as one entity. That definition is similar to sequence family or superfamily, as it is understood today. Those 13 domains accounted for more than two-thirds of the all-beta domains known at the time. Ptitsyn and Finkelstein wanted to go beyond simply enumerating the observed structures and asked for the most plausible explanation of the simultaneous existence of all these structurally similar groups of proteins. They concluded that these groups could not be explained by sequence divergence because there was no evidence for the common ancestor of any 2 of the 13 structural domains [with a trivial exception of two copper-binding proteins, azurin and plastocyanin, which should have been included into one "domain" (i.e., sequence superfamily), as, indeed, they are now]. Neither could these groups be explained by functional convergence because functions were all different, without any molecular commonality such as binding to the same ligand. Ptitsyn and Finkelstein (1980) concluded,

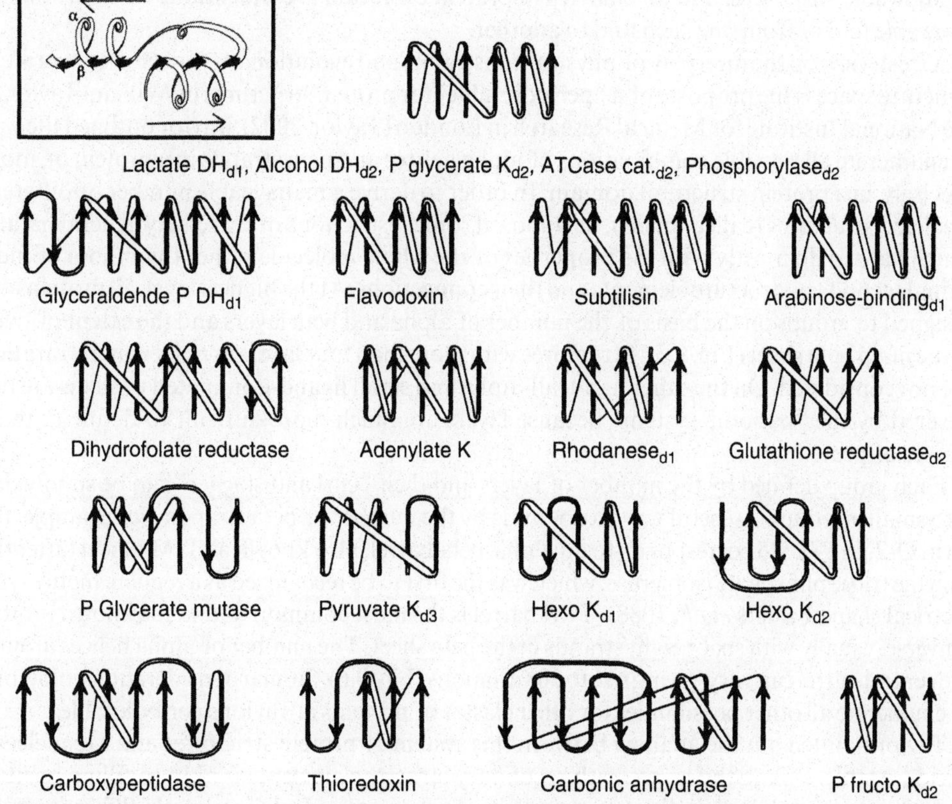

Figure 9.4. Richardson's space of doubly wound alpha/beta folds and related structures. The extent of deviation from the "ideal" structure (*inset*) increases from top to bottom. Reprinted from *Adv Protein Chem.*, 34, Richardson, J. S., The anatomy and taxonomy of protein structure, pp. 167–339, Copyright 1981, with permission from Elsevier.

> *We have to assume that the surprising similarity of beta-protein structures cannot be explained by such biological reasons as evolution and functioning but [is] due to purely physical reasons favoring a very limited class of structures in comparison with all others.*

Today, we may not be satisfied with this conclusion. The three hypotheses explaining structural similarity in the absence of sequence similarity—sequence divergence, function convergence, and satisfaction of structural constraints—are not mutually exclusive. We now know that common function and common ancestry correlate, but imperfectly; in particular, as homologous sequences diverge, functions of paralogs, and even those of some orthologs, change. Therefore, difference in function does not do much for dismissing the hypothesis of the common origin of two structural domains. Furthermore, physical constraints determine the space of possible protein structures, but this does not address the question of divergence and convergence. Even when restricted by these constraints, the existing domains had to evolve to their similar structures either by divergence from a common ancestor (and what is the evolutionary relationship between this ancestor and proteins with other structures?) or by convergence from different shapes.

Thus, if we want to discern the divergence and convergence events in protein evolution, understanding of physical constraints on protein architecture is not sufficient. At the same time, this understanding is necessary because all evolution takes place in this constrained space of allowable structures, and all changes in protein evolution are transitions either within an allowable fold or from one such fold to another.

A fresh view on the interplay of physical constraints and evolutionary trajectories in protein structure space is the proposal of a "periodic table" for protein structures by William Taylor of the National Institute for Medical Research in London (Taylor, 2002). Taylor outlined the way to enumerate all possible combinations of alpha helices and beta strands in a protein or, more precisely, in a protein structural domain. In order to derive a manageable number of different idealized structures (called forms), he imposed some restrictions of extremely general nature, which have to do mostly with the properties of the whole molecule rather than those of individual secondary structure elements and their connections. At the highest level, all proteins are assigned to groups on the basis of the number of alpha and beta layers and the extent of twist in the main beta sheet (Fig. 9.5). Structures with more than four layers are exceedingly rare and are not considered. On the other hand, all-alpha and small ligand-dominated proteins are not covered by the "periodic system" because layers are much more difficult to define in these classes of proteins.

Each group defined by the number of layers and their "curl and stagger" can be subdivided into smaller groups. A useful way to do this is by the number of beta strands. For example, the form O-2_1 in Fig. 9.5 corresponds to an alpha-beta barrel, also known as TIM barrel (after the enzyme triose phosphate isomerase, which was the first to be recognized as a regular, highly symmetrical shape;Lasters *et al.*, 1988). TIM barrel is the most common fold in the known protein universe, usually with six or eight strands in the beta sheet. The number of alpha helices around the central barrel can vary, giving rise to 30 variations (Fig. 9.6). Similar considerations allow one to enumerate all other possibilities for other classes of proteins in Taylor's periodic table.

Taylor pointed out the analogy between this system of protein structures and Mendeleev's periodic table. Filling the layers with secondary structure elements is akin to filling orbitals with electrons, and there is distinct periodicity; for example, ideal forms with different setups of beta layers may nevertheless have the same range of alpha layers. Of course, another analogy with periodic table is that Taylor's system, as well as that of Mendeleev, is designed to be a

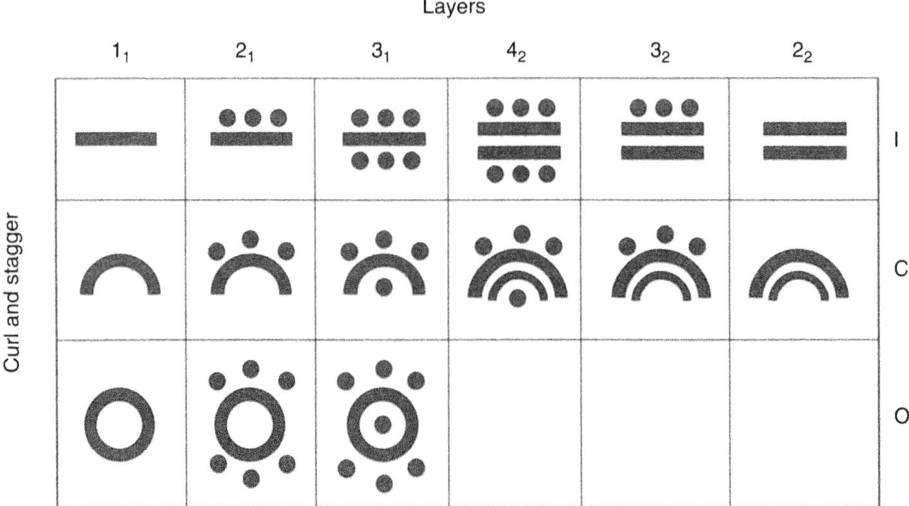

Figure 9.5. Top level of Taylor's (2002) periodic system of protein structures. Beta layers are shown by thick lines or large open circles, and alpha helices are shown as small solid circles. Reprinted by permission of Nature Publishing Group.

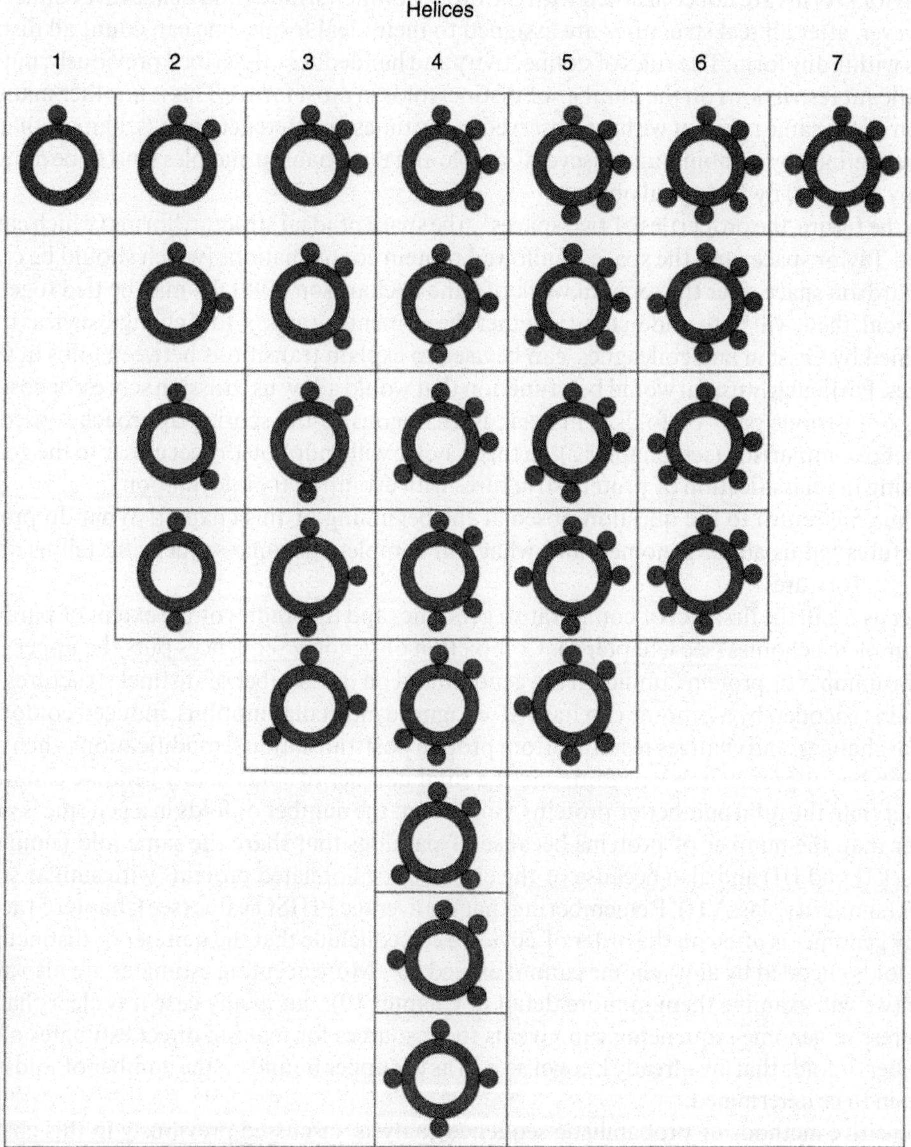

Figure 9.6. Possible arrangements of alpha helices within form O-2_1 of Taylor's (2002) periodic system of protein structures. Reprinted by permission of Nature Publishing Group.

"flat," nonhierarchical system and is not meant to represent the knowledge about evolutionary origins of its elements.

The elegance of both periodic systems is, nonetheless, breathtaking. And one remarkable consequence of Taylor's system for protein classification is that the completely defined space of ideal forms provides us with a novel way to compare any two protein architectures. Namely, we can map two real structures to their closest forms and then find the smallest form that contains both of these forms as its subsets. The number of elements that have to be added, removed, and rearranged to move between all these forms can be taken as the measure of distance between the structures. Algorithms are available to do all this with reasonable efficiency.

Taylor's forms are not concerned with the order in which strands and helices are connected. However, after all real structures are assigned to their ideal forms, one can count all distinct folds within any form. The rules of connectivity and handedness, described previously, impose significant restrictions on the number of distinct folds in most forms. This is another manifestation of the same rule that we have observed many times in different contexts, namely that the classes defined by combination of several simple and reasonable principles tend to be nonrandomly occupied by biological objects.

In the future, the properties of two spaces—the space of ideal structure forms (which can be called Taylor space) and the space of allowed protein conformations [which should be called Richardson space after the seminal work of Jane Richardson (1981)]—may be tied together. The goal, then, will be to understand whether the elementary acts of fold change, such as those outlined by Grishin and colleagues, can be used to explain transitions between folds in these spaces. Particularly useful would be a function that would allow us to assign scores, or costs, to compare various pairs of folds. This task is analogous to the scoring approaches used for sequence comparison (see Chapter 2). In this way, we will undoubtedly get closer to the goal of infusing the classification of protein structures with evolutionary information.

Thus, we return to the question posed at the beginning of this chapter: What do protein structures tell us about genomes, and what can complete genome sequencing tell us about protein structures?

Let us recall the first fact of comparative genomics and the studies of the extent of paralogy in complete genomes (see Chapter 5). Completion of genome sequences puts the upper limit on the number of protein families in the genome and on the number of distinct structures that proteins encoded by a genome can have. If we ignore molecular motions, induced conformational changes, and changes resulting from protein posttranslational modifications, then each protein has approximately one fold, and the total number of folds in any genome cannot be higher than the total number of proteins. But in fact, the number of folds in a genome is much lower than the number of proteins because of paralogs that share the same fold (similarity classes II and III) and also because of the existence of unrelated proteins with similar structures (similarity class VII). Remembering that the average PHISO value (see Chapter 5) across many genomes is often on the order of 50%, we can conclude that the number of distinct protein folds encoded by any genome cannot exceed 10^4. More accurate estimates are also available (we will examine them in more detail in Chapter 10), but in any case it is clear that the progress in genome sequencing can give us the resources for making direct estimates of the number of folds that are already known as well as the upper bound on the number of folds that remain to be determined.

Sensitive methods of probabilistic sequence analysis, discussed previously in this chapter and in Chapter 2, allow us to find homologs with the known three-dimensional structure for a large fraction of proteins encoded by any completely sequenced genome. The most up-to-date analyses indicate that such homologs are tractable for approximately half of all proteins in any genome. This estimate needs several qualifications, however. First, this 50% figure actually refers to proteins that have at least one domain with recognizable fold. Some such proteins may also contain domains that do not have relatives with known structure, so the fraction of total protein length that is covered by structurally characterized homologs may be lower. Second, the percentage is average; the coverage with structurally characterized homologs is higher for compact, globular domains and lower for membrane domains. Third, there are extreme genomes, such as viruses and plasmid-rich genomes of some bacteria, for which the fraction of proteins related to molecules with known structures is lower than 50%. However, as assessment of the progress of structural genomics projects shows, coverage increases with time, although perhaps the rate of such increase would benefit from better planning and coordination of target selection (Chandonia and Brenner, 2006).

Fold distribution in different genomes is also of interest. Proteins belonging to each fold within a given genome can be counted, and these counts in different genomes can be compared (most appropriately, after normalization that takes into account differences in genome sizes). Most often, the justification of these surveys is also given along the lines of structure prediction, modeling, and optimal strategy for selecting representative targets to cover structure space. But in fact, the results are interesting not only because they are monitoring the process of structural sampling from that space but also on their own merit. Indeed, several observations of considerable scientific interest have emerged from these studies.

The first general observation is that the protein structure space as a whole and also each complete proteome (which can be considered a sample from that space) comprise three categories of folds, called "superfolds," "mesofolds," and "unifolds" (see Chapter 10). Superfold in the strict sense means a fold that consists of many sequence superfamilies. Using our definitions, it is a large set of sequences that belong to similarity class VII. In any proteome-scale or database-scale data set, the high end of the distribution of proteins by folds is always represented by one or more such superfolds. The TIM barrel fold is the most common fold in the protein database, and it is also one of the most common folds in most microbial proteomes. On the other hand, a fold may be very common in a proteome, and at the sequence level it may resolve to just one superfamily of clear homologs. This is the case for kinases with serine/threonine/tyrosine/peptide/lipid kinase fold: All proteins in this diverse group of proteins can be shown to share common ancestry by sequence analysis (Cheek *et al.*, 2005), so their shared structure is clearly the result of divergent evolution of one sequence superfamily. In many eukaryotes, this set of kinases is the fold with the largest number of representatives. This is not a superfold in the previous sense, but it is a superfold in a different, simpler sense—a supersized fold that happens to consist of many members of one sequence family (in the case of human proteome, approximately 500; Milanesi *et al.*, 2005). The second result of the fold surveys in complete proteomes is that the usage of folds contains evolutionary and functional signals, but these signals are quite subtle. More than two decades ago, it was noticed that the correlation between structures and broad classes of protein functions is less than perfect. In her groundbreaking review, Jane Richardson (1981) presciently noticed that biochemical activities such as "protease," "peroxidase," "nuclease," or "oxygen carrier," do not define a protein sequence family or fold: Each such activity can be represented by more than one type of protein. This idea reaches its high point in the notion of gene displacement (see Chapter 6), when not just a broad group of activities but, for all practical purposes, exactly the same molecular function is enabled by two or more evolutionarily and structurally unrelated proteins. One exception, however, was noted by Richardson: All nucleotide-binding proteins whose structures were known in 1980 had one and the same fold, namely the three-layer alpha-beta-alpha Rossmann fold (see Fig. 6.2). The Rossmann fold indeed seems to have a special affinity to nucleotides and nucleotide derivatives: The ligands that are bound by different classes of Rossmann fold proteins include ATP, GTP, NAD(P), FAD, *S*-adenosyl methionine, and nucleoside diphosphosugars. It can be said that almost every Rossmann fold protein interacts with some sort of nucleotide derivative. However, the opposite is not true: Not all ATP-binding proteins are Rossmanoids. Indeed, a survey of just one subset of ATP-utilizing enzymes, namely kinases, showed that there are at least 10 distinct structural folds and up to 25 protein superfamilies that have evolved kinase activity (Cheek *et al.*, 2005). Similarly, *S*-adenosyl methionine is used as a ligand and donor of various chemical entities by many Rossmann fold proteins—most notably by a large group of SAM-dependent methyltransferases—but there are at least 12 unrelated groups of SAM-binding proteins with different folds, including at least two groups of non-Rossmanoid methyltransferases (Kozbial and Mushegian, 2005).

A similar picture is seen when fold usage is compared across evolutionary lineages. For example, bacteria and eukaryotes use protein phosphorylation for cellular regulation and signal transduction. Serine/threonine kinases in eukaryotes are well studied and, as already mentioned, are a very large sequence family, one of the largest in any eukaryotic proteome. In bacteria, much of signal transduction is mediated by histidine kinases—also a large family, although relatively smaller than serine/threonine kinases in eukaryotes and not related to them in sequence or structure. For some time before the era of complete genomes, this seemed to be a perfect evolutionary correlation: Protein phosphorylation in bacteria is enabled by histidine kinases, and in eukaryotes serine/threonine kinases play broadly the same role (although protein substrates of these kinases, of course, as a rule are not homologous). Eventually, however, exceptions started to accumulate. First, plant phytochromes and related animal cryptochromes were shown to contain a histidine kinase-like domain (Schneider-Poetsch *et al.*, 1991), and then another histidine kinase was discovered in yeast (Ota and Varshavsky, 1993). On the other hand, serine/threonine-type kinases were discovered in bacteria, first in some human parasites, where they were initially explained (away) by occasional lateral transfer from the host (Chiang *et al.*, 1989; Munoz-Dorado *et al.*, 1991), and then, with the advent of genome-scale sequencing, in almost all bacterial genomes, where their evolutionary history could not be reduced to secondary acquisition from eukaryotes and instead pointed to their ancient origin (Leonard *et al.*, 1998).

Has there been a specific evolutionary advantage for preferential expansion of histidine-type kinases in bacteria, and serine/threonine-type kinases in eukaryotes, or are we seeing stochastic processes at play? Can the size of a family or a fold in a given lineage determine the rate of its own expansion, in some kind of a "rich get richer" rule? Little is known about the evolutionary forces that shape the family and fold usage in various lineages. We examine some of the novel approaches to this problem in Chapter 14.

10

How Many Protein Families are There?

In Chapter 5, we examined the first fact of comparative genomics, which states that protein sequences are well conserved in evolution, or, more specifically, that modern methods of sequence analysis have become sensitive enough to find homologs for the majority of protein sequences in the database and for most proteins encoded by every sequenced genome. In other words, most proteins belong to protein families: Some of these families are ancient, others are more recent; some are large, others are small; some have representatives in many evolutionarily distant species, others are distributed more narrowly—in the extreme case, they may be found in just one genome. So, how many protein families are there?

This is an important question. In Chapters 5–9, we saw that the percentages of proteins that have relatives in the same genome and in other genomes are much more than just genomic trivia: They help us to understand and compare properties of different genomes, to set expectations for predicting function at the genome scale, and to understand genome evolution. As we include increasingly more genomes in our analysis, we will get closer to learning the organizing principles of the protein space as a whole and perhaps to understanding origins and the evolution of life (discussed further in Chapter 13). And one fundamental aspect of the protein space is its partition into protein families.

But how do we define protein family? A natural way to do so is to use homology: A family is such a set of proteins that every protein in it is homologous to all other proteins in the same set, and has no homologous proteins outside of this set. In practice, however, there are at least two difficulties with this definition.

One technical difficulty is that proteins consist of domains that can be fused and split, and in many cases, homologous relationships characterize protein domains rather than whole proteins. Thus, all operations on families have to take the domain organization of proteins into account. Domain partitioning is an important problem that does not have a robust algorithmic solution. Nonetheless, there are several domain databases that are curated by human experts, and they may be seen as a reasonable approximation of the universe of the known domains.

A much larger problem is that, in general, the complete set of homologs of any protein is unknown. For any protein, we can use our favorite state-of-the-art method to find all its homologs. But no method is perfect, and the inference of homology is a statistical one, with associated rates of false negatives and false discoveries. There is always a chance of missing some homologs, especially the highly diverged ones. Obviously, we will also miss those homologous genes that have not been sequenced yet. (If we are concerned with evolution, we have to add to this set all the ancient proteins that existed in ancestral, now extinct species). Thus, every known protein family is really a sample of the true family.

As always with homologs, it is the common ancestry that matters, not the degree of similarity. This is why division of the protein similarities into family and superfamily levels is arbitrary. Sequences within superfamilies are "allowed" to be less similar to each other than in families. The lower bound of a similarity between two homologs within a superfamily is not defined, but pairwise identity of ≥30% is often used as a cutoff to inclusion into a family. This same percentage value is the inflexion point on a graph describing the success of homology modeling: More than 90% of sequences that have homologs with the known structure and sequence identity of ≥30% can be modeled by homology with high accuracy, whereas sequences that have homologs less similar than 30% typically can be modeled only approximately (Orengo *et al.*, 1994; Ginalski *et al.*, 2005). I am not aware of any other significance of the 30% mark, and because the rules of thumb in homology modeling will not concern us much in this chapter, I use "families" in the sense of the known subset of the complete set of homologous proteins. This definition also covers superfamilies.

Comparison of structurally similar proteins is more of the same: Domains need to be taken into account (methods to do so are available, although in general they are even less accurate than the approaches to parsing sequence domains), and each structural family is a sample of what might be the true set of proteins with similar structures. In addition, however, the quantitative theory of structural similarity is usefully defined only for very similar structures, and statistics of structural comparison falls apart for most pairs of distantly related structures. Furthermore, the problem of distinguishing homologous from analogous folds (i.e., separating class III from class VII similarities, discussed in Chapter 9) remains untractable. I call a group of proteins with known structure a "structural family" when there is evidence that all structurally similar proteins in this group are homologous. When the set contains structurally similar proteins that may be analogous, this will be called "fold" (note that upon new evidence a fold may turn out to be a structural family).

Finding sequence families and finding structural families of proteins are tasks of evolutionary inference. At the technical level, however, they are usually presented as problems of clustering by similarity. As for any problem of that type, these clustering tasks include three main components: a measure of similarity between protein sequences or structures, a way of producing clusters, and a statistics evaluating the significance of each cluster.

Frequently, these distinct tasks are lumped together in a somewhat confusing way. For example, Yan and Moult (2005) note,

> *No sequence-based method is able to detect all evolutionary relationships: Experimental structure determinations reveal previously undetectable relationships in many cases. Thus, all sequence-based families are, in some sense, arbitrary, reflecting the effectiveness of current relationship detection algorithms rather than the number of independent evolutionary lines.*

In the same vein, Orengo and Thornton (2005) say:

> *As the number of known structures solved by x-ray crystallography and NMR techniques increased, it became clear that protein structure is much more highly conserved throughout evolution than protein sequence.... In contrast to protein sequence, where in some families relatives have been detected sharing fewer than 5% identical residues, in many protein families at least 50% of the structure, mainly in the core of the protein, is highly conserved . . . and can be used as a fingerprint to detect very distant relatives.*

These quotes and many other similar ones, if read literally, suggest that structural families—by which many authors also mean folds or, at least, they do not distinguish true structural family from a fold—are more reliable estimators of the number of homologous families or "independent evolutionary lineages" than sequence families. But such a statement may not hold a closer scrutiny.

There is no doubt that sequence comparisons, even the most elaborate ones, are not guaranteed to find all homologs. It is also true that for this reason, every partitioning of protein sequence database into families is imperfect. However, as I argued previously, any definition of a sequence family is an estimation of its real size. Different assumptions and statistics may be used for family or fold definition, and as they improve, so too does the accuracy of the estimates. Also, in any given "relationship detection algorithm" the number of independent evolutionary lines will always remain a statistical estimate. In that sense, every partition of the protein space into families is an approximation, unless all existing proteins, all ancestral proteins, and all lines of descent were known in advance (in which case there would be no reason to seek answers for the evolutionary questions).

The same is true of clustering structural families. When we compare structures, we also estimate the size of each structural family. The suggestion that direct comparison of structures may produce a better estimate of real size of sequence families is quite common, but in fact, just the opposite may be true. To illustrate this, let us turn on its head the argument made by Orengo and Thornton (2005). They note that in sequence comparisons we can sometimes detect homologs that have just 5% of identical residues. When comparing structures, however, 5% or even 15% of superimposed residues will not do much for establishing homology. Thus, structure-based clusters of proteins are approximate (but not "arbitrary") in exactly the same sense as sequence families: They are but the estimates of the true clusters.

The number of protein folds is smaller than the number of sequence families—this much is not controversial. This basic fact can be, at the first approximation, derived from the frequencies of different types of events in divergent and convergent evolution. One such event is descent with modification of two protein sequences, which may proceed to the point at which the sequence similarity decreases to the background level, but structural similarity is still preserved. This was described in Chapter 9 as class III similarity. This process, which is most likely the thoroughfare of protein evolution, results in the excess of sequence families over folds. Another event is convergent evolution of folds (class VII similarity), which also results in an excess of sequence families over folds. Finally, abrupt but local sequence change is possible, which changes protein fold but maintains global sequence similarity (class II similarity). In this process, the overall number of folds may increase only when the starting fold was nonunique and the resulting fold is novel. Clearly, this accounts for only a fraction of the relatively rare cases of fold change in divergent evolution (the list of examples of class II similarities is too short to estimate accurately the rate of generation of new folds in this process). If protein evolution is fully described by these processes, the net result is the excess of families over folds, and the ratio of families to folds would keep increasing.

One factor that has not been considered thus far is gene loss. Genes, as well as whole families and folds, are unevenly distributed in different evolutionary lineages. Families can be selectively gained and expanded, but they may also be selectively lost. For example, yeast *Saccharomyces cerevisiae* lacks many genes that are widely distributed in other fungal genomes but must have been deleted in the ancestral lineage leading to hemiascomycetes (Aravind *et al.*, 2000). Even human proteome, thought to be one of the most complex among metazoa, lacks a few genes that other animals have. Not only single genes but also complete families can be lost in the evolution of individual lineages (and some ancestral proteins and protein families must have been wiped out in all surviving lineages so that we may never know that they existed). Here again, however, there is no process to countervail the excess of families over folds. Therefore, and regardless of the exact numbers, an accurate estimation of the number of folds puts the lower bound on the number of sequence families, and accurately estimated number of families puts the upper bound on the number of folds. The problem is derive reasonable estimates for the numbers of families and folds in the protein universe as a whole and in completely sequenced genomes.

The idea of a large excess of sequence families over structural folds has been discussed innumerable times in the literature. In this chapter, however, and also in Chapter 9, we have seen that in order to correctly understand this excess, we have to overcome an apathetic attitude toward evolution (e.g., "divergence versus convergence is difficult to prove and not interesting anyway; what really matters is fold recognition"), as well as simplistic understanding of the generating process (e.g., "structural similarity is the ultimate proof of common origin, even if sequence similarity is not there"). As argued in Chapter 9, a more difficult, yet perhaps more principled, approach is to account, as accurately as possible, for all divergent and convergent events in protein evolution. Enumeration of families and folds, then, can be viewed as a way to organize the protein space in order to unravel these evolutionary events.

Attempts to estimate the number sequence families and folds have been undertaken many times. Interestingly, the number of folds seems to attract more interest than the number of families, probably because of the belief that folds are more informative of the real state of affairs and of the evolutionary history of the protein universe—a belief that may not be well-founded after all. However, there is also a practical reason to estimate the fold numbers, and it has to do with the ongoing needs of target selection for structural genomics projects. In these large-scale efforts, high premium is put on identifying new folds and on increasingly dense coverage of the fold space.

Perhaps surprisingly, the results of different estimates in the past decade vary significantly, by more than an order of magnitude: from 400 to 8000 folds, and from 1000 to 24,000 sequence families. Moreover, the numbers do not seem to converge with time, despite the growing databases of protein sequences, structures, and families, which one would think provide the increasingly representative samples of the protein universe. Sometimes these extremes of the estimated family and fold numbers are thought to be the upper and lower bounds of the actual numbers. In fact, this is probably not true because they have been obtained using different statistical models, with different underlying assumptions, and may not be directly comparable.

Early guesses that the total number of protein families may be on the order of 1000 are the lore of biology. However, as far as I know, the reasons for choosing this number, rather than 500 or 5000, were never presented. It is true, however, that in 1992, Cyrus Chothia at the Medical Research Council in the United Kingdom collected all sequences from the early stages of yeast, worm, and human genome projects and compared them to sequence databases. One-third of protein coding sequences from the genome projects had homologs in the databases (homologs were operationally defined as high-scoring matches; the proportion of sequences with database matches was somewhat underestimated in that protocol; Chothia, 1992). On the other hand, one-fourth of all proteins in the sequence database had similarity to some sequence in the database of protein structures, which, as of 1992, was though to include approximately 120 discrete protein families. Note that this number referred to the sequence families defined on the basis of high sequence similarities. In fact, the database of protein structures contained a smaller number of spatial folds. Chothia examined all cases in which several proteins shared the same fold, and if some of them included dissimilar sequences, Chothia split such folds into a large number of sequence families. This roundabout approach was needed because at the time the classification of structures was even more ad hoc than it is now.

Multiplying $120 \times 4 \times 3$ calculates to approximately 1500 families. Chothia noted, however, that his way of determining whether proteins were related or not (i.e., the alignment of a large fraction of both proteins in each pair, with identity of 25% or more, plus some conserved residues with the known molecular function or other indications of common function) would underestimate the percentage of related proteins in the sample. If only 80% of all sequence similarity signals are detected, the true number of protein families may be closer to 1000. The number of folds should be smaller, but it was not clear by how much.

A crucial contribution of Chothia's work was in his treatment of the three data sets—the database of all known sequences, the database of proteins encoded by the known portions of the model organisms slated for complete genome sequencing, and the set of sequences of each protein whose structure has been determined—as independent samples from the protein universe. Note that this assumption is not entirely correct. For example, families are most likely distributed across genomes in a nonrandom fashion, especially if we consider large families and compare species with vastly different complexity. Databases of protein structures are richer in certain classes of proteins: Notably, families of soluble, globular proteins are overrepresented, and other classes, such as disorganized, nonglobular, and membrane proteins, all of which do not readily produce well-diffracting crystals, are underrepresented. Furthermore, the sample of species represented in the databases is biased. For example, all three data sets used by Chothia were extremely low on proteins from archaea, plants, and protists. However, despite these concerns, the idea of independent sampling was used, and even statistically tested, in later studies.

A few months after Chothia's study, the paper by Philip Green and coauthors on ancient conserved regions (ACRs) was published (Green *et al.*, 1993). We examined it in Chapter 5, mostly focusing on the concentration of ACRs in genomes (PHIDO value). However, in the same work, the total number of ACRs was also estimated. For that purpose, Green *et al.* used the BLOCKS database (see Chapter 2). Green *et al.* selected only those blocks that satisfied the definition of an ACR—that is, those that contained representatives of at least two distantly related eukaryotes or of a prokaryote and an eukaryote. There were 481 such regions. Approximately two-thirds of all database ACRs in yeast and worm proteomes were represented in BLOCKS. Recall that database ACR is such an ACR that also has a match in SWISSPROT. Then, $481 \div 2/3 = 730$ is the number of ACRs—or in other words, ancient protein families—that SWISSPROT should contain. Given their own estimation that approximately 85% of all ACRs already had a representative in SWISSPROT, the total number of ACRs would be near 900. Strictly speaking, BLOCKS and SWISSPROT are not independent; the former has been derived by analysis of conserved families in the latter, so this estimate is rather biased.

Green *et al.*'s and Chothia's numbers may seem in good agreement, but this may be more of a coincidence than true convergence of the estimates. Green *et al.*'s number concerns only ACRs—that is, only such protein families that had sequences from two distant clades. Chothia's estimated set, then, would have only approximately 100 extra families to account for all prokaryote-specific families, families specific to just one metazoan clade, and families found in plants. This clearly does not make sense.

The other extreme of the family count—that is, 23,100 sequence families—comes from the work by Orengo *et al.* (1994). In this pioneering study, several interesting questions were raised, and some assumptions about protein universe were first tried out. Unfortunately, its opening of the article is rather misleading: "It is important to realize that, as with sequence comparisons, structural similarities form a continuum. Cutoff points for acceptable levels of structural similarity for related folds were derived by empirical trials ... on proteins having distinct or common folds. If "distinct or common folds" are purported to form a continuum, there needs to be an external criterion to decide which structures are similar and which are different. Nothing much has been proposed in this arena other than "expert knowledge." Moreover, if this idea about continuum is taken literally, then the purpose of delineating and counting folds and families loses focus and becomes just a tactic of exploratory clustering, not the study of evolutional, functional, and structural signals.

As with everything else in this book, I will hold a different view dear to my heart. The statement of sequence continuum is true only in a trivial sense, namely that we can measure a distance between any two sequences, even between the unrelated ones (see Chapter 2). However, most pairs of sequences in the database are related at a random level, and therefore much of "sequence continuum" is noise. Similarly, we can define a distance measure and compare every two structures; here again, most of the results of these comparisons are very low-scoring similarities, which are not of interest.

Proteins are Pauling and Zuckerkandl's sense-carrying units, and so are protein families. If there are evolutionary, structural, and functional signals in families and folds, and if such signals can be distinguished from the noise, there has to be something discrete about them. Thus, it is the *dis*continuity in the sequence and structure spaces that interests us—as a real phenomenon, not the result of an arbitrary "cutoff of convenience." The counts of families and folds, when done appropriately, should indeed give us an idea of what was going on in evolution, i.e., the number of "independent evolutionary lineages," and what is going on today in structural and functional organization of living species.

Let us return to the estimation of the number of sequence families by Orengo *et al.* Note that despite a later publication date and, presumably, larger sizes of public databases than those available to Chothia and Green, Orengo *et al.*'s study was still performed before the complete genome era. The authors examined 2511 protein structures available at the time, and the protein sequence corresponding to each of these structures was compared to every other protein sequence in this set. Pairs with sequence identity of 25% or higher were joined into clusters. This gave 212 families, 80 of which contained a single protein and 132 contained two or more proteins. At this level of sequence identity, homology is inferred quite reliably, so there were very few false positives or none at all. On the other hand, false negatives could not be excluded. The authors noted that there are known examples of homologous proteins with pairwise identity less than 25%; for example, the lowest percentage of identity between two members of the globin family is 15% (the family is nonetheless confidently defined because each globin sequence is linked to at least some homologs by much higher similarity). The authors' empirical estimates showed that for a pair of proteins with known structures and less than 25% sequence identity, the probability of having a common fold was approximately one-third; in other words, some of the proteins and families perhaps could be lumped into a smaller number of larger groups. Here, the authors made an assumption that two groups of proteins with low similarity and similar fold can be considered homologous if most proteins in one group have similar function to most proteins in another group. With these modifications, the space of proteins with known structures can be partitioned into 131 "hyperfamilies," each consisting of one or more families. Some of the similar folds remain distinct hyperfamilies even in this case, mostly when it can be shown that they have different function.

There are several problems with this approach. First, biological and even molecular function is difficult to define. Many proteins have multiple functions, and a complete set of functions may not be known for any protein (see Chapters 5–7). Often, new functions are discovered for what was thought to be an exhaustively well-studied protein. One of the more famous examples is the role that a citric acid cycle enzyme, aconitase, plays in controlling cellular homeostasis of iron (Rouault *et al.*, 1991), but there is no shortage of "moonlighting proteins" anywhere we look, even among such seemingly dedicated proteins as ribosomal proteins and aminoacyl tRNA synthetases (Copley, 2003; Jeffery, 2005). At the whole-organism level, pleiotropic action of genes is also a rule rather than an exception. Second, functions of proteins evolve in a way that is not completely dependent on the extent of sequence identity. For example, a large group of beta-barrel-like proteins (cupin superfamily) is certainly a homologous family, as evidenced by statistically significant sequence similarities. The list of known functions for the members of this family includes phosphomannose isomerases, CENP-C centromeric proteins, metabolite sensors fused to helix–turn–helix transcription factors, epoxidases and dioxygenases with various specificities, oxalate decarboxylases, seed storage proteins, dTDP-4-dehydrorhamnose 3,5-epimerases, histone deacetylases, and other activities (Dunwell *et al.*, 2001, 2004;). With this diversity of function, even looking at similar structures, we may not be able to use functional criterion to join all these proteins in a superfamily or a hyperfamily, unless sequence similarity is also present. The third and most serious problem is in some sense the opposite of the previous one: Because of possible functional convergence, joining structures by similar function cannot be done by default when there is no sequence similarity. Thus, the hierarchy built by Orengo *et al.* starts as evolutionary classification at a lower level, when sequence similarity is high, and then injects functional criterion into clustering. This may be convenient for many practical purposes, but it is not the way to determine the "number of independent evolutionary lineages."

With this concern in mind, let us examine the rest of the argument. Orengo *et al.* consider the following simple model (in this chapter, I sometimes change the original authors' designations in order to make different models more easily comparable to one another). The most important numbers that we are looking for are F_O and F_A—respectively the total number of folds in the protein structure universe and the total number of families in protein sequence space. We will designate the observed numbers of folds and families F_O' and F_A', respectively. The estimated values, as opposed to actual numbers, will be designated by $\hat{F}_O'$ and $\hat{F}_A'$. F_O and F_A are estimations by definition.

Suppose that each of the F_O folds in the protein structure universe is equally probable. Suppose also that sampling of proteins for structure determination is random. Let us compare only such families that none of the sequences within one of them shares more than 25% sequence identity to any sequence from another family (in Orengo *et al.*'s framework, this means that there is no good sequence-based argument for these families to share similar fold). Under this extremely simple model, the probability of two families having the same fold is $1/F_O'$, and the expected number of folds in the data set can be expressed as the ratio of all possible comparisons to the observed structural matches. With 212 superfamilies, there are $212 \times 211/2 = 22{,}366$ possible structural comparisons. Among those, 583 pairs had significant structural matches. Dividing the numbers, we estimate that the data set should contain 38 folds. If extended hyperfamilies are compared, there are $131 \times 130/2 = 8515$ possible pairs and only 191 matches, which gives the $\hat{F}_O'$ of 45. Both numbers are clearly off base because the F_O' in that data set was at least 80. Clearly, some of Orengo *et al.*'s assumptions must be invalid. Which ones?

The authors concluded that, most important, the equal probability of every fold was not compatible with the data: There were just nine superfolds, which contained between 3 and 11 sequence-unrelated families—56 families in total. On the other hand, 71 folds each contained a single family. Thus, 11% of folds comprised 44% of families; the distribution of folds by the number of families was "one-tailed." The authors also studied the number of sequence families (if pairwise sequence identity was between 30 and 40%, this was called superfamilies, but we agreed not to distinguish them from families). At this similarity level (called 30SEQ), there were approximately 7700 sequence families. Using Chothia's 1992 estimate that approximately one-third of predicted gene products coming from the large-scale sequencing projects have similarities to the database sequences, Orengo *et al.* arrived at the F_A of 23,100. From this, F_O can be reestimated: Given that there are 234 families with the known structure (the slight inconsistency with a previous number of 211 is due to a slightly different inclusion threshold, 25 vs. 30%; we are also ignoring the existence of large proteins that contain more than one structural domain). Then, F_A of all 30SEQ families is 23,100, or almost 10 times higher. This means that the currently known F_O' of approximately 80 expands to the F_O value of ~7900.

Thus, Orengo *et al.*'s main results are the following. First, the distribution of proteins over families, and families over folds, is not random. This, perhaps, is one of the most important facts of computational structural biology. The properties of this one-tailed distribution have been debated and studied ever since.

This distribution had been anticipated by much earlier studies in polymer physics and thermodynamics, which had suggested that laws of physics and chemistry may favor adoption of certain folds by proteins over other folds (see Chapter 9). This has often been discussed under the names of "designability," "optimality," "attraction in the evolutionary space," "robustness," etc. Indeed, we do observe some very large families, as well as hyperfamilies and superfolds. The existence of such extremely populated folds and families is, most likely, a complex phenomenon that has to be explained by a combination of factors acting at different levels. For example, two folds adopted by a variety of enzymes with diverse activities, TIM barrel and Rossmann fold, are both highly symmetrical superfolds that may be viewed as two "most designable" ways of packing a series of repetitive units of a beta strand and an alpha helix.

But let us now count the occurrences of two representative classes of each superfold in different genomes: For example, yeast and archaea *Sulfolobus sulfataricus* have almost the same numbers, but different percentages, of the Rossmann-fold S-adenosyl L-methionine (SAM)-dependent methyltransferases (respectively, 30 and 31, accounting for 0.4 and 1% of all genes in these genomes), and very different numbers and frequencies of another class of SAM-binding proteins, namely TIM barrel SAM-radical enzymes (respectively 5 and 24; Kozbial and Mushegian, 2005). At this level, designability must have very little to do with the observed counts of these protein families; selection for function may have played a more important role. Thus, the reasons for the existence of superfolds may be complex.

The second significant lesson of Orengo *et al.*'s analysis is that the calculations of F_O rely on the F_O' value, which, in their case, was derived from the database of folds. Structural classification of proteins, however, is a difficult problem for both computers and human beings, and there is not much choice with regard to databases of structural families, with only two databases, SCOP (Andreeva *et al.*, 2004) and CATH (Pearl *et al.*, 2005), dominating the field. Most studies, therefore, in effect estimate the total number of folds, as they are defined by SCOP (or CATH). This is good for many practical purposes, but it may not provide a satisfaction of knowing the number of different evolutionary lineages because, as already discussed, these databases do not deal with the problems of structural convergence at the high levels of their hierarchy.

Third, it became clear that the earlier estimate of 1000 for F_A was an inspired, but not very accurate, guess. The F_O value is bound to be smaller than F_A (and, as more recent studies show, may indeed, with some assumptions, be close to 1000), but the F_A value may be an order of magnitude higher than 1000.

The fourth conclusion was that estimation of F_A and F_O may benefit from the rapid growth of databases of sequences and sequence families. Two sides of this dynamic are of interest: (1) the picture of families and their sizes at any given moment and (2) the information about growth of these numbers with time. Orengo *et al.* did not make much use of this information in their 1994 study, but they calculated the number of sequences and 30SEQ families in the database for each year from 1960 to 1992 (Fig. 4 in their article). Although both numbers were steadily growing, the family discovery rate (i.e., the ratio of 30SEQ families to sequences) reached a plateau and even started to decline slightly sometime between 1981 and 1986.

More than a decade later, we are still counting families and folds. The new studies are based on more involved statistical models than simple calculations of the early 1990s and also on much larger databases of folds, families, and complete proteomes. But the estimates of F_A and F_O continue to vary widely.

Chun-Ting Zhang of Tianjin University in China was one of the first to explicitly model the relationship between sequences, families, and folds (Zhang, 1997). He used essentially the same data as Orengo *et al.* but examined them from a different angle. He formulated some desirable properties that the inference of F_A and F_O should have. First, the growth of families and folds over time should be an important aspect of the model. Second, the way to define sequence families (e.g., which score threshold to use for including a protein into a family) should enter the model as a constant. Zhang derived an equation that shows a log-linear dependence between the newly identified families and the increase in number of the database sequences. Zhang did not give the F_A and F_O values for the protein universe, but he provided estimates for four species: *Escherichia coli*, yeast, worm, and humans. All these numbers were on the higher side (e.g., 17,000 families and 5200 folds for human proteome).

In 1998, Chao Zhang and Charles DeLisi of Boston University further studied the distribution of folds by the number of families (Zhang and DeLisi, 1998). They proposed a random sampling model that suggested the geometric distribution for the family numbers over fold numbers. The estimate of fold number was given by the following equation (in our notation):

$$F = \frac{F'_{AS}\ F'_O}{F'_{AS} - (1 - \frac{F'_{AS}}{F_A})F'_O}$$

where all designations are as usual, and F'_{AS} is the observed number of protein families with at least one representative with known structure. Zhang and DeLisi noted that when $F_A >> F'_{AS}$, the formula is not very sensitive to the F_A value. But F'_{AS} and F'_O had to be substituted from somewhere, and the values (736 and 361, respectively) were taken from the June 1997 release of the SCOP database. Solving the equation, we have $F_O = 687$, and the number remains close to 700 even if F_A is allowed to grow as high as 10^5. The F_O of ~700, obviously, is an order of magnitude less than the estimate of 7900 given by Orengo *et al.*

In 1999, Sridhar Govinjaran, Ruben Recabarren, and Richard Goldstein of the University of Michigan pointed out that most of the previous statistical theories of fold and family sampling were based on simplified assumptions, and that the fit of the derived distributions to the data was not very good. They noted that the estimation of the number of folds is the instance of a long-known "species problem," dating back to the father of 20th-century statistics (and one of the fathers of modern genetics), R. A. Fisher. In this problem, the shape of the distribution and the number of entities need to be estimated from a random sample drawn from the population. Govinjaran and coauthors produced their own, more sophisticated model in which the counts of folds over families were approximated by stretched exponential distribution, which appeared to give the best fit to the data in a likelihood ratio test. They found that their distribution modeled reasonably well the folds that had extremely high, medium, and extremely low numbers of families in them. In contrast, Zhang and DeLisi's geometric distribution fails to predict the excess of families in the superfolds, and it shifts the density toward the middle of distribution. Govinjaran and co-workers give the F_O estimate of approximately 4000, with the 90% confidence interval between 2105 and 8069. They also note that because of the extreme rarity of many folds that consist of just one sequence family, perhaps 2000 of the existing folds are unlikely to ever be observed. The F_A value was not estimated.

Andrew Coulson at the University of Edinburgh and John Moult at the University of Maryland Biotechnology Institute studied these distributions further and concluded that there may be no way to approximate the shape of the distribution by one analytical formula (Coulson and Moult, 2002). They showed that Zhang and DeLisi's model resulted not only in undercounting of families in superfolds but also underestimation of the number of folds that consist of just one family. Because the latter category of families contributes the most to the fold count, Zhang ad DeLisi's model is also an underestimation of F_O, and in fact, on the updated data sets, Zhang and DeLisi's model even underestimates F'_O. At the same time, Govinjaran *et al.*'s model overestimated the number of protein folds that had just one family in them. Coulson and Mount proposed a three-zone model, in which all folds are partitioned into unifolds, mesofolds, and superfolds. Unifolds are all folds that consist of just one sequence family each, superfolds are previously noted nine folds that contain more than 10 families each, and mesofolds are folds with 2–10 families in them.

Coulson and Mount suggested that three types of folds have different properties, and that any statistical estimation has to deal with three separate models, one for each zone. They derived such a model, fitted parameters using SCOP release 1.37, and then asked whether thus adjusted model would correctly describe the distribution of folds by the number of families in a later, larger release of SCOP (1.48). The results seemed to be satisfactory, which allowed the authors to estimate F_O. The abstract of their publication states that "the total number of folds is at least 10,000." In the text, however, they note that the number is obtained if F_A is taken to be 50,000. The proportion of unifolds is estimated indirectly by assuming that there are nine superfolds and by inferring the fraction of mesofolds using the "well-behaved" region of distribution. The fraction of mesofolds f_m, in turn, is given by the expression

$$f_m = \frac{N_m}{D' \cdot F_A},$$

where N_m is the number of mesofolds, and D' is the mean number of families per mesofold. Most of the parameter fitting, however, is done using Orengo *et al.*'s F_A of 23,100, and at that point the total number of folds is ~4600. Clearly, the value of F_A is of the utmost importance for all other computation. However, I suspect that 23,100 is an overestimation of F_A, and the brief explanation by Coulson and Moult as to why they switched to a twofold higher F_A is confusing to me.

In 2000, Yuri Wolf, Nick Grishin, and Eugene Koonin of the National Center for Biotechnology Information produced their own estimates for both F_A and F_O. This work contained two important novel ideas, both having to do with sampling of families and folds. (Coulson and Moult only mentioned this work in passing.) All previous efforts were essentially confined to one and the same sample, in which families were defined in some way and then assigned to SCOP folds. The resulting sample can be fitted to a favorite analytical distribution, but the ways to evaluate the correctness of the model are limited to resampling (as was done by Govinjaran *et al.*) or to evaluation of performance on the growing fold database (as was done by Coulson and Moult). In contrast, Wolf and co-workers decided to take the suggestion of Chothia (1992) and Green *et al.* (1993) seriously and used the information on the completely sequenced genomes as an essentially independent sample of families and folds. As shown in Chapter 9, a substantial fraction of proteins in all completely sequenced genomes belong to families, and many proteins encoded by each genome have homologs of known structure; thus, each proteome contains 10^3–10^4 proteins that can be assigned to families and 10^2–10^3 proteins that can be assigned to folds. Although each particular proteome may be enriched in particular folds and depleted of the others, the union of many phylogenetically diverse proteomes is probably not strongly biased compared to the universal population. Then, the distribution may be fit to analytical form using families defined and folds predicted for complete proteomes and then extrapolated to the universal population.

Wolf *et al.* employed logarithmic distribution, which is often used to model biological phenomena that involve hierarchies and populations. For example, distribution of species by the average number of individuals in populations appears to be well described by logarithmic series (Pielou, 1969). Here, as in most other studies, the likelihood of the model was not systematically evaluated but, instead, Wolf *et al.* optimized partitioning of proteins into sequence families by focusing on the lower 90% of all families (i.e., ignoring superfolds) and finding such a threshold for clustering sequences into families that provided a good fit of the data to the logarithmic distribution. Interestingly, there was a sharp peak in the quality-of-fit function at the threshold of 0.3 bits per aligned position. With the aid of these inventions, two things could be done. First, the number of families and folds in the universal population was estimated. F_A was in the range of ~4400–7300, and F_O was between 900 and 1400. Second, families and predicted folds in complete proteomes could be counted. The percentage of families with known or predicted fold varied from 5% in archaea and yeast (the only eukaryote in the data set) to 10% in most bacteria and 20% in the universal population. In each case, this translates to a much larger proportion of predicted folds for individual proteins because many families contain large numbers of proteins.

Most recently, Moult's group provided a direct count of F_A by clustering all proteins encoded by 67 completely sequenced prokaryotic genomes (Yan and Moult, 2005). They compared several methods of family definition using SCOP families and folds and parsing proteins into domains in order to avoid chimeric clusters. A total of 178,310 protein sequences were dissected into 249,574 domains and then clustered into 31,874 sequence families. Distribution of families by size is dramatic: 20,992 families are singletons, 4810 families are "doubletons," and 6072 families contain three or more members. At the same time, singletons and doubletons account for a small fraction of all proteins—respectively, 8.4 and 3.9% of the sequence space. These results suggest that complete coverage of the sequence space by structural genomics may be an unattainable goal, but almost complete coverage of 80–90% of structural space is achievable with the current high-throughput structural biology approaches. For example, 88% of proteins in 67 genomes fall in just 6072 families, and projection for 1000 genomes indicates that 8000 experimentally determined structures may afford 70% domain coverage (ignoring the complications associated with nonglobular and transmembrane proteins).

The idea that structural genomics will reach the point of diminishing returns is becoming widely accepted; however, it is clear that at that point, a very substantial fraction of proteins, on the order of 80%, will have a template for structural modeling. At the same time, it is evident that for any *completely sequenced* genome, the complete coverage of proteome by structural templates will be limited not by the high number of singletons but by the recalcitrance of some proteins to current methods of structure determination. Indeed, if Yan and Moult's numbers are representative of the prokaryotic world, then a medium-sized proteome of 5000 proteins will cluster to approximately 1000 sequence families, including all singletons, and for a large fraction of those (at least one-third, I think) the template is already in the structure databases.

The fundamental question of the F_A value, however, remains unresolved. I believe that Moult's study overestimates the proportion of singletons. I compared the proportion of larger (tripletons and up) families in the set of 67 genomes with the coverage of the individual genomes by the COG database. For example, the genome division of GenBank lists 1729 proteins in *Methanocaldococcus jannaschii*, and 1514 of them (87.5%) are found in COGs. In plague bacterium *Yersinia pestis* CO92, there are 3885 proteins, 3342 of which (86%) are in COGs. At first glance, this seems like a remarkable corroboration of Yan and Moult's approach, and it is tempting to explain this agreement by pointing out that COGs, too, are required to contain three or more proteins, and that substantial work on manual domain dissection is part of COG definition. However, closer analysis suggests a mere coincidence. First, the training set of Yan and Moult included many pairs and triplets of closely related organisms, and homologs of each protein could come from any of these organisms, including the one in which the query was found. In contrast, COGs cannot be made of any three matches but are initiated by three proteins found in three species separated by certain evolutionary distance (see Chapter 5). If paralogs in one organism or homologs in strains and subspecies of one species, or in species within one genus, were all allowed to form a COG, the coverage of genome by COGs would be much higher. Second, Yan and Moult matched only proteins encoded by a large yet finite selection of completely sequenced genomes. Using all homologs found in the NR database for construction of family models would increase sensitivity of matching and move a fraction of singletons and doubletons into larger families. Finally, the PSI-BLAST protocol that involved three iterations and threshold for model inclusion of 10^{-4} may be optimal for keeping the false positives-to-true positives ratio under 1%, but in my experience this underreports many matches.

These three factors, all contributing to overestimation of the number of very small families, may be offset, to an unknown degree, by domain parsing. However, I believe that when all is said and done, the number of small families will decline, perhaps by one-third.

The excess of proteins with paralogs over unique proteins is a complex affair. There are several dozen very large sequence families, and together they account for a very large fraction of proteins. On the other hand, when the excess of families over singletons is averaged over all families, the number of proteins per family may turn out to be a small number, perhaps close to 2.

Attempts to estimate F_A and F_O will, undoubtedly, continue. Increasing coverage of sequence and structure space will result in more accurate numbers than are currently available. One of the most interesting questions that will be answered in the not-so-distant future is whether the excess of folds over families is, likewise, a relatively trivial phenomenon. Could it be that despite the existence of nine superfolds (are there only nine of them?), the average number of families per fold is also less than two?

11

Phylogenetic Inference and the Era of Complete Genomes

In the past 10 chapters, I tried as hard as I could to stress the importance of the evolutionary inference for everything that is going on in computational biology. From inference of homology to complex interplay of divergence and convergence in protein folds and complete pathways, nothing indeed makes sense except in light of evolutionary comparison of sequences. But I barely mentioned the main workhorse of evolutionary inference, namely phylogenetic trees. I now discuss this theme, but not without trepidation.

Only a few years have passed since the publication of the instant classic, *Inferring Phylogenies*, by Joseph Felsenstein (2003). That book is the ultimate source of all things related to phylogenies, from statistics and algorithms to idiosyncrasies of various schools of evolutionary thought and the aesthetics of tree drawing. Add to it Felsenstein's beautiful prose, and the only question is, Why even bother to say anything more about phylogenetic approaches in biology?

A few things, however, are not in Felsenstein's book. First, very little is said there about actual phylogenetic history of the known life-forms. Felsenstein knows more than anybody else about the methods of phylogenetic inference, so I can imagine that application of these methods, even with the goal of reconstructing the history of life on Earth, might look to him, well, applied. Moreover, being aware of—indeed, keenly interested in—the limitations of each method of phylogenetic inference, Felsenstein probably sees better than most that with the current state of the data and algorithms, too many difficult phylogenetic questions, especially concerning the ancient speciation events, do not have good answers. Continuing heated debate about specific phylogenies, two of which are reviewed in Chapter 12, can be seen as indication of exactly that—the problems may be still too difficult to settle even with the most advanced approaches.

Insufficiency of methods and data, of course, never prevented researchers from trying to answer their favorite questions. However, the scientists of the 1960s and 1970s, and all the way to the early 1990s, had to focus on a relatively small set of molecular sequences, but in the past dozen years the picture became qualitatively different—complete genome sequences entered the scene. And this era of complete genomes is another theme mentioned only in passing in Felsenstein's book. But in my opinion, the impact of completely sequenced genomes on our understanding of phylogenetic problems and of the history of life is significant, and it only grows as the space of genome sequences and deciphered gene functions becomes denser. We always wanted to use molecules to reconstruct the evolutionary history of life—just as Pauling and Zuckerkandl told us to (see Chapter 1)—and now we are closer to this goal than ever before.

Phylogenetic approaches in the era of complete genomes should not be viewed as just "more of the same": The challenge is not just to handle many more molecular characters by scaling up our algorithmic approaches and computer hardware. As with everything else in biology, complete genomes bring large, perhaps qualitative changes to our perspective of the phylogeny of life and to the methods of phylogenetic analysis. With that, let us now review some basics of trees and biological phylogenies.

A tree is an object of mathematics, which can be defined using some notations of graph theory. A graph is the set of points, called vertices or nodes, where some pairs of vertices are connected by lines, called edges. A degree of a vertice is the number of other vertices in this graph to which this vertice connects. An unrooted tree T is a connected graph of n vertices of degree 1 or 3 with no cycles and with $2n-3$ edges (branches); a nonnegative real number (branch length) may be assigned to every branch. Defined in such a way, trees are purely formal constructs, and they do not convey any biological information.

Living forms evolve by descent with modification, and one neage may be split in two, with two descendants having different modifications in their sense-carrying units. This is why the ancestor–descendant relationships, phylogenies, can be modeled as tree-like graphs. In a species tree, nodes with degree 3 are speciation events. In a gene tree, these nodes can also represent speciations, or they can be intraspecies gene duplications—resulting in orthologs and paralogs, respectively (see text and figures in Chapter 3, and note that speciations and duplications are depicted differently in Fig. 3.1 because we know in advance which is which. In Fig. 3.2, however, the nodes corresponding to duplications and to speciations are indistinguishable).

The nodes in biological phylogenies are also called operational taxonomic units (OTUs). OTUs are not always genes or species: They can be anything that reproduces with modification, from kingdoms to single nucleotides. The internal nodes usually represent ancestral OTUs, which are often unavailable to direct observation (but not always: when comparing morphological characters, we may have paleontological evidence of the state of ancestral OTUs, and when studying DNA and protein sequences, we may be able to recover ancestral molecules from ancient specimens and sequence them).

Phylogenetic tree seems to be the most obvious way to represent evolutionary history. The Tree of Life is taught in every life science class, from kindergarten to college, and the power of metaphor makes it a favorite in popular literature and art. Even biologists sometimes are convinced that life history is equivalent to a tree. But two things have to be remembered here.

First, not every evolutionary history can be represented by a tree. In a most obvious example, genealogy of individuals is certainly an example of an evolutionary history, but if an individual has two parents, there is no tree. If evolutionary history of two species involves hybridization, there is also no tree, and if evolutionary history of a gene involves horizontal transfers, there is no tree either. These histories can be depicted as trees only if we agree to change the aforementioned definition of a tree, or to allow some simplifications. For example, human genealogy can be represented as a tree if only one gender is considered, which is indeed done in mitochondrial genome-based phylogenies of humans (mitochondria are inherited maternally by us). Phylogeny can be presented in a tree-like form if we assume, or conclude from the data, that horizontal gene transfers can be ignored because they were so rare as to play no significant role in this phylogeny. In addition, in a tree as previously defined, all splits produce two OTUs from one, but there is no reason why this would always be so in evolution: Splits into more than two species or nearly instantaneous generation of more than two gene copies are possible. This results in patterns that are called stars or bushes and also in nodes with degrees other than 1 or 3.

Second, genetic and genomic data can be represented in the form of a tree, even though they do not represent evolutionary information. Consider the quantitative picture of gene expression (often called "microarray data," although, of course, information on gene expression can be obtained by any kind of mRNA assay, from DNA array hybridization to massive signature sequencing, or measured at the protein, not RNA, level by quantitative proteomics). The patterns of gene expression are commonly organized as hierarchical clusters (Eisen *et al.*, 1998), which satisfy the definition of a tree, and yet contain no evolutionary signal; expression patterns of genes within single organism are not OTUs and, unlike genes, they do not share a common ancestor nor did they evolve from one another. This tree, however, contains a biologically informative functional signal, for example in the form of groups of coexpressed genes. Thus, not every tree built from genomewide measurements is evolutionary, and not every evolutionary history is a tree.

We now discuss those trees that do convey evolutionary information, and we start with a brief introduction of the methods of inferring phylogenies (for an in-depth review, the reader cannot do much better than reading Felsenstein's book). We have a set of OTUs as the input and would like to have a tree on which each OTU is placed at one node (usually, although not always, a tip) in such a way that it reflects the evolutionary relationship between the OTUs. Even for a handful of OTUs there exists a surprisingly large number of different trees. The true tree has to be constructed by quantitative comparison of OTUs and optimization of some parameters that come out of this comparison. For now, we will defer the question of what to optimize and will describe one class of methods for phylogenetic inference, which relies on construction of distance matrices.

Let $d_{ij} = d(i, j)$ be a nonnegative real function d: $X \infty X \varnothing R^+$, satisfying (1) $d_{ij} > 0$ for $i \neq j$; (2) $d_{ij} = 0$ for $i = j$; and (3) $d_{ij} = d_{ji}$ for all i, j. Then d is a distance measure, and $D = \{d_{ij}\}$ is a distance matrix. In the preceding chapters, we discussed comparative analysis of genes, proteins, and genomes, but there was almost no discussion of distance measures that are used to compare them. Instead, we were talking about similarity and different ways of measuring it. It is not difficult to see that similarity and distance can be converted into one another. The space of distance measures, however, is better studied and easier to treat mathematically than the space of similarities. Some of the known properties of the distance measures are discussed next.

If for a given $D = \{d_{ij}\}$ there exists a tree T such that the sum of branch lengths along the shortest path between any pair of terminal vertices i, j is equal to d_{ij} for all i, j, D is said to be additive. If, in addition, d_{ij} satisfies triangle inequality $d_{ij} \leq d_{ik} + d_{kj}$, then d is called a metric. A necessary and sufficient condition for additivity of D (or, in other words, for d to be a tree metric) is the four-points condition (Zaretsky, 1965; Buneman, 1974): For all sets of four elements there exists some labeling i, j, k, l X such that $d_{ij} + d_{kl} = d_{jl} + d_{ki} \geq d_{il} + d_{jk}$. Furthermore, D is said to be ultrametric if the three-points condition holds: For any three elements i, j, k, the two closest elements i, j are at the same distance from the third element—that is, $d_{ij} \leq d_{ik} = d_{kj}$. In other words, distance matrix is ultrametric if, for each triplet of elements, there is a tie for the maximum of pairwise distances between them, and distance matrix is additive if, for each four elements, there is a tie for the maximum of pairwise sums of distances between them. Ultrametric condition is stronger than additivity, and additivity is stronger than metric property: If distance matrix is ultrametric, it is also additive, and if it is additive, it is metric, but the opposites are not true.

These properties of distance matrices are useful for constructing trees from them. For example, it has been proven that if the distance matrix is additive, there exists a unique tree corresponding to it, and this tree can be constructed from the matrix in time that grows as a squared number of the OTUs. The problem of reconstructing tree from a nonadditive distance matrix has not been solved in polynomial time, although approximate methods, such as neighbor joining, can infer the tree in cubic time. Similar theorems have been proven for ultrametric distance matrices (Gusfield, 1997). These facts are described in computer science textbooks and in many books on computational biology.

All this is helpful when the set of distances have been already obtained, but these properties do not tell us how to get them. In Chapter 2, we reviewed the history of measuring similarity between amino acid and nucleotide sequences. Perhaps the main conclusion was that the *a priori* models need to be modified using the observations on the real data. Also, in the case of scoring function for protein sequence alignments, the mathematical properties of the distance measures, such as additivity, did not play a major role in the process. Moreover, it is not quite clear that all genomewide distances could or should be ultrametric, additive, or possess any other property. Every justification why this should be so goes back to computational tractability, not to any fundamental biological reason. Therefore, I propose the following as the "unnumbered fact of comparative genomics" or perhaps its "first Zen observation":

> *There exist a great many ways to measure distances between any entities in comparative genomics, be it gene sequences, distributions of genes in genomes, patterns of gene expression, or something else. The choice of good distance measure is extremely important for uncovering biological signals, but it cannot be dictated solely by useful mathematical properties of distances.*

Distance matrix methods are not the only way of reconstructing phylogenies. Most introductions to phylogenetic approaches start with another group of methods, namely those that involve maximum parsimony. Distance matrix methods and parsimony methods try to find the "best" tree by optimizing different properties of the tree. Roughly speaking, the former methods try to minimize the distortion between the matrix of observed distances and the set of distances that is induced when all OTUs are assigned to specific nodes in the tree, whereas the latter methods try to minimize the amount of evolutionary change that is needed to explain a particular tree.

Many authors believe that parsimony approaches are more intuitive or more valuable as a didactic tool. Moreover, the past few decades have witnessed a passionate debate in which one side posed that parsimony is the only viable approach to phylogenetic inference, that parsimony is in the most basic way opposite to statistical inference, and that this is good. For example, parsimony is purported to descend directly from Karl Popper's teachings or from the Okham's razor. However, Felsenstein and others argue that both parsimony and distance matrix methods are, in most cases, different incarnations of a fundamental statistical approach—that is, maximum likelihood. Interestingly, Felsenstein notes that all major classes of approaches to phylogenic inference can be traced back to a series of papers by A. Edwards and L. Cavalli-Sforza, published at the same time as Pauling and Zuckerkand's work (Edwards and Cavalli-Sforza, 1963, 1964, 1965). [For discussion of nonstatistical justification of phylogenetic inference, see Chapter 10 of Felsenstein (2003). For a philosophical perspective, see Sober (1991, 2004), and for lively discussion of strengths and pitfalls of cladistics, see Ridley (1986)].

Other methods of building trees and inferring phylogenies, based on various algorithmic and combinatorial ideas, continue to appear. We will not examine them here, but it important to remember that the field is not lacking new ideas. And with the abundance of methods for tree reconstruction, it is often assumed that a tree-like relationship between any OTUs exists and is recoverable. But I already noted that in genomics we may have reasons to build and analyze trees that lack evolutionary content, and we also encounter evolutionary histories that do not look like trees. In this chapter, we are particularly interested in biological processes that can be modeled as edges connecting branches on the tree. A graph that contains these connections, or reticulations, is formally not a tree: Reticulations produce cycles, and some paths from the OTU to the root are no longer unique (Fig. 11.1, top). Such graph may still be "tree-like" if the number of cycles is moderate and it is clear that they are but an embellishment of a clearly

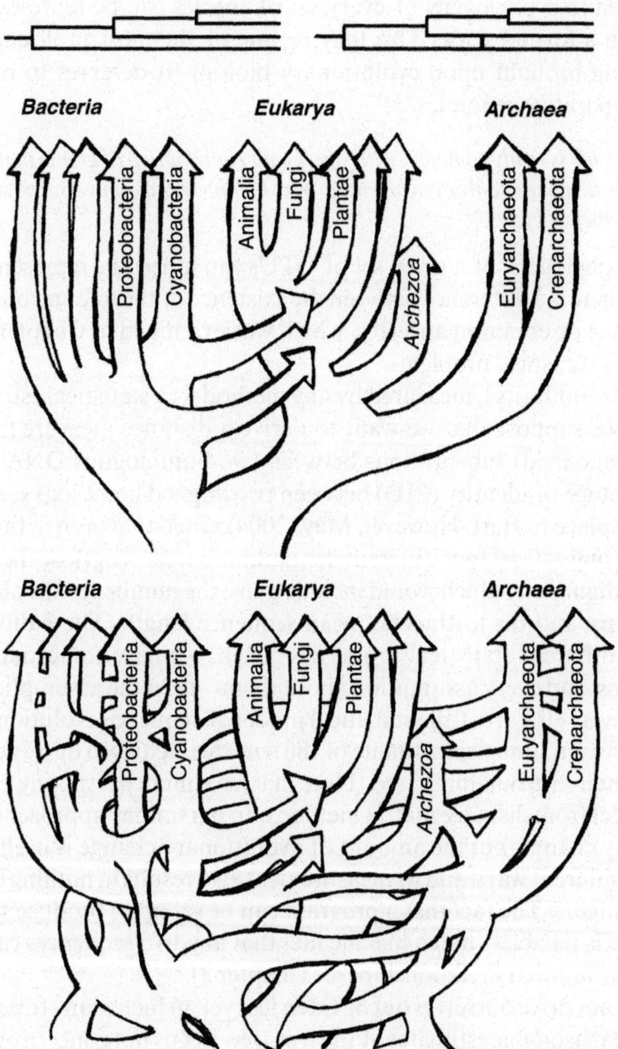

Figure 11.1. A tree or not a tree? (**Top**) The graph on the left is a tree, and the graph on the right is not a tree. (**Middle**) The tree of life as imagined before the era of complete genomes. Two symbiogenetic transfers of genes from organellular to nuclear genomes were thought to be the only horizontal gene transfer events of any consequence (note that these two reticulations mean that the Tree of Life is no longer a tree from the graph theory point of view). (**Bottom**) The era of complete genomes may provide so many examples of horizontal gene transfer that the Tree of Life is not at all like a tree, at least in its deepest branches. Note that the picture at the bottom is a metaphor. In fact, a pessimistic conclusion that the high rate of horizontal transfer makes it all but impossible to infer the ancient evolutionary events with any certainty may not be true, and further improvements in phylogenetic methods may allow us to reconstruct these early events in sufficient detail. The middle and bottom panels are reprinted with permission from Doolittle (1999), copyright 1999, American Association for the Advancement of Science.

discernible tree structure. But, of course, it is quite possible to imagine an evolutionary history, rich in reticulations, so that its tree-like aspect is barely visible (Fig. 11.1, middle and bottom).

Comparison of bacterial and archaeal genomes, especially those that live in the same habitat, has produced much evidence for unusual evolutionary relationships between some of these genomes—basically, the excess of bacteria-like genes in certain archaea and archaea-like genes in certain bacteria. We examine this evidence in more detail later in this chapter and also in Chapter 12. What is important to note now is that these relationships upend, in the most direct way, the idea that phylogeny of every set of species can be represented by an acyclic graph, or a tree in a formal sense. This may be one of the most profound discoveries that genome sequencing brought upon evolutionary biology. It deserves to be recorded as the "third fact of comparative genomics":

> *Natural history of life is really complex, with significant contribution of events that cannot be represented on a conventional, cycle-free phylogenetic tree. In other words, universal phylogeny contains a nonnegligible number of cycles.*

But what if the phylogeny of a given set of OTUs can indeed be represented as a true tree? Even in this case, there is a difference between the existence of the tree and our ability to recover it. As with inference of common ancestry, which was examined in Chapters 2, 3, and 9, this, fundamentally, is a statistical problem.

Any distance (or similarity), measured by any method, is a statistical estimate of a true distance. For example, suppose that we want to derive a distance measure from the counts of nucleotide (or amino acid) substitutions between two homologous DNA (or protein) fragments. The percentage of identity (PID) between two aligned homologs seems a well-defined measure—a good place to start. However, May (2004) noted that even in this simple case, one can express PID using at least four different denominators: the length of the shorter sequence; the length of the alignment, which would include gaps; the number of the aligned pairs, which would exclude gaps; and the arithmetic mean sequence length. These differently expressed PID values have different statistical properties. Then, we have to account for reverse and repeated mutations, and every assumption about them—even the assumption that they occur infrequently or never—is part of the statistical model of sequence evolution.

Likewise, any tree is a statistical estimate of the true tree, achieved by way of modeling many parameters that enter into building a tree. (Note that maximum parsimony methods are no different in this respect from distance matrix methods: In parsimony approaches, we seek to minimize evolutionary change, but the amount of evolutionary change is itself a parameter that can be defined in different ways, and its measurement also results in nothing more than the estimate of "true change.") The fact that a program can be set up to produce some tree does not mean that the tree is reliable—much like the fact that any two sequences can be aligned does not ensure their evolutionary relationship (see Chapter 2).

However, let us not drive ourselves out of Eden just yet; in fact, many trees built from molecular sequences are reasonable estimates of the true tree. Let us not concern ourselves with "bad trees" and instead ask what "good trees" might be good for.

In 1995, Walter Fitch published one of the most comprehensive lists of uses for evolutionary trees in the pre-genomic era. The most obvious purpose of a tree, of course, is to represent the evolutionary history of a set of homologous traits. However, there are many other things for which the tree-like representations are helpful; Fitch showed 22 examples of different uses. The following are the most important types of questions that can be answered with the aid of evolutionary trees:

1. Trees are tools to study gene duplications. To distinguish duplications from speciations (and paralogs from orthologs), one usually needs at least two trees—one for genes and

another for species that contain them. We examined this theme in considerable detail in Chapter 3.

2. Duplications is one example of a larger class of cases in which histories of individual genes/proteins (or protein domains) are not the same as the history of the species in which these genes reside. Other such cases include gene recombination of various kinds, as well as gene conversion. Evidence of all these events can be noticed in the trees. Here again, comparison between two trees is of significant help.

3. Yet another example of difference in species history and gene history, which can be studied by examination of evolutionary trees, is xenology (see Chapter 3) and other types of horizontal gene transfer. This will be further discussed later in this chapter and in Chapter 12.

4. Trees are tools to study evolutionary models and parameters of sequence evolution. Some of it can be done on the basis of just one tree: For example, one can infer a phylogeny of a group of homologous sequences and then use the knowledge of tree topology to find the relative rates of evolution of several sequences within an in-group using comparison of these sequences to the out-group (Kumar, 2005). In a more involved experiment, several detailed, multiparameter models of evolution can be compared by asking how likely they are to produce the observed data. These ideas are applicable not only for molecular sequences: Trees can also be used, for example, to estimate the rates of gene gain and loss in genomes and to study the behavior of other characters as well.

5. Trees are tools to infer the ancestral states of the OTUs, as proposed by Zuckerkandl and Pauling (1965; see Chapter 1). Most classes of methods for tree construction can be modified to be used for that purpose, and all sorts of OTUs are of interest in this respect. There is growing literature on reconstructing and then synthesizing the ancestral molecules based on alignments and trees of their present-day descendants (see the next example), and I dedicate a significant part of Chapter 13 to the problem of reconstruction of ancestral gene content.

6. Trees are also tools of building up biological hypotheses by connecting different types of data. For example, Thomson *et al.* (2005) examined the evolutionary origins of alcohol metabolism in yeast of the *Saccharomyces* group. Many yeast species live in fleshy fruits or in sugar-rich plant sap. There, they convert sugars to pyruvate by glycolysis. Pyruvate is further converted to acetaldehyde, which is reduced by alcohol dehydrogenase 1 (Adh1) to ethanol. This is the ability for which yeasts have been domesticated by humans. Yeasts also have Adh2, a paralog of Adh1, which works mostly in the opposite direction (i.e., to consume the accumulated ethanol). Alignment of Adh1 and Adh2 is 348 amino acids long, with 24 substitutions. Thomson and co-workers sequenced many Adh homologs from hemiascomycete yeasts, constructed a maximum likelihood evolutionary tree, and inferred the sequence of common ancestors, including AdhA—the putative common ancestor of Adh1 and Adh2 of *S. cerevisiae*. Using one of the models for codon and amino acid evolution (PAML11; Yang, 1997), they reconstructed almost all positions unambiguously, except for three sites where equal support was given to two or three candidate residues. Twelve possible combinations of amino acids were synthesized, and the kinetic behavior of candidate AdhA enzymes showed that all of them (except for a single nonfunctional variant) were more like Adh1 than Adh2, suggesting that the ancestor was mostly making ethanol rather than consuming it. Thus, before the Adh1–Adh2 duplication, production of ethanol was most likely not the strategy of food hoarding but, rather, served some other purpose, perhaps recycling of NADH generated in glycolysis or poisoning the competitors that cannot live in ethanol.

Another question is whether the Adh1–Adh2 duplication can be explained by a specific selective pressure. The comparison of topologies of the Adh family tree and hemiascomycetes consensus species tree suggests that the Adh1–Adh2 duplication occurred before the divergence of the species of *Saccharomyces* sensu stricto but after the divergence of *Saccharomyces* and *Kluyveromyces*. The latter event might have occurred approximately 80 million years ago (this is an estimation from molecular sequences—another use for a tree—since paleontology is of no help here). Interestingly, at least eight other genes in the ancestor of *Saccharomyces* seem to have undergone duplication at approximately the same time, and six of these duplications involved proteins that participate in the conversion of glucose to ethanol. With all the reservations about the accuracy of the assumed molecular clock, the time of these duplications seems to be close (enough) to the age when fleshy fruit arose in the Cretaceous, during the age of the dinosaurs. We do not know why fleshy fruit habitat would be conducive to having two Adh enzymes with preference for two opposite reactions, but at least this question can now be studied experimentally by inoculating some peaches with yeast strains of defined phenotypes. Furthermore, trees can be studied in conjunction with geospatial, demographic, and almost any other data that contain some sort of distance measurement or records changes in character states.

The common thread running through all examples is that a tree is cross-referenced with other data, which, in essence, provides labels to the nodes or branches in a tree. Then, distribution of these labels over the tree is studied in order to extract biologically important signals. Fitch concluded his review (1995) thus: "While a nimble mind might well discover the results inferred from these tree studies without recourse to a tree, the use of the tree simplifies and speeds the road to understanding evolutionary processes."

This comment is worth contemplating. For example, if I accept distance matrix methods as a legitimate way to build a tree, many of the problems described previously can be solved by making computations on such a matrix directly. This idea of talking about evolutionary problems "without recourse to a tree" is not about banning trees; it is about better understanding what trees are for. However useful trees may be for presenting biological information, it is often possible that they are not necessary for obtaining this information in the first place.

I believe that another list, complementary to Fitch's, should also be of interest to biologists. Let us call it "misuses of (or, at least, honest mistakes in using) evolutionary trees."

The first type of misuse of evolutionary trees involves the wrong choice of characters. With regard to sequences, the gravest misuse is to infer the tree from the alignment of characters that should not have been aligned in the first place. In Chapter 2, I argued that sequence alignment has two most basic but distinct uses: to screen the space of pairwise similarities and to find statistically significant signals in the space of all possible alignments. Any two sequences can be "aligned" in the sense of the Gibbs–McIntyre square diagram or Needleman–Wunsch approach; that is, some subsequences from each sequence can be written on top of one another, and a score can be associated with such a match. As we have seen, computation of such a score does not prove the existence of any signal, including an evolutionary one—a separate statistical theory is needed for that.

I also warned in Chapter 2 that any computer algorithm of pairwise or multiple sequence alignments will build an "alignment" from almost any bunch of sequences. Some programs will warn the user about random-level similarity in some of the sequences that it tries to align, but I do not know of programs that would exit if similarity is too low. Likewise, programs for tree inference are set up to build trees from the data, more or less regardless of what the data actually represent. As long as alignment is in the right format, the tree-building program will process it. In truth, some programs will complain if the values of some parameters are out of whack: For example, NEIGHBOR from the popular PHYLIP package (Felsenstein, 2005) will post the errors about infinite distances (which are roughly equivalent to background-level similarity between sequences) but will output the tree nonetheless. However, just as alignments of sequences with random-level similarity may lack biological meaning, the same will be true of a tree built on such a basis.

Biological interpretations of meaningless alignments, leading to meaningless trees, can still be encountered in, and need to be chased out of, scientific literature (Iyer *et al.*, 2003). For example, a secreted factor from human macrophages, called macrophage migration inhibitory factor (MIF), had been thought to conjugate with glutathione *in vitro* (Blocki *et al.*, 1992). A cautious interpretation of this observation would be that MIF has a cysteine residue that can participate in redox reaction(s) under some experimental conditions. Instead, the authors of the observation became enamored by a hypothesis that glutathione *S*-transferases (GSTs), already known to play a role in protection of cells from certain toxic molecules, might also be part of a cellular-level resistance of organism to pathogens. On that premise, they produced an alignment of MIFs and glutathione *S*-transferases GSTs and then a "family tree" (Blocki *et al.*, 1993). William Pearson, who had spend a considerable part of his career studying GSTs, showed, using several statistical tests, that the purported connection between MIFs and GSTs was not supported by any sequence signal (Pearson, 1994). The tree only showed the relationships within each of the two classes of proteins, but the edge drawn between MIFs and GSTs was arbitrary. Later research proved the correctness of Pearson's analysis: MIFs turned out to have their own catalytic activity—not GST nor any transferase, but 4-oxocrotonate tautomerase (Rosengren *et al.*, 1997). The now known three-dimensional structure of MIFs is also different from that of GSTs (Sugimoto *et al.*, 1999). Glutathione conjugation has never been shown to play any role in MIF biology, and cysteine residue that ostensibly conjugated glutathione in mouse and human MIF is not conserved in many other homologous proteins.

Thus, homology between sequences is a necessary condition of any analysis that involves a tree, and it has to be inferred from statistical assessment of sequence similarity and, perhaps, similarity in protein structure. Homology cannot be inferred from the tree.

Another kind of misuse of a tree may happen even when the characters are fundamentally sound (i.e., their homology is determined correctly). Modern methods of sequence analysis, such as those examined in Chapters 2 and 5, are very sensitive. They can establish homology between sequences when the absolute values of sequence identity/similarity are extremely low. This is excellent for a qualitative conclusion—that is, that there is an evolutionary relationship between two sequences—but most of the methods of phylogenetic inference are not well suited to decide on the best tree in these circumstances. (A maximum likelihood model, which takes account of all relevant evolutionary parameters, could be a general solution to a problem, but such detailed models are rarely available in practice, and working with them may be computationally prohibitive.) However, as just discussed, most tree-building algorithms attempt to build the best tree, no matter what the data are.

This is where assessment of statistical significance of a tree comes into play. Methods to do so have been around for more than 20 years, beginning with different versions of data resampling, such as bootstrap or jackknife tests, and supplemented more recently by Bayesian approaches. A tree that does not show the degree of statistical support for each internal node is not good for any biological inference.

In effect, this is the problem of discerning signal from statistical noise, by finding those branching events in a tree that are well supported and those that are not. In a review of Felsenstein's book, David Penny (2004) noted that systemic biases in the data may be a much more significant problem than statistical noise. There are indeed many types of systematic bias in the data—most obvious, in the frequency of different characters and in the model of transition between different states. However, the difference between these biases and "noise" is mostly in the degree of our understanding of the process. When we do not know that our data are biased, we use simple models and simpler null hypotheses, and it is all noise to us. When we become aware of the bias, and start taking it into account either by learning from the data or by inventing a more sophisticated *a priori* model, then the bias becomes part of the model. The rest remains noise until we learn even more.

One kind of bias, however, deserves special discussion. In 1978, Felsenstein examined one of the statistical properties of the tree inference methods, called consistency. An estimator (in this case, phylogenetic inference) is statistically consistent when, upon the addition of more characters, the best tree estimate converges on the true tree. Felsenstein studied simple models of rooted phylogeny of three OTUs, in which a character has state 0 at the root and can change to state 1, with certain probability along each branch, but can never change back. Probability of change in character state can take only two values, for example, *P* along some branches and *Q* along the other branches. Felsenstein asked whether certain methods of tree inference, namely, a character-compatibility method and two different parsimony methods, are consistent under various values of *P* and *Q*. He also examined unrooted phylogeny of four OTUs, with similar properties of *P*, *Q*, and branch lengths. The main result was that, in these simple models, there were always zones of *P* and *Q* values for which these methods will estimate wrong trees, and the addition of more characters would not result in any improvement. Details aside, this zone is defined by all cases in which *P* is much larger or much smaller than *Q*.

The insight from this work was summarized in the article abstract as follows (Felsenstein, 1978): "In all cases the conditions for this failure (which is the failure to be statistically consistent) are essentially that parallel changes exceed informative, nonparallel changes." In other words, statistical inconsistency is found in trees, particularly those inferred by parsimony and compatibility methods, when there are some (in Felsenstein's example, two) relatively long branches. A relatively small amount of parallel change in these long branches may artificially draw the two branches together.

Felsenstein noted that unequal branch lengths may be a result of either noncontemporaneous OTUs (he had in mind the cases of morphological characters, some of which might come from ancient organisms and others from existing organisms, and now we also have molecular sequences related in the same way), or different rates of evolution along the branches leading to several present-day OTUs. He also noted that a proper maximum likelihood estimate of a tree would not be prone to such an artifact. On the other hand, the knowledge about the inconsistency zone in the *P*/*Q* parameter space [called the "Felsenstein zone" by John Huelsenbeck (1997)] is useful for examining the properties of phylogenetic approaches: One can simulate sets of sequences evolving under different values of *P* and *Q* and use these sequences to test the consistency of any tree-building method.

Thus, there are three important ideas in Felsenstein's paper: the observation that long branches may be attracted to one another, the warning that this may cause problems in evolutionary inference, and the proposal to use this effect for assessing consistency of phylogenetic inference.

In the 1980s and early 1990s, several investigators studied the *P*/*Q* parameter space further, under various evolutionary assumptions and within different frameworks of phylogenetic inference, and interesting theoretical behaviors were noticed with simulated sequences. However, it took almost two decades until the next milestone. In 1997, John Huelsenbeck, then at the University of California at Berkeley and currently at the University of California at San Diego, studied a real phylogeny that may be sensitive to the long branch attraction (LBA). Two years earlier, a tree of insect orders had been constructed on the basis of 18S rRNA. *Diptera* (flies and mosquitoes) was grouped in this tree with *Strepsiptera*, twisted-winged parasitic insects that, according to morphology and physiology, have long been placed closer to *Coleoptera* (beetles; Carmean and Crespi, 1995). Carmean and Crespi did not believe the *Diptera*/*Strepsiptera* clade was correct and thought it was an LBA artifact. This hypothesis had to be proven, so Huelsenbeck made extensive simulations and compared tree topologies under different evolutionary models (using mostly maximum parsimony and maximum likelihood methods of tree construction). He obtained significant supportive evidence in favor of fast evolution of both taxa and the artifactual origin of the clade *Diptera*/*Strepsiptera* (Huelsenbeck, 1997).

Thus, by the late 1990s, LBA entered not just theory but also practice of phylogenetic inference. Other comments made by Huelsenbeck (1997), however, remained not sufficiently appreciated. In my opinion, they should be better known, especially the following (pp. 69–70):

> *(1) The branches leading to Strepsiptera and Diptera are both very long and (2) the support for this grouping [i.e., the clade of these two taxa] is moderately high according to the bootstrap method. Unfortunately, these criteria for long-branch attraction are weak because they fail to identify whether the branches are long enough to attract each other in a parsimony analysis. According to these criteria, if the longest branches of a tree happen to be linked together, then long-branch attraction or method inconsistency can be invoked. Yet, using this criteria, it is impossible to ascertain whether (1) long branches do, in fact, belong together or whether (2) the long branches should be separated by short branches but were linked together because of long-branch attraction.*
>
> *I argue that two more tests must be passed before long-branched attraction can be invoked: It should also be shown (1) that the branches are long enough to attract (i.e., if the long branches were separated, that the maximum parsimony method [i.e., the same method as was used in the first place] would link them together in the estimated phylogeny) and (2) that a method that is less sensitive to the long-branch attraction problem gives a phylogenetic estimate in which the long branches are separated.*

These criteria continue to be important, especially because in the past few years, several labs, notably Herve Philippe's (at Paris-Sud and, recently, at the University of Ottawa), demonstrated that unequality of evolutionary rates in different taxa (commonly, in the OTUs at the rank of family or above) is widespread, that an evolutionary history of even one clade may include periods with significantly different rates, that popular distance matrix methods are as prone to LBA artifact as parsimony and compatibility, and that even some of the maximum likelihood methods are in practice inconsistent.

Much of this is serious theoretical phylogenetics. But it also had a side effect of introducing some postmodern discourse in evolutionary biology, which can be summed up as yet another nonnumbered fact of comparative genomics:

> *In a discussion about phylogeny of any group of organisms, the probability of bringing up unequal evolutionary rates and ensuing LBA artifacts approaches 100% with time.*

In order to avoid an ad hoc introduction of LBA into scientific discussion, and to find it when it really happens, we should follow Huelsenbeck's advice. As he pointed out, the ultimate question for practicing phylogeneticists is usually not whether there are long branches in a given tree and not whether they are attracted to each other. What is most often of interest is the correctness of phylogenetic inference. With this in mind, and remembering the "ten commandments of detection," which define what is permissible for an author of a mystery novel (Knox, 1928), I propose the "five rules of (long branch) attraction in phylogenetics," or, rather, the rules of invoking the LBA artifacts. Rules 1–4 are taken, with modification, from Huelsenbeck's (1997) work. Rule 5, as far as I know, has not been explicitly introduced, although, when this manuscript was submitted to the publisher, I became aware of a thoughtful review by Bergsten (2005) that discusses essentially the same rule and many related issues.

Rule 1: In order to invoke LBA, it is necessary but not sufficient to have both long branches and short branches in a tree.

Rule 2: In order to invoke LBA, it is necessary but not sufficient that relatively long branches in a tree indeed attract each other.

Rule 3: In order to invoke LBA, it is necessary to show that two or more branches are grouped together, but the same is not sufficient because these branches may be a proper clade.

Rule 4: LBA cannot be invoked automatically if phylogeny inferred by a more LBA-resistant approach displays the same clades with comparable statistical support.
Rule 5: LBA cannot be invoked automatically if there are more than two long branches in a tree.

In order to better understand rule 5, let us examine an unrooted tree with four branches, three of which are "long enough to significantly attract each other" and one is short (Fig. 11.2 and Table 11.1). The true topology of the tree is (AB)(CD). There are three pairs of long branches in this tree and, accordingly, three ways in which two branches can attract one another: (BC), (CD), and (BD). Note that in the second of these cases, LBA may mislead us into thinking that C and D are closer to each other than they really are, but the clades (AB) and (CD) are still inferred correctly (this has been shown to happen more often in maximum parsimony inference, and the parameters that favor this particular kind of LBA are called "Farris zone" after famous evolutionist and proponent of parsimony methods John Farris; see Swofford *et al.*, 2001). Furthermore, depending on the relative strength of each pairwise attraction, there are four formal possibilities of "net attraction," two of which do not change

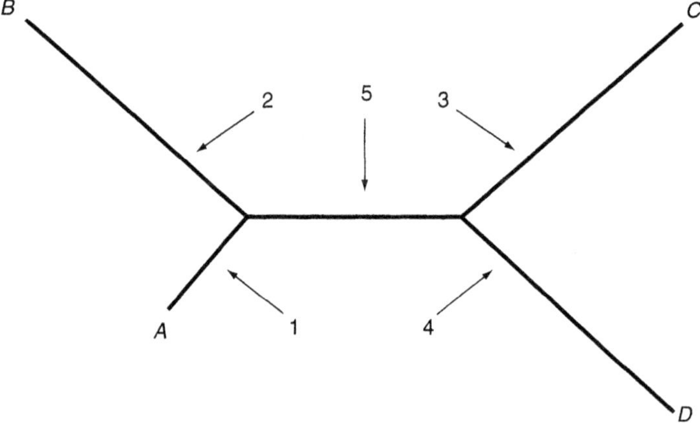

Figure 11.2. A phylogenetic tree with four taxa, A–D, and five possible places for the root of the tree (1–5). The correct topology of the tree is (AB)(CD).

Table 11.1. Possible Artifacts of Attraction of Two Long Branches in Such a Tree as Shown in Fig. 11.2[a]

	Net LBA result					
	(BC)		(CD)		(CD)	
True root position	Wrong clade	Wrong rooting	Wrong clade	Wrong rooting	Wrong clade	Wrong rooting
1	Yes	No	No	No	Yes	No
2	Yes	Yes	No	No	Yes	Yes
3	Yes	Yes	No	Yes	Yes	No
4	Yes	No	No	Yes	Yes	Yes
5	Yes	Yes	No	No	Yes	Yes

[a]Artifactual clades and wrong positions of the root are possible in some, but not all, cases.

the topology of the tree (see Fig. 11.2). In addition to these varied effects on tree topology and on apparent branch length, there may also be consequences of LBA on rooting because many methods of root inference involve computation of branch lengths, whereas rooting on the out-group brings into the picture a new branch (the said out-group) that is long by definition.

From these preliminary qualitative considerations, it is clear that a detailed study of LBA on more than two branches and a theoretical model of behavior for an arbitrary number of long branches are needed. In Chapter 12, we examine two cases in which LBA has been invoked without regard to some of the rules outline previously, leading, in my opinion, to erroneous conclusions.

LBA is a special case of a broader class of events, which have been known in phylogenetics under the name homoplasy. Homoplasy, for all practical purposes, is a synonym of convergence, the term that we examined in detail in Chapter 6.

Let us now discuss how the discovery of one such major inconsistency led to the third fact of comparative genomics, which was introduced earlier in this chapter. The first archaeal genome, *Methanococcus (Methanocaldococcus) janaschii*, was sequenced in 1995. The original report stated that most proteins in this species were unique (PHIDO of 44%; see Chapter 5). Craig Venter, then of The Institute for Genomic Research and later of Celera and human genome sequencing fame, stated that "two-thirds of its genes are unlike anything we've seen in biology before" (quoted in Department of Energy press releases, for example, www.pnl.gov/er_news/10_96/down_ar.txt; accessed August 11, 2006). However, Eugene Koonin and co-workers, including myself, reanalyzed the genome using gapped BLAST, the most sensitive database search program at the time, and found that statistically significant similarities, coupled with the analysis of conserved sequence motifs, increased the fraction of evolutionarily nonunique proteins in *Methanococcus* to slightly more than 72%. This is a large difference, which had brought archaeal PHIDO fully in line with other species (see Chapter 5). However, average and median highest scores between *Methanococcus* proteins and their closest homologs were significantly lower than for the other microbes sequenced at the time. This difference must be explained by much better sampling of the evolutionary neighborhoods of these other bacteria at the time (i.e., many sequenced *Proteobacteria* for *Haemophilus influenzae*, many Gram-positive bacteria for *Mycoplasma genitalium*, and bacteria at-large for *Synechocystis* sp.).

Approximately 5% of *Methanococcus* genes had detectable homologs only in archaea. The remaining ~68% of nonunique genes had detectable homologs in bacteria, eukaryotes, or both. This is where the real evolutionary surprise was awaiting. But to explain what it was, I have to give an extremely brief summary of the leading theory of the three domains of life. This theory, the debates surrounding it, and how it survived and became accepted deserve better than what I reduce it to in the next three paragraphs. Fortunately, the firsthand accounts are available, most notably from Carl Woese and co-workers: The interested reader should consult Olsen *et al.* (1994), Woese (2004), and additional references that will be given in the rest of this chapter as well as Chapter 12.

A hypothesis that was decidedly unorthodox in the 1960s and 1970s when first proposed by Carl Woese, but by the 1990s enjoyed considerable support and respect, with only a few detractors, postulated that Archaea are the third major kingdom of life, along with Bacteria and Eukarya. The relationship between three kingdoms was under debate for many years, but two main facts seemed to be pinned down. First, the phylogeny based on ribosomes (started in the 1960s and 1970s as comparison of oligonucleotide and peptide maps of ribosomal RNA and proteins from different prokaryotes and followed in the 1980s by sequencing and comparison of rRNA and ribosomal protein genes) indicated two ancient splits. One was separation between bacteria and the rest of living forms, and another was separation between archaea and eukarya. This topology of tree of life can be written as ((AE)B) in the standard Newick format.

Partial genome sequencing in the 1980s and early 1990s sampled many prokaryotic and eukaryotic genomes, and it was found that many protein coding genes also supported the split into (AE) and B. This was confirmed at two levels. If a protein had orthologs in all three kingdoms A, E, and B (phyletic pattern 111), the phylogenetic tree of such orthologs tended to display the ((AE)B) topology. However, if a protein was not omnipresent, it usually had phyletic pattern 110 or 001; that is, it was found either only in AE or only in B. This was mostly true for proteins involved in translation (ribosomal proteins, initiation and elongation factors, and some aminoacyl tRNA synthetases), transcription (multisubunit RNA polymerase and several eukaryote-like transcription factors), and replication (most notably, Archaea/Eukarya had type B replicative polymerase and ATP-dependent DNA ligase, whereas bacteria had type C polymerase and NAD-dependent DNA ligase; see Chapters 6, 12 and 13). Thus, based at least on this large collection of essential genes, it was concluded that prokaryotes are most likely two different kingdoms, not one.

The second important fact had to do with the position of the root. ((AE)B) is a rootless tree of life, and theoretically there may be three places to insert the root in it, one on each branch. The two most popular methods of rooting evolutionary trees were not applicable in this case: Rooting on the out-group was out of question because there is no out-group in the tree that encompasses all life, and rooting on the midpoint between the major clades is also technically impossible when there are only three clades. The new idea was to search for a particular class of genes, namely those that were represented by pairs of paralogs in all living forms. Such genes must have been duplicated prior to divergence of A, E, and B, and each ancestral member of such a pair gave rise to its own set of orthologs. The complete tree produced by the progeny of one ancestral paralog will then be an out-group for the tree given by the descendants of the other paralog. Thus, trees for two ancestral paralogs can be rooted on one another, if we can determine which genes have been duplicated prior to the divergence of A, E, and B.

Reconstruction of gene content in an ancestral genome has to combine traditional methods of phylogenetic inference with some novel algorithmic approaches, which help to account for gene losses and displacements, and to be checked for internal consistency (i.e., that the inferred metabolism makes sense) and for compatibility with paleontological, biochemical, and other planetary evidence. We have seen one example of a partial historical reconstruction of one short pathway earlier in this chapter, and I devote much of the next two chapters to reconstruction of ancestral cells and metabolisms. And it will become evident that many genes in the genome of the last common ancestor of A, E, and B had paralogs. In 1989, Iwabe and co-authors at Kyoto University used one such pair, two GTPases involved in translation initiation, to root the tree of life, and shortly thereafter several other authors attempted to do the same with additional pairs of paralogs, including V- and F-type ATPase subunits, aminoacyl-tRNA synthetases, and a few others. Details aside, most trees agree on the same root placement between AE and B.

Thus, by the mid-1990s, the most straightforward interpretation of the tree of life was that the ancestral lineage evolved by two major splits: first into (AE) and B, and then by bifurcating (AE) into A and E. Of course, major questions remained, particularly concerning the identity of the common ancestor (was it more like modern bacteria, more like archaea, or yet something else?) and the origin of eukaryotes. However, the expectation of the complete genome sequencing was that the *Methanococcus* proteome would broadly support the ((AE)B) topology of the three kingdoms, in the same way as the translation-related subset of archaeal genes did: either by the ((AE)B) topology of the gene trees or, if genes are lineage specific, by phyletic pattern 110 (A E -). This, however, was not what we observed.

We examined the taxonomic identity of the closest database homologs of each protein of *M. jannaschii*, which were collected using the gapped BLAST software, and found the following. For 44% of proteins, the closest bacterial homolog had significantly higher similarity to the *M. jannaschii* protein than the closest homolog from Eukarya (the difference of at least

5 percentage points between similarity levels determined by WUBLAST) or, in many cases, no eukaryotic homolog of the archaeal protein could be detected at all. The percentage of *M. jannaschii* proteins for which the opposite was true (i.e., the closest homolog from Eukarya was significantly closer to the query than the nearest homolog from Bacteria) was much lower at only 13%. The remaining 43% of proteins had exclusively archaeal homologs, none at all, or they were approximately equidistant from their bacterial and eukaryotic homologs. In the years since our initial study, with improved sequence comparison and increase in dense coverage of the sequence space by genome projects, the majority of this group was parsed into either Bacteria-like or Eukarya-like genes, and Bacteria-like proteins are still in excess after these later recalculations.

Thus, essentially only 13% of archaeal proteins supported the Woesean ((AE)B) topology, which, by the mid-1990s, had seemed to be finally settled by the rRNA and other evidence. In contrast, 44% of proteins supported the ((AB)E) topology. We also saw the same trends in a large collection of proteins from a distantly related archaeon, *Sulfolobus solfataricus*, whose genome sequencing was ongoing at the time. Thus, the distribution of Bacteria-like and Eukarya-like genes was not peculiar to *Methanococcus* but seemed to be a genuine property of all Archaea.

There was another clear trend in the data: The "Bacteria-like" genes in *Methanococcus* tended to be concerned with energy supply and with biosyntheses of amino acids, nucleotides, coenzymes, and polysaccharides. In contrast, the "Eukarya-like" genes mostly had to do with genome maintenance and expression. There were some exceptions to this rule; for example, we saw a fragment of eukaryote-style mevalonate pathway for isoprenoid biosynthesis and, on the other hand, bacterial-style primase of the DnaG family. In any case, proteome of Archaea appeared to consist of two components with different evolutionary histories. Hence the suggestion of "chimeric origin for the Archaea" in our article (Koonin *et al.*, 1997).

Several months later, James Lake's group at UCLA published an article with the title that tells essentially the same story: "Genomic Evidence for Two Functionally Distinct Gene Classes" (Rivera, 1998). Notwithstanding some technical differences with our work (they used ungapped BLAST for similarity detection and transformed the raw scores in a different way), their observations agreed with ours: Archaeal genes involved in metabolism (called "operational" by Lake and colleagues) tended to be Bacteria-like, and genes involved in replication, repair, transcription, and translation ("informational genes") were more Eukarya-like.

The main difference between the two articles was in the evolutionary scenarios that were offered to explain the data. We speculated that the origin of Archaea included a massive horizontal gene transfer between two types of ancient organisms—the ancestor of eukaryotes, which donated informational genes, and the ancestor of Bacteria, providing operational genes. A rather technical question is whether this process is more appropriately described as horizontal gene transfer or as a wholesale fusion of two organisms, perhaps followed by massive gene losses (to explain the "missing subsets" of ancestral genes). Lake's interpretation also involved massive gene fusion and loss, but he thought that the donors of genes were ancient Archaea, on the one hand, and Bacteria, on the other hand, whereas the resulting clade was the ancestor of modern eukaryotes (Fig. 11.3). As we will see later, the most important, and new, point on which both theories agree is that the evolution of life involves massive gene transfer/genome fusion, and therefore the tree of life is not a tree: It contains at least one cycle. The difference between our and Lake's scenarios was mostly in the position of the root of the tree.

Since this work, there has been much debate about the role of horizontal gene transfer in the early evolution of life. In my opinion, the discussion of the deep branches in the tree of life boils down to the following three problems:

1. Whether Woesean topology ((AE)B), with two splits and a root between bacteria and all the others, is sufficient to explain true phylogeny of all life.

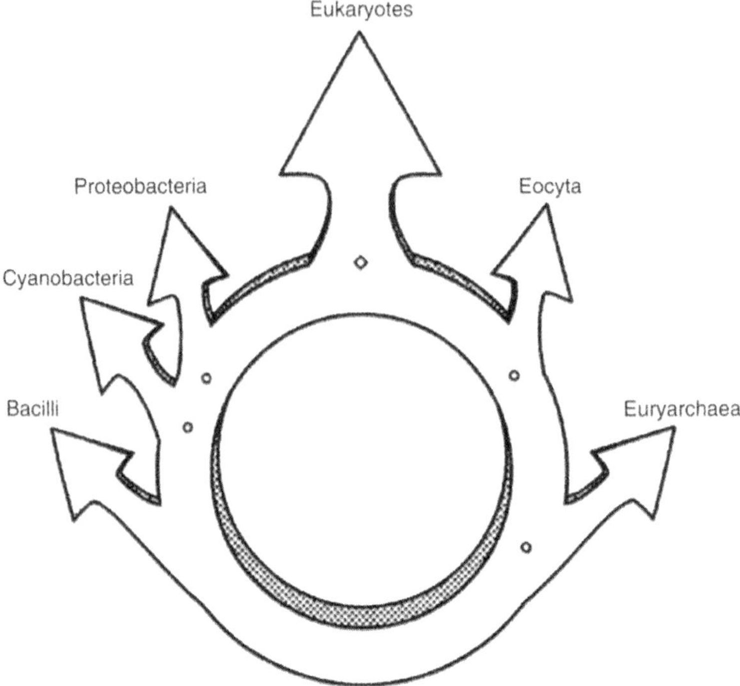

Figure 11.3. The "ring of life." Eocytes is a name given by James Lake to a subset of Archaea; neither this name nor the notion of the high taxonomic rank of this group are widely accepted. The circle at the bottom of the tree of life is accepted by most and in fact is not much different from the reticulations shown in Fig. 11.1. The identity of genomes that have been fused in early evolution of life remains to be thoroughly established. Reprinted from Rivera and Lake (2004) by permission of Nature Publishing Group.

2. If it is not, and a better explanation of the evidence is needed, whether such better explanation involves massive horizontal gene transfer (with variations such as merger of complete genomes and massive gene loss).
3. If horizontal gene transfer (HGT) played a role, what was the identity of the donor(s) and recipient(s).

In my opinion, the debate regarding the first two questions is, for all practical purposes, over, and the answers are, respectively, negative and affirmative. At least one important cycle close to the root of the tree of life has to be postulated in order to explain the evidence. The answer to the third question, however, is still open, although the decade of accumulating molecular data seems to lend more support to Lake's scenario.

Here, I have to digress and to discuss a somewhat technical matter. BLAST similarity score is a standard surrogate measure of similarity between sequences that can be obtained automatically and relatively quickly using a popular, well-supported suite of software. Probability of chance observation of score equal or higher (see Chapter 2) is simultaneously computed, and the rank of similarity (e.g., the highest or second highest score) is yet another measure easily derived from BLAST analysis. But we are often warned against using BLAST scores when talking about evolution.

I know of three main grievances here. First, to some investigators, discussion of evolution has to be over a phylogenetic tree, whereas BLAST comparisons emphasize distance/similarity, score ranks, and other parameters that are seen as mere "raw material" for building a tree.

This is a verbal jousting: Good information about evolution may be obtained "without recourse to a tree" (Fitch, 1995). Thus, "BLAST is not a program for studying evolution" is not a valid argument. A specific reason needs to be provided why BLAST similarities are unsuitable for answering a specific evolutionary question.

The second concern is about incompleteness of BLAST results. One famous example is the hypothesis of the bacterial origin of a subset of proteins encoded by human genome (Lander *et al.*, 2001). This explanation, which involved xenology, was offered when BLAST searches of the National Center for Biotechnology Information NR database failed to report any orthologs of these human proteins in other vertebrates or even in two completely sequenced metazoan animals, nematode and fruit fly, but showed many bacterial orthologs. A comparison of the same human sequences to the EST databases (also by BLAST), however, found orthologs of human proteins in many EST libraries from various metazoa (Salzberg *et al.*, 2001). Thus, instead of the single scenario of direct xenologous transfer of bacterial gene into a human genome, one has to consider others, such as horizontal transfer from bacteria much earlier in evolution, perhaps in the *Deuterostomia* clade, or even a phylogeny that includes hardly any xenology and instead explains the phyletic pattern of these genes by losses in worm and fly lineages (Andersson *et al.*, 2001). But even though this and similar stories often presented as an example of insufficiency of BLAST analysis for evolutionary inferences, this is really a problem of access to sequence and has nothing to do with the BLAST program: The fault of initial observation was that EST databases were not searched, not that the search method biased the evolutionary estimates. (Interestingly, at least one group of enzymes in *Metazoa*, namely those involved in biosynthesis of neuromediators from amino acids, nonetheless may have been xenologously transferred from various groups of bacteria to the metazoan lineage on at least five separate occasions; for discussion of the advantages of this scenario over most alternatives, see Iyer *et al.*, 2004).

The third type of complaint states that BLAST scores may be too crude of a distance measure between sequences. This objection is heard even from the authors who otherwise support the use of distance matrix methods in phylogenetic analysis. For example, Ludwig *et al.* (1998) warns,

> *Only careful data analysis starting with a proper alignment, followed by the analysis of positional variability, rates and character of change, testing various data selections, applying alternative treeing methods and, finally, performing confidence tests, allows reasonable utilization of the limited phylogenetic information.*

Technical sophistication and understanding of limitations of our methods are good. If there is anything inherently careless or sloppy about BLAST scores, we should not use them—but is there? Indeed, these scores are derived using sensitive probabilistic models of sequence similarity, which utilize significant evolutionary information in the form of substitution matrices (see Chapter 2). When transformed into bits, similarity scores are not arbitrary units but reflect specificity of signal detection at the background of complete sequence database (Altschul, 1991). High similarity scores are indicative of evolutionary and functional connections between sequences—this is the first fact of computational molecular biology (see Chapter 1). Moreover, transformed BLAST distances allow "testing various data selections, applying alternative treeing methods and, finally, performing confidence tests." All this is not controversial. So, what is the reason to distrust the phylogenetic inference based on the BLAST data?

In fact, all distances, whether derived from BLAST, multiple sequence alignments, or anywhere else, have their limitations. The key fact in phylogenetic inference is that every distance between two molecules is a statistical estimate of the true evolutionary distance; same as in parsimony methods, every minimization of character state changes is an estimated value of the true minimum. Likewise, every reconstructed phylogenetic tree is an estimate of the true phylogeny. The advantage of the multistep protocol outlined by Ludwig *et al.* is that incorporating more knowledge into the method may result in a better statistical estimate. However, if BLAST scores or their simple transformations are equally good statistical estimators as the distances induced on the trees, we would certainly want to know about it in this era of complete genomes and phylogenomics. In other words, no approach should be dismissed solely because it is too simple; each method of computational experiment needs to be evaluated in its own right. This is not different from wet lab: I may contend that only DNA purified in CsCl gradient is the pure DNA, but I may not need that pure of a sample in order to clone a gene. After all is said and done, there are few specifics regarding what may be wrong with BLAST scores in any sense.

One of the very few reports in which the problem has been clearly defined is that by Koski and Golding (2001), with the memorable title, "The Closest BLAST Hit Is Often Not the Nearest Neighbor." The basic phenomenon is what the title says: Suppose I use a protein sequence A_A from organism *A* to search a collection of other proteomes *B*, *C*, *D*, etc. with the BLAST program and find that a homolog A_B in the proteome *B* is the highest scoring BLAST match of A_A. I then build a tree of all detected homologs and ask what is the nearest neighbor of A_A in this tree. The nearest tree neighbor is defined, for example, as the protein that I can access from A_A by passing the smallest number of internal nodes in the tree (the tie in the number of nodes may be broken, e.g., by the shorter length of traversed branches). It turns out that in a number of cases, the nearest tree neighbor is not the same as the BLAST neighbor A_B but, rather, a homolog from some other species. Thus, the BLAST neighbor and the tree neighbor may not be one and the same.

We studied this effect, which we call Koski–Golding incompatibility (KGI), in more detail (G. Glazko, M. Goel, and A. Mushegian, unpublished). First, we examined the occurrence of KGI at the genome scale. For most interspecies comparisons, BLAST neighbor and tree neighbor properties are very good statistical predictors of one another. For two proteomes specifically studied by Koski and Golding, bacterium *Escherichia coli* and archaeon *Aeropyrum pernix*, KGI was observed in less than 18 and 45% of all cases, respectively. Second, the relatively large gap between the rate of KGIs in *E. coli* and *A. pernix* narrowed significantly if instead of one-way BLAST matches, the symmetric BeTs are considered. SymBeTs are excellent rejectors of the null hypothesis that BLAST neighbor property was uninformative with regard to the tree neighbor property. The percentage of KGI among the proteins involved in SymBeTs decreases to 9 and 18%, respectively. Approximately half of these KGIs go away with better normalization of BLAST scores and when the statistical ties between the first few top-ranking matches in the case of complex co-orthologous relationships are taken into account. The fraction of KGIs that remains unexplained is therefore less than 10% in most species.

Thus, Koski and Golding made an important contribution to comparative genomics by defining a specific technical problem with BLAST scores. They think that the only way to know whether a gene suffers from a KGI is to build a tree. We have observed, however, that at the genomic scale the rate of KGIs can be reduced without recourse to a tree by implementing some automated filters on the BLAST neighbor sets. This is analogous to removing non-informative or incompatible characters from sequence alignments, which is a standard practice in many methods of phylogenetic inference. Much of the following analysis in this chapter and Chapters 12 and 13 concerns similarity and orthologous relationships between SymBeTs.

So, was there massive HGT in early evolution of prokaryotes, and could there be significant extent of HGT in more recent times? This question remains contested. Lawrence and Hendrickson (2003) wrote,

It ... seems that complete genome sequences have generated more debate, speculation, discussion, and publication of works—both those presenting objective analyses of new data and extrapolation of data according to one" point of view—regarding horizontal (lateral) gene transfer (HGT) than any other subject.

Let us review the evidence.One could approach the phenomenon of HGT, for example, from the mechanistic angle, studying the enzymatic activities and molecular complexes that enable gene passage between species. Another angle is to focus on the adaptation value—that is, the utility of HGT to an individual organism or evolving species (including comparison of the adaptationist hypothesis with various "selfish DNA" models)—and yet another perspective is that of population genetics—the dynamics of allele transfer and spread in populations. My interest here, however, will be in the evidence that comes from comparative genomics and that tells us how HGT shapes the gene content of living species.

I believe that there are essentially nine groups of relevant observations:

1. Perhaps the earliest type of evidence that is naturally explained by HGT is the abundance of highly similar genes in diverse bacteria—genes encoded by plasmids and other mobile elements. Various types of "selfish genetic element" narratives have been used to explain dissemination of viruses, insertion sequences, transposons, and other such elements that are of no immediate use to bacteria. But many plasmid-borne genes provide hosts with immediate advantages, such as detoxification of antibiotics, bacteriocins, or resistance to phages. There is also ample evidence of plasmid-borne groups of genes that provide novel metabolic functions, notably utilization of novel sources of food, but also essential biosynthetic functions. In such cases, "selfishness" is beside the point.

2. This set of observations is closely paralleled by the evidence at the nucleotide level. In most microbial genomes, there are reasons with unusual nucleotide frequency distributions, which differs from average base composition, and often contain genes that encode proteins more closely related to gene products in distant bacteria or in mobile genetic elements, such as phages, insertion sequences, or broad host range plasmids. Nucleotide evidence of gene mobility, however, seems to be eradicated (or "ameliorated") relatively quickly because of apparent, but not well-understood, constraints on local base composition in many bacteria. The rate of amelioration appears to be lower in the amino acid sequence, which may be useful for dating the HGT events (Lawrence and Ochman, 1997).

3. Eukaryotes have acquired two large batches of bacterial genes by symbiogenesis. First, an ancient alpha-proteobacterium gave rise to mitochondria after being engulfed by an ancestor of eukaryotes (perhaps the ancestor of all known eukaryotes, since the currently known amitochondrial eukaryotes may be the result of a secondary loss of mitochondria). Second, some eukaryotes have additionally acquired ancient cyanobacteria, which gave rise to chloroplasts. This may have happened on more than one occasion. Genes that used to be encoded by genomes of these previously autonomous bacteria have been mostly transferred into the nuclear genomes of the hosts. This is another massive horizontal transfer with which no one seems to have any major disagreements.

4. As discussed previously, proteomes of Archaea and Eukarya can be partitioned into two distinct and large classes of proteins with apparently different evolutionary affinities: Some protein sequences display ((AE)B) topology, whereas others show topology ((AB)E). All sensible phylogenetic explanations of how this came to be include horizontal transfer at a large scale (tens of percent of all genes) or whole-genome fusion and subsequent gene losses. One radically dissenting opinion on the origin of archaea has been offered by Thomas Cavalier-Smith (2002a); it is of great interest (at least to me) despite its eccentricity, and I will discuss and, I hope, rebuff it in Chapter 12.

5. Complete genomes of distinct microorganisms sharing the same habitat provide further evidence. A first thermophilic bacterium with completely sequenced genome, *Aquifex*

aeolicus, has a set of proteins that are commonly found in thermophilic and hyperthermophilic archaea but are rarely found in other bacteria, including proteobacteria to which *Aquifex* is related (Aravind *et al.*, 1998). Nelson *et al.* (1999) have shown that *Thermotoga maritima* also has a fraction of proteins related to homologs from thermophilic archaea but rarely found in mesophilic bacteria, including low-GC Gram-positive bacteria (most likely the sister clade of *Thermotoga*, as judged by analysis of many molecular characters). *Methanosarcina mazei*, a methanogenic archaeon that, unlike most of its close relatives, is mesophilic, has a larger proportion of genes with only bacterial homologs compared to other archaea: Of 3371 protein coding open reading frames, as many as 544 had statistically significant matches only in bacteria, but none in other archaea (Galagan *et al.*, 2002). Note that the meaning of this latter number was quite different in 2002 than it was in 1996; when we reported that a *M. jannaschii* protein had matches only in bacteria, this may have reflected, among other things, insufficient information about archaeal genomes. But two dozen completed archaeal genomes later, a *M. mazei* protein with recognizable homologs only in bacteria, but not in genomes from its own taxonomic neighborhood, provides a much stronger suggestion of the HGT scenario. Finally, a bacterium that lives in crystallizing concentrations of sodium chloride, *Salinibacter* sp., was found to share a large fraction of its genes with halophilic archaea (Mongodin *et al.*, 2005).

6. The phylogenetic tree of a gene family may not always be the same as the tree of the species in which these genes reside. Anomalies of this sort are more commonly observed in operational genes, but informational genes also display such evidence, and for some of them, strong cases have been made for horizontal transfer. Aminoacyl-tRNA synthetases seem to be especially prone to having family trees distinct from those of their host species (Aravind *et al.*, 1998), and strong disagreement has also been observed between bacterial tree and the family tree of ribosomal protein S14 (Brochier *et al.*, 2000).

7. Evidence that is compatible with HGT can be obtained by comparing trees of different genes from the same set of species, without recourse to a species tree. For example, there are 188 genes that are specifically shared by five phylogenetically diverse photosynthetic bacteria (Raymond *et al.*, 2002). When trees are built from the sequence alignments of these genes, there are 15 different topologies, 4 of which are supported by at least 10% of all family trees. A qualitatively similar picture is observed with the trees of archaea-specific gene families (Makarova and Koonin, 2005).

8. In Chapter 6, we discussed DOGs—displacements of orthologous genes, or functional convergences at the molecular level. Every gene displacement is based on recruitment of two or more genes to perform the same function. There are only two possibilities of recruitment: A new gene can be recruited either from genes present in the same genome or from another genome. In the latter case, the DOG is produced by HGT.

9. In Chapter 8, we discussed phyletic patterns, which are binary vectors coding the presence and absence of proteins/COGs in different species. Currently, the COG database includes more than 110 microbial species and almost 14,000 COGs. By definition, COG is a set of orthologous groups found in at least 3 species, so a phyletic vector of a COG may have three or more coordinates set at 1. Only 46 COGs have every coordinate equal to 1, whereas 88% of all COGs are found in less than 30 species. Thus, an overwhelming majority of phyletic vectors are dominated by gene absences: Out of 10 broadly conserved proteins, 9 will not be found in a randomly picked bacterial or archaeal species. One explanation for such sparse distribution of genes in genomes, especially when these genomes are evolutionarily distant from each other, is that genes are passed between genomes by HGT.

These nine lines of evidence are summarized in Table 11.2. The correctness of each observation is not in doubt—for example, no one argues with the existence of two gene classes in archaea or with the notion of gene displacement—it is the evolutionary history that leads to each observation that needs to be explained. Thus, let us call them the "nine facts that have to

Table 11.2. Nine Facts That Argue Strongly in Favor of Ancient and Ongoing Horizontal Gene Transfer

Evidence	Evolutionary distance between donor and recipient	Rate of HGT: % of all genes in the recipient genome	Explanations that do not involve HGT?
1. Similarity of genes encoded by mobile elements	Various	~10	No
2. Nucleotide frequency statistics	Various	~10	No
3. Symbiotic origin of organelles	Large	>10	No
4. Two gene classes in Archaea and Eukarya	Large	>>10	Confluence of artifacts (e.g., unequal evolutionary rates)
5. Closely related genes in species that share the habitat	Large	>10	Retained ancestral genes
6. Discordance between the consensus genome tree and protein family tree	Large	~10	Differential gene losses and unequal evolutionary rates
7. Discordance between the individual family trees in the same set of genomes	Large	>10	Differential gene losses and unequal evolutionary rates
8. Displacement of orthologous genes	Large	5–10	Differential gene losses
9. Dominance of sparse phyletic patterns	Various	Unknown	Differential gene losses

be explained." Each of the nine facts can be seen as the effect of HGT, either in ancient or in more recent times. However, other explanations, that do not involve HGT, can be put forward to account for some of the nine facts. It is the strength and implications of these alternative explanations that will concern us now.

We have already had a look at the first line of counterargument, which claims technical artifacts, such as incompleteness of the sequence databases or poor choice of distance measures. It is clear that each of the nine facts is quite robust to these methodological problems. In a careful phylogenetic study—for example, if BLAST-derived distances are replaced by JTT evolutionary model and maximum likelihood analysis—the HGT hypothesis may be refuted for a fraction of genes, but none of the nine facts can be dismissed: Each of them remains supported by many different genes in many different genomes. There has to be a more substantive explanation.

The second type of argument makes no specific refutation of evolutionary scenarios for specific genes. Rather, general functional and evolutionary considerations are brought up. For example, there are known functional barriers, such as site-specific and nonspecific DNA degradation systems, that may be guarding cells against invasion of parasitic and pathogenic DNA. Of course, on the other hand, some bacterial species are naturally competent, which means that they can take up large fragments of foreign DNA (the first completely sequenced bacterium, *H. influenzae*, is one of the best studied examples). Moreover, it is well documented that many genes are transferred between cells not by serendipitous fragments of genomic DNA but, rather, by viruses and plasmids that are already engineered to resist host defenses.

The next set of arguments examines the implications of gene transfer at the level of evolving populations. In particular, Charles Kurland of the University of Lund has described many factors that he believes to be formidable evolutionary obstacles to "rampant horizontal gene transfer" (Kurland *et al.*, 2003; Kurland, 2005). He essentially sees four main barriers. First, there is a negative effect of large population size: Any new gene is likely to be acquired by only a few recipient cells, and mathematical modeling indicates that in the absence of very strong selective advantage, the transferred gene will be lost before it can spread in the population. Second, many gene products belong to co-adapted cellular systems, where proteins, nucleic acids, membranes, polysaccharides, and other components are "optimally fitting together." Such optimality is thought to be difficult to achieve when a recipient gains some but not all components of a larger complex. Third, even the most benign foreign gene incurs costs to the recipient, such as the expense of replicating extra DNA and, potentially, of mutation load if the gene is inserted in a functional region of the recipient's genome. Fourth, selection is never distributed evenly across the habitat. Hence, even if a new gene initially spreads because of positive selection, it will have additional opportunities to be lost when selection is removed.

However, each of these obstacles is relative. For example, the existence of the first barrier is equivalent to the suggestion that HGT should be favored in small populations, in which the density of both donor and recipient is sufficiently high. This immediately brings to mind habitats such as hydrothermal vents, mats, and, as the ultimate case of dense cohabitation, symbiogenesis. These ideas are fully compatible with the nine facts, especially with facts 3–5. The third and fourth barriers essentially state that the transferred gene has to offer immediate and strong selective advantage. This may be fulfilled in the case of transferred operons, which can bring in whole new pathways, but even a singly transferred gene can confer a strong selective advantage, such as resistance to antibiotics produced by some of the cohabiting microorganisms. This would be compatible not only with fact 1 but also with fact 8. In any case, patchiness of the environment and changing selection pressures affect all genes in the genome, and there is nothing particular here about horizontally transferred genes. Regarding the third barrier, "optimal fit" essentially seems to be the same as the "complexity hypothesis," which states that it is more difficult to transfer a gene if its product needs to interact physically with many molecules at the same time (Jain *et al.*, 1999). But we know that DOGs do happen, although they are less frequent among the informational than the operational genes. Thus, all four of Kurland's barriers are worth thinking about, but more than anything else, they help us to understand the circumstances in which HGTs are more likely to occur. Examination of Kurland's arguments leads me to suspect that his disagreements are not as much with the facts as they are with the characterization of HGT as "rampant" or "massive." We will return to the semantics later.

The next line of defense against the prominent role of HGT is the suspicion of tree artifacts, most prominently LBA. It is true that most of the branches in the phylogenetic trees of prokaryotic lineages are long. However, LBA has to be demonstrated in each case separately, for example, using the rules outlined previously. The alternative ("at this branch length, we will not believe any tree, no matter what it shows") can lead to eccentric theories, as exemplified in Chapter 12. In any case, LBA is all but useless for explaining the issues of gene content, when some genes are shared by disparate groups of genomes, as in facts 3, 8, and 9.

But perhaps at the heart of the debate of the HGT status is the argument crisply stated by Koonin (2003b). It points out that any phylogenetic evidence of HGT can also be attributed to a combination of gene duplications and gene losses. In Fig. 11.4, this is illustrated for a gene with patchy phyletic distribution. The only way in which this distribution can be explained without any HGT at all is if the ancestral gene was present in the genome in the last node from which all depicted species descend—in this case, the last universal common ancestor. But under the same hypothesis, we also have to accept 10 gene losses in different lineages. An alternative explanation is one act of HGT, most likely from bacteria to halophilic archaea.

One could argue that the consensus species tree is reconstructed incorrectly and that in a true tree, the phyletic pattern of this gene would have been more compact and explainable without HGT, or with a much smaller number of gene losses. This argument does not hold: The tree topology shown in Table 11.2 is well supported by many molecular characters of all kinds and

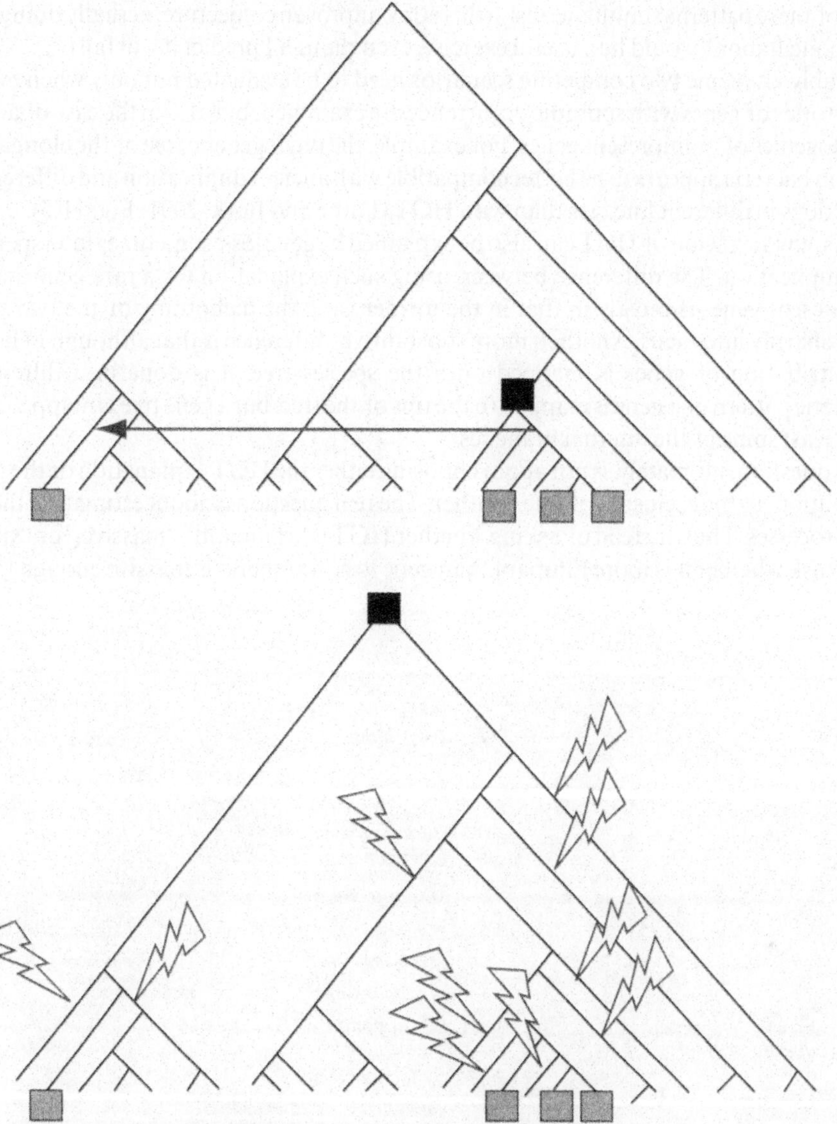

Figure 11.4. How many gene losses are we prepared to accept in order to reject a single act of horizontal gene transfer? Two scenarios of evolution for anaerobic glycerol 3-phosphate dehydrogenase GlpB. The consensus phylogenetic tree is shown, and the species that have this protein—(left to right) *Halobacterium* sp. (archaea), *Vibrio cholerae*, *H. influenzae*, and *Escherichia coli* (all gammaproteobacteria)—are marked by gray boxes at the tips of the tree. (**Top**) Scenario with horizontal transfer: Emergence of the gene in the gammaproteobacterial ancestor (*black box*) and one act of transfer to halobacteria explain the tree. (**Bottom**) Scenario without horizontal transfers: *Lightning bolts* indicate gene losses that need to be invoked to explain phyletic distribution of GlpB in this case. Modified from Koonin (2003b) by permission of Blackwell Publishers.

is perhaps as close an approximation of the (non-reticulated) evolutionary history of life as we can get. Moreover, in the last release of the COG database, there are approximately 10^3 phyletic patterns altogether, and 90% of them are "patchy" (fact 9 of HGT). We would be hard-pressed to produce a tree with such rearrangement of branches that could satisfy an HGT-less scenario for all of these patterns simultaneously (this is my unproven conjecture; actually doing this sort of "reconciliation" would be a useful exercise, even though I predict it will fail).

Notably, the same two competing scenarios need to be evaluated not only when evolutionary histories of genes with sporadic occurrence are examined, but also in the case of anomalies in phylogenies of omnipresent genes. For example, the phylogenetic tree of the elongation factor Tu in bacteria appears to be better compatible with ancient duplication and differential loss of paralogs in different lineages than with HGT (Lathe and Bork, 2001; Fig. 11.5).

Thus, any suspicion of HGT can also be explained by gene loss, sometimes interspersed with gene duplication. The difference between using such explanation for a rare gene and for the omnipresent genes is mostly in that in the former case, the dichotomy of the two scenarios comes sharply into focus. Another, more substantive, difference is that although in both cases the distribution of genes is mapped onto the species tree, it is done in a different way: A phyletic pattern of a gene is mapped to the tips of the tree, but a gene tree is mapped onto tips and at least some of the internal branches.

The question, ultimately, is not about choosing either the HGT explanation or the gene-loss explanation, to the exclusion of one another. The real question is about estimating the rates of both processes. Thus, instead of asking whether HGT is "rampant," "massive," or "sporadic," we can ask whether it is more rampant than gene loss, whether it is massive enough to explain

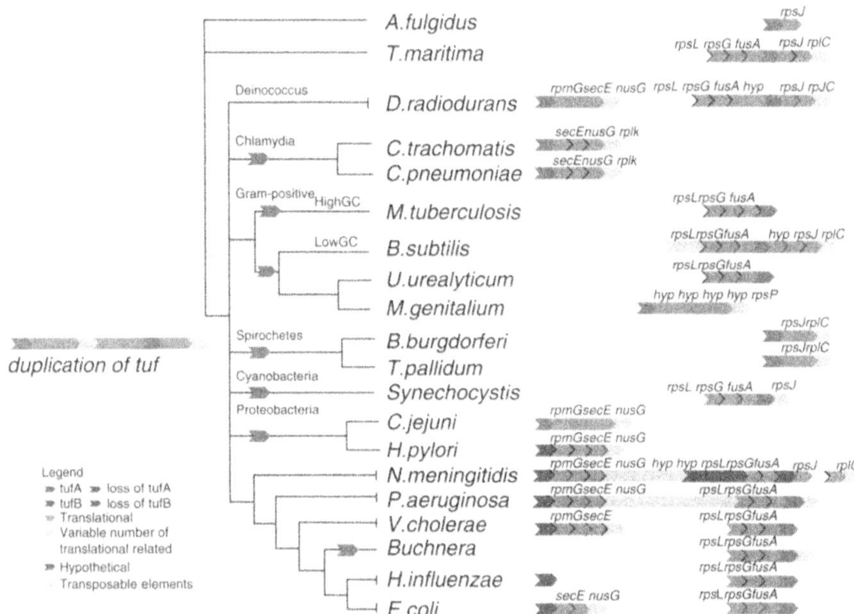

Figure 11.5. Gene duplication and loss in the evolution of elongation factor Tu. An ancient duplication of the *tuf* gene (*tufA*) occurred early in the divergence of eubacteria with the addition of a second *tuf* gene (*tufB*) upstream of the *rpmG* gene, the first gene in the proposed ancient transcriptional unit. As inferred from the current genome location, *tufA* has been lost in both the *Chlamydia* and epsilonproteobacteria, whereas *tufB* is lost in *Buchnera*, *Synechocystis*, the spirochetes, and low- and high-GC Gram-positive bacteria. Clostridia, a Gram-positive low-GC taxon, has maintained the *tuf* duplication, suggesting the other low-GC Gram-positive taxa lost *tufB* in a separate event from the high-GC Gram-positive clade. The duplication is maintained in the remaining proteobacteria and *Deinococcus radiodurans*. Reprinted by permission of Federation of the European Biochemical Societies from Evolution of tuf genes: ancient duplication, differential loss and gene conversion, by Lathe, W. C., 3rd, and Bork, P., *FEBS Letters*, 502(3), 113–116, copyright 2001, Elsevier.

the history of a substantial portion of genes in the genome, whether it is too sporadic to perturb evolutionary signal in a species tree built on the basis of concatenated sequence alignments of multiple genes, and so on.

The question of rates of HGT and gene loss, or about the form of the function that assigns costs to both processes (and also to gene births, which, obviously, had to occur in some ancestral life-form), is one of the central questions of gene content-based phylogeny, and indeed of comparative genomics. One type of such function is unweighted parsimony, in which we simply count the number of events that explain the distribution of genes given the species tree. The scenario with the smallest number of events is declared a winner. This, clearly, is not an especially realistic model for several reasons, the most immediate of which is the relative ease with which genes seem to be lost upon changes in selection. Many examples of that can be seen in the genomes of pathogenic microorganisms, which were among the earliest completely sequenced species. These bacteria live in human and animal body cavities—environments that are rich in several classes of nutrients, for example, amino acids and nucleobases (although, on the contrary, some essential microelements, such as iron, seem to be depleted in such environments, and many human pathogens have specific adaptations for scavenging iron and gene switches that sense iron concentration). Genes in the pathways of *de novo* biosynthesis of these metabolites are lost quite frequently (Fig. 11.6). Thus, it is tempting to say that loss of a gene is much more common than gene gain, either by birth or by HGT. But here again, the crucial question is, by how much?

In the absence of direct studies of gene loss and gain—studies that have not been done in the laboratory yet, even though it should be possible to set up such an experiment with the existing technology—we can only estimate these rates from the gene content of the existing genomes. One way to do so is to compare many genomes of closely related species. For example, one of the best sampled clades of bacteria are gammaproteobacteria. Here, we have a large number of closely related orthologous genes, which can be used to estimate the relative times since divergence of strains, species, and genera. In the same set of genomes, we can count all cases of gene gains and gene losses, especially those that are seen in just one or a small number of lineages. This provides a direct estimation of the instantaneous rate of gene loss and gain in an evolutionary lineage, which can be extrapolated to larger evolutionary distances (Lerat *et al.*, 2005). The main result of this and other studies work is that there are both rampant and rare aspects to HGT: Most important, although the absolute number of genes that have been horizontally transferred at least once in their lifetime appears to be high (on the order of hundreds), the HGT events are rare in the sense that most genes (>90%) have been transferred not more than once in their lifetime or, commonly, not at all.

Another, indirect way to define the cost function for gene loss and HGT is to assess the effect of different loss-to-HGT ratios on gene content of ancestral genomes. The point here is that if we prohibit HGT altogether, the result will be either a grotesque "mother of all genomes"—the last common ancestor that had all imaginable genes and evolved only by differential gene loss—or an implausible scenario of gene birth in multiple lineages (i.e., parallel and/or convergent sequence evolution on an unprecedented scale). Alternatively, we can put a constraint on the size of the common ancestor of a group of species. For example, we can request that the number of genes in the common ancestor not exceed the genome size of a present-day species that lives in a similar environment. We can also require that the ancestral set of genes contains a relatively coherent set of genes, sufficient to build most of the pathways that enable functioning of a primitive organism. Then, we can use information about the distribution of genes/COGs in today's species to infer the status of each gene in the common ancestor and to find the values of gene gain and loss rate that satisfy the constraints imposed on the ancestor. This search for the weight function has been performed by Snel *et al.* (2002) and by Mirkin *et al.*

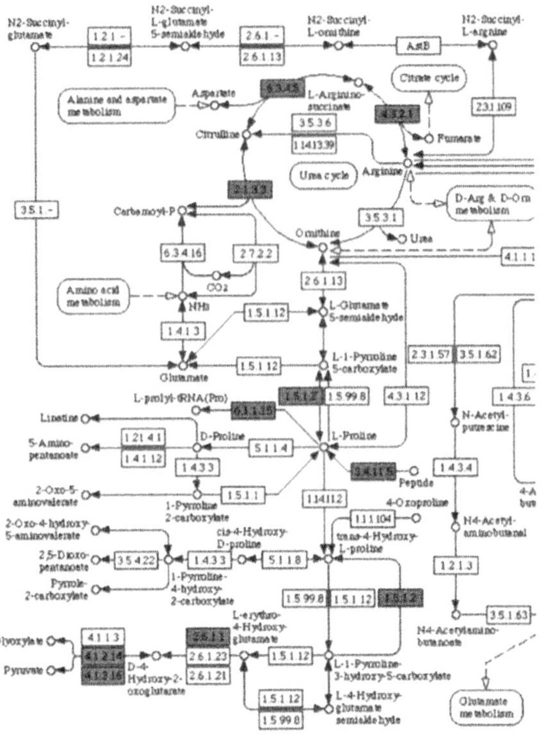

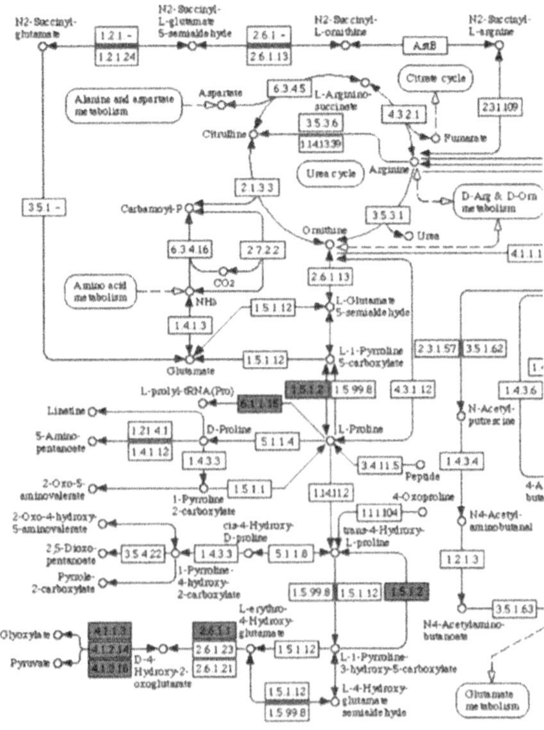

Figure 11.6. Gene losses in parasitic bacteria. Fragment of the metabolic map, representing biosynthesis of amino acids proline and arginine. *Open boxes* indicate enzymes present in *E. coli*, where these pathways were defined originally. *Gray boxes* indicate genes that have been retained in *Neisseria gonorrhoea* (**top**) and *Treponema pallidum* (**bottom**). From *Kyoto Encyclopedia of Genes and Genomes* Web site (www.genome.jp/kegg; accessed May 15, 2006); Copyright 1995–2006, Kanehisa Laboratories.

(2003), and the most consistent results were obtained when the rates of gene loss and HGT were comparable (see Chapter 13).

Thus, is HGT rare, common, or rampant? Perhaps, in the end, this is not the correct question to ask. Different aspects of HGT may be quantitatively different. So, inspired by Chapter 17 of James Joyce's *Ulysses*:

What is rare in horizontal gene transfer?

The transfer over the lifetime of an individual gene. Most genes are transferred only once during their lifetime or not at all (Lerat *et al.*, 2005).

Most likely, fixation of a transferred gene in the population of recipient. Most genes are fixed only if they confer strong selective advantage, and perhaps even then only if the population of recipient is small enough (Kurland *et al.*, 2003; Kurland, 2005).

What is neither very rare nor very common in horizontal gene transfer?

The DOGs in two similarly sized microbial genomes. Perhaps only 5–10% of genes are DOGs, and not all of them are the result of HGT (see Chapter 6).

Genes that have been horizontally transferred in the recent history of a given lineage (Price *et al.*, 2005).

What is common in horizontal gene transfer?

The fraction of genes in archaeal and eukaryotic genomes that have been acquired by ancient HGT or a whole genome fusion.

The fraction of genes/COGs that have been horizontally transferred at least once in their lifetime.

What is of unknown frequency in horizontal gene transfer?

The relative rate of losses and horizontal transfers in the general population and in the evolutionary histories of individual genes. The best current estimate is that they are not strongly different for prokaryotic genes.

The proportion of reticulations that need to be added to the tree of life to fully explain what was going on in evolution. I believe that the ratio of reticulate to "conventional" branches will be not higher than 20–30%.

12

Two Stories about Evolution

This chapter is about two episodes in the history of Life and about human efforts to make sense out of complex data.

The first story is about the origins of Archaea and their relationship to Bacteria and Eukarya. The discovery that Archaea are a unique form of life belongs to Carl Woese. He postulated that an unusual relationship exists between Archaea and the other two divisions of Life almost 30 years ago. This is what we also think today, but during the past three decades there have been several shifts in understanding of what this relationship really is. Scientific discussions of this question continue, but something resembling a mainstream opinion has eventually emerged. Computational analysis of molecular sequences and gene content in various completely sequenced genomes mostly supports the opinion that can be summarized as follows:

1. Bacteria and the common ancestor of Archaea/Eukarya are ancient sister groups.
2. The common ancestor of Archaea/Eukarya most likely was a prokaryote of archaeal type.
3. Early evolution of life included a period of massive horizontal gene transfer; therefore, the tree of life has one or more cycles close to its root.
4. The direction of transfer and the identity of the partners that exchanged genes are not completely clear: They probably were all prokaryotes, and massive exchange of genes and gene modules may have played a role in the emergence of the eukaryotic cell.
5. There are two components of the cell that display physical continuity between generations: lipid plasma membrane and DNA genome. Whether coincidental or not, the protein modules that are required for their maintenance—respectively, enzymes of lipid biosynthesis and the system of DNA replication—are fundamentally different in Archaea and Bacteria. The situation with Eukarya is even more complex: Their DNA replication is archaeal, but lipid biosynthesis is similar to that in Bacteria. All this suggests ancient DOGs (see Chapter 6), which need to be understood and worked into any evolutionary scenario.

This is a broad framework for the understanding of early phylogeny of life. But long-time dissenter Thomas Cavalier-Smith, at Oxford University, takes issue with most of these statements and puts forward an elaborate hypothesis, which proposes that Archaea and Eukarya are extremely recent, extensively derived descendants of Gram-positive bacteria (Cavalier-Smith, 2002a) that should be called "neomura" for their derived membrane structure. My intent here is to show that Cavalier-Smith's hypothesis deserves serious evaluation and criticism. I will also argue that despite its appealing features, this theory is not as well supported by the evidence as Cavalier-Smith would like us to believe.

The second story is about invertebrate animals. Textbooks describe many lineages of invertebrates, but scientists cannot agree on Metazoan phylogeny. Analysis of shared traits gives a puzzling result: Almost every subset of invertebrate animals shares a few common anatomical and embryological traits. It is difficult to determine which of these traits are true synapomorphies—that is, shared innovations specific to the clade in which they are found [according to cladistic schools of systematics, synapomorphies are the only reliable basis of phylogenetic inference (Ridley, 1986), but they are useful even for those who do not strictly adhere to the cladistic canon]. For two centuries, the main route of invertebrate evolution was thought to be marked by the growing complexity of body plan, and several clades of animals were defined on the basis of morphological innovations having to do with gross morphology and anatomy. True body cavity (coelom) was thought to be one such synapomorphy, missing from flatworms and roundworms (including nematodes) and present in more complex animals. Within animals with true coelom, Coelomata, a clade of animals with segmented bodies (sometimes called Articulata) consisting of annelids, arthropods, and some minor groups, had been defined.

In 1997, this two-centuries-old understanding was shaken. Sequencing of rRNA from diverse invertebrates was undertaken, and Ecdysozoa, "a clade of molting animals" that consists of nematodes and arthropods, to the exclusion of plathelminthes, annelids, and other protostomes, was proposed (Aguinaldo *et al.*, 1997). Moreover, rRNA evidence appeared to support the monophily of most of those other protostomes, under the name of Lophotrochozoa, in contrast with the more traditional view, in which plathelminthes that lacked any body cavity and true mesoderm were considered far more primitive than three-layered, coelomic mollusks or annelids. Under this theory, Ecdysozoa and Lophotrochozoa were true clades, and Coelomata was not. Some support of these views could be found in older work too.

Intense discussion of invertebrate taxonomy continues. Ecdysozoan and lophotrochozoan clades are seen in phylogenetic trees, although mostly under special circumstances. But not every kind of molecular character supports these clades—many characters continue to support the coelomate phylogeny. In my second story, I review the Coelomata–Ecdysozoa standoff and argue that inconsistency between metazoan trees built on different sets of molecular characters is a real and important phenomenon, but that it cannot be ruled in favor of Ecdysozoa with the existing evidence. Here again, I examine the results of genome-scale analysis. I also show, using very recent data on molting-related genes in the nematode, that only a diminutive proportion of genes required for molting are synapomorphic in insects and worms.

In 2002, Cavalier-Smith published two long treatises, one focusing on the evolution of prokaryotes and early eukaryotes and the other mainly dealing with radiation of different groups of protists (Cavalier-Smith, 2002a,b). I focus on the first of these papers, which was concluded by the author thusly: "I invite the strongest possible reasoned criticisms of this synthesis."

The dissenting opinions of Cavalier-Smith on various evolutionary questions have been published throughout the years, and the history of science figures quite prominently in his narrative. However interested in history, I will not concern myself with the debates of the 1970s through the 1990s. This chapter pertains to Cavalier-Smith's synthesis as published in 2002, as well as to relevant facts that came to light since then. [When the manuscript for this chapter was almost finished, Cavalier-Smith published a new work on the placement of the root of Bacteria (Cavalier-Smith, 2006). This interesting question, to which I believe Cavalier-Smith continues to make a genuinely important contribution, is beyond my current scope. However, he appears to presume in the latest work that his 2002 paper has largely settled the question of the evolutionary novelty of Archaea, which is why I believe that strong "reasoned criticisms" of that paper are all the more important.]

Cavalier-Smith (2002a) summarizes his critique of the current views of the origin of Archaea and Eukarya by "concentrating on six things" that form the core of his argument. Here is my summary of these six things:

1. Almost all neomuran characters are adaptations to thermophily and "nearly all neomuran characters can be used to polarize unambiguously the direction of evolution from posibacteria to neomura, not the reverse".
2. After the emergence of neomura, archaea went on to adapt to hyperthermophily, and all characters unique to modern archaea can be explained by such adaptations (or else by adaptations to high salt).
3. Paleontological evidence shows the same, and it also shows that the Archaea/Eukarya clade is four times younger than Bacteria.
4. Evolution of prokaryotes and early eukaryotes was "quantum and mosaic"—that is, long periods of relatively slow evolutionary change were interrupted by bursts of extremely rapid evolution. "Rapid evolution" here has two meanings: acceleration of the rate with which molecular characters change within a lineage, and a burst of divergence that gives rise to new lineages. Such patterns of evolution ridicule the idea of the molecular clock and distort all phylogenetic trees, particularly by making shorter branches look artifactually long and by obscuring the true branching order. In addition, these effects prevent correct rooting of the tree of life because the only more or less reliable way to do so relies on analysis of trees made of pairs of ancestral paralogs, and such trees would also be distorted by quantum and mosaic evolution.
5. The root of all life is within Gram-negative bacteria, most likely green nonsulfur bacteria.
6. Horizontal gene transfer (HGT) is frequent in life but not as frequent as to distort the evolutionary signal. We just have to focus on coadapted cellular systems, and not overemphasize any single molecule.

Thus spake Cavalier-Smith. I will now give my brief response to the "six things," and then will explain my disagreements with some of them in more detail.

I have no quarrel with Thing number 6. HGT is indeed frequent, but it is nonetheless amenable to quantitative analysis. The evidence for HGT was examined in Chapter 11. Calling HGT "rare" or "frequent" may rub some people the wrong way, but the real question, of course, is which genes were horizontally transferred, and with what consequences for genetic makeup, structure, function, and further evolution of living forms.

I believe that Thing number 4, the notion of "quantum and mozaic evolution" in prokaryotes, at present lacks unequivocal quantitative support, and the burden of proof is on Cavalier-Smith. Thus far, such signal has not been teased out from either molecules or paleontology, at least as far as paleomicroorganisms are concerned. Rather, these evolutionary patterns are invoked by Cavalier-Smith as an ad hoc hypothesis, without which he would be unable to reconcile his theory with conflicting evidence. The alternative evolutionary scenario, which I support and which Cavalier-Smith mentions, only to dismiss without much evaluation, does not require quantum and mozaic evolution, although it does not reject such evolutionary pattern either. Unequal evolutionary rates are like HGT: They certainly perturb the evolutionary signal, but not to the extent that we cannot trust our own judgment.

I accept Thing number 1 as far as thermophily is concerned, and parts of Thing number 2 that have to do with adaptation to hyperthermophily. At the same time, not all traits specific to all Archaea are relevant to hyperthermophily, and hypersalinity is probably of minor importance in emergence of these traits. Finally, and going to the core of my disagreement with Cavalier-Smith, I discuss genomic evidence supporting a different explanation of all those things—an explanation that Cavalier-Smith is aware of but that he does not take seriously—namely that bacteria and archaea are ancient sister groups, and only eukarya are relatively recent (although probably not as recent as he thinks).

Several general premises of Cavalier-Smith's work are valuable. He reminds us that treating living species as "bags of genes" (i.e., as simply lists of genes or COGs) is not sufficient for understanding evolution. Tracking evolutionary trajectories of individual genes has to be supplemented with cell biology and with analysis of functional coherence of gene ensembles. Many researchers try to emphasize exactly this in their work, looking not just at the individual genes but also at biochemical pathways and structural assemblies formed by genes, or at groups of genes whose correlation only became known as a result of complete genome sequencing (recall the discussion in Chapters 7 and 8 of phyletic patterns and conserved clusters of genes on the chromosomes). Cavalier-Smith argues for even more comprehensive synthesis, which would account for spatial organization of pathways and systems in the living cell. He is particularly concerned with the importance of cellular membrane—an organelle that, like DNA and unlike typical protein, is long-lived and physically passed between generations. He is interested in geological, paleontological, and paleoclimatological facts and is not shy to build hypotheses that take all these facts into consideration, together with molecular record. Much of this championship is exactly in line with biology of today, in which we try to capture the spirit of such a research program by calling it "systems biology" and even "planetary biology."

Cavalier-Smith is a feisty writer. Compared to the generally monotonous style of scientific discourse, some of his invectives sound almost Bardean. The following quotes from his 2002 treatise give an idea: "despite repeated vociferous denial of this basic fact [i.e., fundamental similarity of cell organization of eubacteria and archaebacteria] by a few influential biochemists"; "an obsession with gene expression has prevented molecular biologists from understanding cell evolution, for which novel properties of gene products are fundamentally more important"; "as has long been evident to anyone not seduced by the false dogma of the molecular clock"; "GenBank ignorantly uses the term 'crown eukaryotes' for an arbitrary subset of eukaryotes that have short branches on rRNA trees"; and "archaebacteria are just somewhat unusual bacteria" (which their unwarranted and undesirable renaming as archaea... attempted to conceal).

Cavalier-Smith is serious about words, especially about names of taxa. Not only does he criticize names that he finds unfortunate or misleading but also he coins new Latinisms, giving dozens of new formal names, or sometimes renaming to formalize earlier suggestions, viz. alpha subdivision of proteobacteria into alphabacteria. I agree that clear and informative names are important. But Cavalier-Smith prefers "archaebacteria" over "archaea," and this, I think, is figured out backwards. When Carl Woese first came up with "archaebacteria," his meaning was clear: "ancient bacteria," or, most likely, "the oldest bacteria of them all," with the implication that they possess the features of the common ancestor of all bacteria. This meaning was almost certainly wrong because current evidence does not support origin of bacteria from within archaea; Cavalier-Smith and everyone else can agree on this without a fight. When Woese proposed to change "archaebacteria" to "archaea," it was mostly to reflect that fact—"ancient, and distinct from bacteria" (Woese, 2004). One may disagree with this statement and put forward some alternative theory, such as Cavalier-Smith does, making the case that archaea are derived, not ancient, bacteria. However, "archaebacteria," currently favored by Cavalier-Smith, are neither here nor there: The term appears to state the bacterial nature of "archaebacteria," but the root *archae* runs against Cavalier-Smith's own notion of their evolutionary novelty.

One can have legitimate grievances with Woese and, perhaps, hold a different type of conceit against him. Indeed, Woese presided over sequencing of the first archaeal genome *Methanococcus* (*Methanocaldococcus*) *jannaschii* (Bult *et al.*, 1996); appears to have not noticed the unprecedented breakdown of its genes into two large categories with distinct evolutionary affinities, which was noted by others (Koonin *et al.*, 1997; Rivera *et al.*, 1998); and then discoursed as if this evolutionary pattern has been evident all along, or perhaps as if it was novel but minor observation compared to the really important question of what happened earlier in evolution (Olsen and Woese, 1997; Woese, 1998a,b, 2000, 2002). But as far as opposing the name "Archaea" goes, I would argue that it is the "bacterial" part of the latter that we have more reason to drop, not the "archaic" part.

Cavalier-Smith's substantive arguments are from four angles: cell biology, physical environment of the ancient life-forms, paleontology, and molecular characters. As already said, central to his explanations is the notion of neomura ("the new-walled organisms"). According to his hypothesis, neomura is a bacterial clade that had emerged from within the gram-positive bacteria approximately 850 million years (My) ago and gave rise to strongly derived bacteria, including "archaebacteria"– (what most people call Archaea) as well as perhaps other bacteria that are either extinct or not available for our analysis, and also to eukaryotes. Sometimes, however, "neomuran" refers to the existing branches of this clade (i.e., only present-day neomura). Remaining cladistically and linguistically consistent about these meanings (only tips of the tree, or also all the stems) requires some concentration.

Cavalier-Smith starts with 19 molecular traits defining "neomura." Here, I cluster these traits into functional systems. Trait numbers are as given by Cavalier-Smith, except that I prepend them with "AE" (Archaea + Eukarya). I am not sure why Cavalier-Smith numbered them in the order he did. Trait descriptions are nearly verbatim from Cavalier-Smith, and my commentaries are in square brackets.

1. Genome replication and maintenance

   Trait AE5: Replicative DNA polymerase of the B family (palm-like fold), inhibited by aphidicolin. Replicative sliding clamp is PCNA-type, not part of type C DNA polymerase holoenzyme [in fact, sliding clamps in all living forms belong to "PCNA-type," which is not only a fold type but also a monophyletic sequence family; a detailed comparison of shared and unique components of replication systems can be found in Leipe *et al.* (1999), which Cavalier-Smith selectively cites but not on this topic].

   Trait AE4: Core histones are present, with characteristic fold ("histone fold"). In some species, they are secondarily lost.

   Trait AE6: Flap endonucleases and RAD2 repair enzymes are common. [They are not necessarily an AE innovation: Recent analysis of sequence and structure similarity connects FLAPs to two other superfamilies of nucleases, bacterial YacP and widespread PIN, and the common ancestor of these nucleases can be traced back to the common ancestor of all prokaryotes (Anantharman and Aravind, 2006)].

   Trait AE17: DNA topoisomerase VI (known in Eukarya as meiotic protein Mre11).

   Trait AE19: DNA initiation helicase is represented by hexameric Mcm instead of bacterial-type DnaB.
2. RNA metabolism

   Trait AE7: mRNA is transcribed by RNA polymerase consisting of more than seven subunits, not four as in bacteria [in fact, core RNA polymerase in bacteria has five-subunit composition, i.e., alpha$_2$-beta-beta′-omega].

   Trait AE3: Pseudouridylates in rRNA, at least some of which are inserted by a mechanism that requires C/D box snoRNAs.

   Trait AE10: Some unique modifications in tRNA.

   Trait AE11: Exosomes.

   Trait AE14: Triplets CCA at the 3′ ends of tRNAs are not encoded by the tRNA genes and added posttranscriptionally [in fact, substantial diversity of the mechanisms of tRNA 3′ end formation are observed, from precise encoding of CCA in the genome to consistent adding of all 3′ terminal adenines with intermediate mechanism that involves complementary synthesis of 3′ terminal nucleotides followed by their removal and repair. Analysis of taxonomic distribution of these strategies does not identify strong synapomorphies].

Trait AE16: 5′-OH/3p protein-spliced tRNA introns with homologous endonucleases. [This section is an incomplete list of synapomorphies. More comprehensive lists can be found in Anantharaman *et al.*(2002) and Mushegian (2004). For example, Archaea and Eukarya share two RNA-binding domains, Peter Pan/Brx1/Ssf1 and PAZ, which play prominent roles in Eukarya in ribosome maturation and posttranslational gene silencing, respectively, although both of these domains are derived versions of more commonly distributed folds, similar to the case of FLAP nuclease].

3. Protein synthesis

Trait AE8: "Many similarities in rRNA and proteins." [Most of those are genuine synapomorphies, which have suggested evolutionary importance of Archaea in the first place (Woese *et al.*, 1990). It is this importance that Cavalier-Smith thinks is overblown. For analysis of these synapomorphies, see Anantharaman *et al.* (2002), Klein *et al.* (2004), and Mushegian (2005)].

Trait AE15: Translation initiation: The initiatory amino acid is methionine, not formyl-methionine. Larger repertoire of initiation factors, including eIF-, 2A, 2B, 5A [and 6].

Trait AE12: Translation elongation: More similar elongation factors, sharing some biochemical properties not found in bacteria, such as sensitivity to diphtheria toxin.

Trait AE13: Similar rules cotranslational selenocysteine insertion into proteins: Well-defined RNA element (SECIS) that directs insertion, and specific SECIS-binding protein.

4. Protein sorting and turnover in the cell

Trait AE1: Signal recognition particle contains extra components—7S RNA has a synapomorphic helix 6 that interacts with the SRP19 protein. Secretion in "neomura" is mostly cotranslational. Bacterial SecA, which binds to secreted proteins in cytoplasm and carries them to SRP, is missing.

Trait AE2: Cotranslational glycosylation of proteins that occurs by transfer of acetylglucosamine and mannose from an isoprenoid (dolichol) carrier to *N*-asparagine and is mediated by synapomorphic oligosaccharyl transferase.

Trait AE9: Unique set of chaperones, including CCT-type group II chaperonins with eightfold symmetry (not sevenfold as in bacterial Hsp60). Built-in cap that replaces co-chaperonin Hsp10 [Our tentative assignment of the Hsp10 homology and function to *Methanococcus* protein MJ0073 (Koonin *et al.*, 1997) was an error]. Prefoldin, a chaperone complex that delivers proteins to the lumen of chaperonin.

5. Miscellaneous

Trait AE18: Specific insertion in the catalytic subunit of V-ATPase.

According to Cavalier-Smith, there are also Archaea-specific, "unique archaebacterial" molecular traits, but there are relatively few of them. Here they are, again with his numbering:

Trait A1: Side chains of membrane lipids are made of prenyl ethers, not of acyl esters.

Trait A2: Flagellar shaft of archaea is made of acid-insoluble glycoproteins related to pilin, not of acid-soluble flagellins.

Trait A3: DNA-binding protein 10b.

Trait A4: Unique tRNA modifications, such as archaeosine in the D-loop.

Trait A5: Tiny protein LX in the large ribosomal subunit.

Trait A6: No HSP90 chaperone.

Trait A7: RNA polymerase A subunit split into two proteins.

Trait A8: Glutamate synthase split into three proteins.

The point of these two lists is to illustrate the notion that all neomuran traits, and also most Archaeal traits, are adaptations of one or more of the following three cellular systems: (1) cell envelope; (2) the way the ribosomes interact with cell envelope, especially with regard to protein secretion; and (3) chromatin. Cavalier-Smith makes the case that all this is best understood as concerted adaptation to thermophily, with further adaptation to hyperthermophily in Archaea. He also states that none of these involve changes in metabolism. Moreover, all these changes, Cavalier-Smith says, can be unambiguously polarized—from ancestral mesophilic bacterial-like form to derived Neomura. We cannot help noticing that all the AE traits are preserved in eukaryotes, which have been largely mesophilic throughout their evolutionary history, but to that Cavalier-Smith says that these adaptations are fixed because they have turned out to also be superior at cool temperatures. Moreover, "since none of them [traits AE1–AE11 and A1–A6] are reversed in secondary mesophiles, they are 'valves' that can be used to polarize from mesophilic bacteria to hyperthermophilic archaebacteria."

It is interesting, and at times inspiring, to witness creative thinking of Cavalier-Smith, as he links, one by one, most of the traits AE and A to thermophily (although differences between "simply" thermophily and hyperthermophily are not consistently explained). Some of his explanations are plausible, such as the hypothesis that bacterial predecessor of proteasomes, the heat shock-inducible HslUV protease, must have been replaced by constitutive proteasome in thermophiles, where removal of denatured proteins from the cytoplasm might be more of a nagging problem.

Other constructs of Cavalier-Smith are more strained. Consider, for example, the contorted story about evolution of protein secretion—a story that, in fact, is central to Cavalier-Smith's synthesis. In bacteria, he says, membrane proteins are inserted in plasma membrane cotranslationally by the SecYEG complex. Secreted and periplasmic proteins, on the contrary, are expressed in cytoplasm and need to bind to SecA (in some bacteria, there is also a backup subunit SecB, which may have been invented later) in order to be delivered for insertion into the membrane. In "neomura," all secreted and membrane proteins experience translational arrest after synthesis of the leader peptide, and they resume translation upon ribosome binding to evolved signal recognition particle. This, according to Cavalier-Smith, is adaptation for thermophily in order for protein to not be denatured in cytoplasm. This streamlined protein secretion, Cavalier-Smith says, happens to be as good for mesophiles as it is for thermophiles, and that is why it remains in place even after secondary switch of most "neomurans" to mesophily.

However compelling this story may sound, it is not difficult to construct an alternative. Could translational arrest and cotranslational insertion of the growing protein chain into the membrane have evolved in mesophiles? There is at lease one plausible reason to suggest that, indeed, it could. What had to be survived and conquered was not thermal or other stress-induced denaturation of proteins in the cytoplasm. For that, all living forms, thermophiles and mesophiles alike, have cytoplasm teeming with molecular chaperones that work on renaturing misfolded proteins and with proteases to degrade the proteins that could not be refolded. Later in evolution, these nonspecific proteases are supplemented in the cytoplasm by specialized HslUV and proteasome. In contrast, in periplasmic and membrane compartments, similar problems of repair and garbage removal also exist, but the chaperones are scarce. Ostensibly, (hyper)thermophiles are affected by this problem even more than mesophiles. The mode of protein secretion is not of much relevance here; however the cell did it in the cytoplasm, the problems outside the cell are exacerbated at high temperature. Cotranslational protein secretion does not do much to protect cytoplasm from misfolded proteins upon moving into the hotter habitats.

A factor that would force a solution similar to the cotranslational secretion mechanism, however, exists, and it is not strongly dependent on the temperature of the habitat. This factor is the increase in complexity of interactions with the environment, which requires a larger repertoire of membrane-bound and secreted proteins. These classes of proteins account for a larger proportion of proteomes in heterotrophs than in autotrophs, in agreement with the need for heterotrophs to seek and acquire a diverse array of chemical compounds, which requires more proteins that serve as sensors, carriers, and transporters (Galperin, 2005). An intrinsic property of any exported or membrane-associated protein is the presence of a signal peptide, an (usually) N-terminal hydrophobic protein segment, which is inserted into the membrane at the first step of secretion. Nature has not come up with any other major way of targeting proteins into plasma membrane in prokaryotes (in eukaryotes, protein prenylation is one notable alternative, but it does not concern us here). Thus, if a cell needs to insert a protein into membranes, this protein usually contains a hydrophobic segment, even if the rest of the sequence is not strongly, or not at all, hydrophobic. This segment either remains unfolded, which is energetically disfavored and makes it a prey of cytoplasmic proteases, or it may try to minimize potential energy and shield itself from the solution by other parts of the same protein, thus preventing these other parts from folding correctly, with the same detrimental effect on the protein function. It follows that the main advantage of translational arrest and cotranslational membrane insertion is to prevent the leader peptides from interfering with the correct folding of the remainder of the protein, in order to keep the cytoplasmic burden of misfolded proteins under control. Similar burden in extramembraneous space is avoided by cleaving the signal peptide off as it traverses the membrane. The temperature of the environment plays at best only a secondary role in all this. I believe this casts serious doubt on Cavalier-Smith's proposed tight connection between the origin of cotranslation secretion mechanism and thermophily; indeed, my proposal appears to rely on a more robust selection pressure, which is independent of the hot environment.

In Cavalier-Smith's mind, many other AE traits are consequences of the cotranslational protein insertion mechanism. This seems to be his explanation of all ribosomal synapomorphies, such as specific subsets of AE ribosomal proteins, which are either very strongly diverged in bacteria or, more typically, not found there at all (at least with the existing methods of sequence and structure comparison). This explanation has several weak aspects. First, as discussed previously, increased complexity of secretion may be explained "isothermically" by increased access to different sources of food in heterotrophs, with concomitant dedication of an increasingly larger fraction of proteome to secreted proteins. Second, there are approximately 25 AE-specific ribosomal proteins, only 3 or 4 of which are directly involved in interaction of the ribosome exit channel with membrane (Harms *et al.*, 2001; Klein *et al.*, 2004; Mushegian, 2005). Thus, to say that all ribosomal proteins polarize the direction of evolution from mesophilic bacteria from thermophilic "neomura" would be an exaggeration. Viewing the loss of the Hsp90 chaperone in archaea as hyperthermophily related, in the meantime, is outright strange.

Occasionally, Cavalier-Smith has to invoke other extreme habitats in order to explain the AE/"neomuran" traits, such as high acidity as the factor selecting for acid-resistant archaeal flagella. High salinity is also mentioned as a bacteria-to-neomura polarization factor but is not explicated. This is good because, in my opinion, salinity is of minor, if any, importance for explaining archaeal origin (and sequencing of *Salinibacter*, a bacterium that lives in crystallizing conditions of NaCl, indicates that despite ample opportunity of horizontal gene transfer between this bacterium and cohabiting halophilic archaea, *Salinibacter*'s archaea-like genes are involved mostly in energy transduction and ion transport, not in protein secretion or membrane maintenance; Mongodin *et al.*, 2005).

One more protein complex, which, in Cavalier-Smith's opinion, is a crucially important synapomorphy—namely, the system of biosynthesis of N-linked glycoproteins—appears, according to his own assessment, to be adaptive not to hyperthermophily "but to lysozymes and antibiotics secreted by posibacterial ancestors" of neomura. Indeed, this synapomorphy is an important evolutionary marker, and utility of this biochemical pathway for protection against lysozymes and antibiotics may well be the reason for its provenance. But lysozymes, microcides, and antibiotics are produced by all kinds of bacteria, wherever researchers look for them, and most bacteria have evolved multiple strategies to defend themselves against these toxins. Posibacterial ancestral relationships to Archaea, meanwhile, is a hypothesis that needs to be proven separately.

I am far from saying that none of the archaeo-eukaryal synapomorphies are adaptations to thermophily. On the contrary, I agree with Cavalier-Smith about several of them. The aforementioned replacement of HslUV by constitutive proteasome may be one. Also likely to be true, although perhaps not uniquely contributed to science by Cavalier-Smith, are the cases that can be made for the emergence of histones and reverse gyrase, which are viewed by most authors as adaptations for negative supercoiling of DNA, useful when melting of the double-stranded molecule has to be resisted (Forterre, 2002). Cavalier-Smith may also be correct that some other peculiarities of AE-type replication can be seen as ways to handle complications imposed by histones.

On the crucial distinction of the main DNA replicative enzymes in Bacteria versus Archaea/Eukarya, however, Cavalier-Smith has nothing much to say, except for advising not to obsess over them and asserting that bacterial, evolutionary unique PolC/DnaE polymerase is ancestral [when this book was at the final revision stage, this structure became known (Lamers *et al.*, 2006); it is similar to polymerase beta and to various nucleotidyltransferases, but its ancestral status with regard to Archaea has not been proven]. Archaeal/eukaryotic type B polymerase, he holds, is derived from bacterial repair polymerase, but specific selective factor facilitating this is not identified.

Up to this point, we have seen that thermophily, which Cavalier-Smith promised to be the overriding theme for all "neomuran" and even more so for the Archaeal traits (hyperthermophily in this case), in fact is not sufficient to explain even a fraction of them. For a remaining subset of synapomorphies, the evidence is also shaky. Pseudouridylation, says Cavalier-Smith, rigidifies RNA. Physical measurements *in vitro* support this notion, as do the indications that at least some pseudouridines are found in "loose regions" of rRNA. On the other hand, ribosomal large subunit in mesophilic *Escherichia coli* contains 10 pseudouridines, whereas the numbers in the studied archaea vary from 3 to 6 [see Ofengand *et al.* (2001) and Del Campo *et al.* (2005), in which the number of pseudouridines in rRNA of archaeon *Haloarcula* was revised downward)], which does not follow from the predicted role of pseudouridylation. Neither can I make heads nor tails of Cavailer-Smith convoluted argument about the terminal nucleotides of tRNAs, which seems to end with doubt as to its relationship to thermophily. Cavalier-Smith concedes that not each trait on his list was selected directly as adaptation to heat, but selective sweeps and gene hitch-hiking must have also been involved. I have no argument with this general statement, but of course it can be invoked to explain any evolutionary scenario. Finally, splits of genes (traits A7 and A8) are good archaeal synapomorphies, but, as with splits and fusions of other genes in various other prokaryotic lineages, there is no robust evidence of selective advantage of these events; as discussed in Chapter 8, clustering on the chromosome seems to be the defining adaptation in prokaryotes, after which the precise type of fusion (coordinated expression, polycistronic mRNA, or multidomain protein) seems to be a relatively easy, perhaps in some cases almost neutral, modification. Until a specific role of these splits in archaea is discerned, all they argue for is monophyly of Archaea, which is not in doubt, at least within the frame of this debate.

Let us now look at the same evidence of "neomuran" synapomorphies, as it is presented to us by comparative genomics, starting from the reviled bags-of-genes approach and working our way toward a higher-level view. One of the best organized bags of genes is the COG database, which was discussed in detail in several previous chapters. Currently, the database includes 110 prokaryotic species (16 archaea and 94 bacteria) and almost 14,000 COGs. This data set allows us to identify a mix of archaeo/eukaryal and archaeal gene-gain synapomorphies automatically and then separate AE from A by comparison with a separate set of eukaryotic othologous groups (KOGs, also developed at NCBI).

Synapomorphies in this approach are detected by analysis of phyletic patterns. It can be either deterministic, when the state of COG is set at "1" in all archaea and at "0" in all bacteria, or probabilistic, when certain flexibility is allowed to account for secondary gene loss and horizontal transfer—two processes that were discussed in Chapter 11. I used the psi-square program for probabilistic matching of binary vectors (Glazko *et al.*, 2006) to derive the set of A + AE synapomorphies. These COGs tell a story that is quite similar to what I described previously: There are between 300 and 400 COGs largely specific to archaea as opposed to bacteria, more than half of them with confidently predicted functions, the exact number depending on the parameters of the search. This list includes many COGs that indeed belong to the functional systems emphasized by Cavalier-Smith. For example, a relatively conservative set of search parameters results in 24 genes making up most of the replication complex and nonbacterial machinery for DNA recombination and repair, 71 genes whose products are involved in RNA biosynthesis and turnover, 22 transcription factors, seven components of flagella, and nine factors of protein folding and secretion machinery. At the same time, at least 23 proteins on the list are metabolic enzymes—mostly archaea-specific analogs of bacterial enzymes with similar activities—and one-third of A/AE synapomorphies are completely uncharacterized proteins, some of which are likely to be involved in intermediary metabolism as well. Thus, the list of archaeal synapomorphies is a much broader palette than a tight bunch of coinherited functional modules. Polarization from bacterial to archaeal traits, hinged on transition from mesophily to (hyper)thermophily, has to be more or less arbitrarily hand-picked from the extended set of other synapomorphies, many of which may be more amenable to other explanations.

Thus, only some, as opposed to all, "neomuran" synapomorphies may indeed have been adaptations to hyperthermophily. In reality, however, all this is a mere prerequisite for the debate on the evolutionary position of "neomura." The crucial question is when these synapomorphies, and the clades defined by them, emerged. The earlier hypothesis of the cenancestral position of archaea and the suggestion of ancestral position of eukaryotes, with prokaryotes being secondarily simplified, need not be examined here. The real debate is about three points:

1. Whether archaea-like prokaryotes are ancient or recently derived: Many people, including myself, say "ancient," whereas Cavalier-Smith says "recently derived." This can be called the question of long vs. short Archaea stem. (In effect, this is also the question of the identity of the last universal common ancestor—whether it was a form most similar to a modern-type bacterium, as Cavalier-Smith would have it, or some other type of organism, as I will argue in more detail in Chapter 13).

2. Whether eukaryotes are more recent than archaea: This can be called the question of long vs. short Eukarya stem.

Four logical combinations of answers to the above two questions are possible. Cavalier-Smith supports "short A, short E," whereas many others think "long A, long E." I tentatively think (in part convinced by some of Cavalier-Smith's arguments) that the answer is "long A, shorter E" ("long E, short A" need not be considered seriously). Absolute dating is important here.

3. Whether the universal tree of life involves massive horizontal gene transfer and, perhaps, even a wholesale fusion of genomes: This can be called the question of cyclic vs. acyclic graph of life. The emerging consensus view, that it is cyclic, with some uncertainty as to the placement of the reticulate branches and the root, was discussed in Chapter 11. Cavalier-Smith agrees

that HGT occurs in nature, but this should not prevent us from reconstructing the history of life. We disagree on the ratio of gene losses to gene gains (which include HGT events), and in the previous chapter I followed Eugene Koonin's argument that any model that allows large excess of gene losses over HGT would produce absurdly large ancestral genomes. Cavalier-Smith, in the meantime, appears to have no problem with a large, perhaps up to tenfold, excess of gene losses.

I now discuss Cavalier-Smith's argument that "neomuran" origin is a recent event, coming from within actinomycetes, and conclude that this is not likely to be the case. Instead, Archaea originated, perhaps indeed in large part as adaptation to thermophily, in more ancient times. Therefore, Archaea are a sister group of most, perhaps all, of contemporary groups of Bacteria. As already shown, the arguments about concerted evolution of neomuran synapomorphies as adaptations to thermophily are not as airtight as Cavalier-Smith would like us to believe. This means that we have to turn for supporting evidence to his other arguments, namely paleontology and trees based on molecular characters, and the picture here is not much brighter for Cavalier-Smith's theory.

What follows is the summary of my understanding of the paleontological evidence, and "the strongest reasoned criticism" is invited from practicing paleontologists, which I am not. Morphology of both bacteria and archaea is too simple to reliably distinguish their fossils with the available methods, either by eye or by microscope, and conclusions have to be supplemented by analysis of chemical markers that are produced only by living organisms. There are two major types of such markers: One consists of specific molecules and the other of specific patterns of shift in relative abundance of isotopes of certain elements. Protists, with their more complex morphology and, in many cases, sculpted cell envelopes, provide a larger array of fossilized morphological traits.

Biological fixation of CO_2 by the Rubisco enzyme is biased against ^{13}C and enriches the biomass in ^{12}C. Carbon isotope ratio that may be indicative of Rubisco-like fixation can be dated at 3.7 Gy ago (Sirevag *et al.*, 1977). Biological sulfate reduction depletes ^{34}S compared to ^{32}S, and this depletion can be traced to perhaps 3.47 Gy ago (Shen *et al.*, 2001). From this pair of datings comes the first of the paleomicrobiological one–two punches of Cavalier-Smith synthesis: Since the sulfate reducer-suggesting deposits (Warrawoona) are in gypsum, which is unstable above 60°C, the deposits must have been left behind by a mesophile. All currently known archaeal sulfate reducers are hyperthermophiles, and the known mesophilic sulfate reducers are found among Gram-positive bacteria or proteobacteria. Cavalier-Smith thinks this leaves proteobacteria as the culprit of the 3.5 Gy-old deposit. So far, so good; I am not certain that these were modern-type proteobacteria rather than some of their ancestors, but I can tentatively agree with the 3.5 Gy dating as a reasonably early estimate of bacterial life. Deposits from mesophilic environments, of course, tell us nothing about what was going on with thermophiles at the time.

The second punch of Cavalier-Smith is morphology. Eukaryotes began to diversify in shapes and surface sculpture approximately 850 My ago, and Cavalier-Smith suggests that this should be a robust estimate of the origin of eukaryotes. He argues that "any assignment of a 2.0 Gy fossil to eukaryotes needs to explain why eukaryotes went so long without diversification," and on that topic he is indignant: "If no sound suggestion as to why this should be so [i.e., no fossils before 850 My, many fossils after that time], we should regard this as antiscientific special pleading of the worst kind." This is bold from someone who firmly believes in "quantum and mosaic evolution" in archaeal and eukaryal stems but teaches that no evidence of such evolution should be expected to have survived in the completely sequenced genomes (see later).

There is another piece of evidence that Cavalier-Smith needs to explain. Most Eukarya, and only Eukarya, make steranes with a modified carbon atom in the 24th position. Such molecules are preserved in the fossils and are found at least in 1.64 Gy Barney Creek formation, if not earlier (Pearson *et al.*, 2003). This is twice as old as Cavalier-Smith would have it. He handles the problem as follows: Instead of dealing with specific synapomorphy of ^{24}C-modified sterans, Cavalier-Smith tells us that several groups of bacteria have recently been shown to synthesize sterols and, therefore, sterans are totally useless as evolutionary markers. This is a verbal trick. In fact, genes coding for two enzymes of sterol biosynthesis, squalene monooxygenase and oxysqualene synthase, are found in three bacterial lineages—*Plactomycetales, Myxobacteriales*, and *Methylococcales*—and the latter enzyme is additionally found in mycobacteria. These clades are distant from each other, and whenever both enzymes are present, their genes are found as a pair adjacent on the chromosome, which is compatible with dissemination across distant clades by horizontal transfer. Moreover, sterol modifications in mycobacteria are almost certainly secondary adaptations, facilitating interactions with sterol-containing cellular membranes of their animal hosts (Gatfield and Pieters, 2000), and the pathway of sterol biosynthesis in mycobacteria appears to be incomplete and possibly working in the catabolic direction (Bellamine *et al.*, 1999). Most important, none of the bacterial species makes ^{24}C-modified sterans, which remains a eukaryotic synapomorphy, useful for fossil identification.

So much for paleontology: Neither Cavalier-Smith nor anyone else has good dating for Archaea; the late dating of Eukarya still needs to be reconciled with the more ancient evidence of the unrefuted biochemical marker; and so this leg of the argument for recent origin of Neomura is shaky too, especially as it concerns Archaea. This leaves us with molecular characters and phylogenetic trees inferred on their basis.

Cavalier-Smith suggests that the idea about extreme antiquity of Archaea comes from three sources: methanogenic lifestyle arguably compatible with early atmosphere; split on rRNA; and "apparently large but biologically trivial differences in gene expression molecules." This is disingenuous: The truth of the matter is that virtually every sequence-based tree supports a deep split between Bacteria and Archaea. Cavalier-Smith, of course, is aware of all this, but he is unfazed. Most of his argumentation falls back on rRNA trees anyway:

> *I pointed out earlier that rRNA cannot possibly be a molecular clock, since nuclear rRNA, plastid and mitochondrial RNA must have evolved at different rates... It is now abundantly clear that all three types of rRNA evolve at two or three orders of magnitude different rates in different evolutionary lineages.*

This, of course, is not directly relevant to the rates of rRNA evolution observed in free-living organisms. More to the point, Cavalier-Smith quotes an example of the heterogeneous rate of rRNA evolution in *Foraminifera*, reminds us of the theoretical considerations indicating that the molecular clock may have insufficient mechanistic or theoretical basis (Ayala, 1997, 1999), brings up a classic observation by G. G. Simpson that the origin of new groups is often marked by rapid evolution, and correctly points out that unequal rates of rRNA evolution are often confirmed when the fossil record is available.

It must be noted that methods that assume no molecular clock do exist, and they give qualitatively the same result for bacterial and archaeal origins. Moreover, a deep split between two groups is observed not only for the rRNA trees but also for the trees built on the basis of the distance matrices derived from concatenated sets of orthologous proteins, which are mostly involved in transcription and translation; from pairwise percentage of sequence identity in orthologs; from the proportion of shared genes; from co-occurrence of orthologous gene pairs; and from conservation of local content of orthologous genes, allowing for permutation of gene order. These are discussed in more detail in Wolf *et al.* (2001a), Dutilh *et al.* (2004) and Snel *et al.* (2005), except for the tree based on local neighborhood conservation, which can be found on-line at my Web site (http://research.stowers-institute.org/bioinfo). Note that the last three types of trees capture evolutionary signal from many functional classes of proteins.

Never in these trees are archaea nested within actinobacteria, from which, according to Cavalier-Smith, all "neomura" have originated.

All this requires a potent counterargument. Cavalier-Smith finds one in the radical idea that most, if not all, phylogenetic trees of bacteria and archaea, built on the basis of most, if not all, molecules, are ridden with a king of all artifacts, the most vicious case of the long branch attraction—the attraction that may occur regardless of the observed branch lengths. Enter the long stem attraction, which can disguise every aspect of the tree, including branch lengths (short branches will appear artifactually long), branching order, and root position (no prediction regarding the specific way in which the latter two would be distorted).

The difference between long branch and long stem is the following: In the long-branch scenario, fast evolution goes on, whereas in the long-stem scenario evolutionary rates are high for awhile and then they become low again and continue at a low rate after speciation. Another difference is that a long branch can be observed directly, at least in principle, whereas a long stem may be forever hidden from view.

The long-stemmed affair does not have to be supported by the evidence. On the contrary, the whole point of Cavalier-Smith is that it would be naive to expect the evidence, because of the conniving artifacts. This line of argument is best seen when Cavalier-Smith re-interprets the data of Iwabe *et al.* (1989) on rooting of the universal tree of life. In that classic work, two paralogous proteins present in every genome and more than likely produced by gene duplication before the last universal common ancestor were used as each other's outgroups to root them on one another. This gives the root of life between Bacteria and Archaea, supported by several other similar analyses. Cavalier-Smith points out various weaknesses of these other analyses, and some of these critiques may have a point. However, what he really needs to explain (away) is the most robust rooting from Iwabe and co-authors, and this is how it is done: In the EF2/G part of the tree, the (AE)B topology is said to be wrong because AE evolves too fast and, being a long branch, attracts to another long branch—that is, to the EF1a/Tu subtree as a whole. On the other hand, that latter subtree is also a long branch, but, according to Cavalier-Smith, at the same time it somehow manages to evolve not fast enough and therefore produces the same (AE)B topology for a completely different reason—as an artifact of too few synapomorphies. I suppose we should call this frivolous.

Whether there is an unfortunate confluence of artifacts or not, can we find an independent way to evaluate Cavalier-Smith's claims? The answer is yes, and Cavalier-Smith posits that the help might come from the complete genome sequences. Cavalier-Smith proposed to deduce species' position in a tree on the basis of their cellular organization: For example, examination of morphological and cytological features allowed him to place thermophiles *Aquifex aeolicus* and *Thermotoga maritima* into proteobacterial and Gram-positive clades of Bacteria, respectively. However, this is exactly the same placement that is observed for these species in most proteome-based trees (Wolf *et al.*, 2001a; Dutilh *et al.*, 2004). One would think that the remarkable congruence between these objectively inferred, statistically supported phylogenies and Cavalier-Smith's expert assessment would be seen, in his eye, as the validation of these tree-building methods. But this is to no avail, apparently because these trees are of the disreputable bag-of-genes persuasion.

The other comparative genomics tests that Cavalier-Smith proposes are as follows. According to his own revision of bacterial systematics (which is interesting in its own right and discussed in great detail in the same paper, but it is not evaluated here), there are seven deep phyla of bacteria. Cavalier-Smith says that if a well-supported tree showed that an archaeal gene is shared by all bacterial phylae, this would disprove the novelty of Archaea. To assess the distribution of genes in clades, one could use phyletic patterns of gene families/COGs, together with phylogenetic trees built on the basis of sequences of these same COGs. However, the particular test proposed by Cavalier-Smith would not be conclusive.

Indeed, if all clades (or at least five or six out of seven, allowing for some gene losses, which are common according to Cavalier-Smith and everyone else) have a gene also found in archaea, this means that this gene is widely distributed in all prokaryotes; what would then be a basis for calling it "archaeal" in the first place? A more refined quantitative version of this test may be perhaps devised, but Cavalier-Smith does not offer any guidance here.

Another test suggested by Cavalier-Smith is to show that all cyanobacteria, for which reliable paleontological dating of 2.5 Gy is available, have acquired an archaeal gene. This is a sensible test that should be performed when broader diversity of cyanobacterial genomes is sequenced.

In that same spirit, I tried another test, which can be viewed as a modification of Cavalier-Smith's first test. If actinobacteria were ancestral to neomura, then archaea should be sharing more genes with actinobacteria than with other bacterial phylae. (Perhaps preemptively, Cavalier-Smith dictates that there should be no such expectation of specific sequence similarity between neomuran and actinobacterial proteins, even though, in his theory, the latter are the closest bacterial relatives of the former—all because of the long branch/stem attraction artifacts. My test, however, is different, because it deals not with the rate of sequence evolution but with the relative rate of gene retention; see also Chapter 5 for explanation why COGs are relatively resistant to unequal rates of sequence evolution.) As a first approach to the problem, I measured, for each COG, its archaeal, actinobacterial, and proteobacterial propensities, expressed as the decimal fraction of all species within the corresponding clade that have this COG. Each COG becomes a vector in the Euclidean space defined by these three propensities. It turns out that almost all data scatter in this space is explained by three principal components (PCs), the first of which is parallel to the line defined by points (0;0;0) and (1;1;1). This PC has to do with gene rarity and is not of immediate interest to us. The second PC describes the separation between bacteria-specific and archaea-specific COGs and is also not directly relevant. Finally, the third PC contrasts COGs shared by archaea with actinomycetes to those shared between archaea and proteobacteria. Data projection on this PC indicates that for those genes that are found in most species of their respective lineage, the number of archaeo-actinobacterial exceeds the number of archaeo-proteobacterial ones. In other words, if a gene is found in most archaea, this is a good predictor of whether it will also be found in most actinomycetes and vice versa. The same cannot be said about archaea and proteobacteria. This result needs to be examined further because it points out a possible specific relationship between Archaea and actinomycetes (although it does not say what this relationship is: recent origin of the former from the latter, as Cavalier-Smith would have it, is one, but not the only, possibility).

In Cavalier-Smith's most recent paper (2006), the conclusions published in 2002 are repeated as the settled truth. Some novel tropes are also introduced, the most relevant to the "novelty of neomura" being the evolutionary history of proteasome-like protease HslUV and of proteasome itself. To wit, Cavalier-Smith proposes, "I argue here that the proteasome 20S core particle evolved from the simpler HslV, not the reverse. If this evolutionary polarization is correct, it excludes the root of the universal tree from a clade comprising neomura and actinomycete actinobacteria"—because only Archaea + Eukarya (neomura to Cavalier-Smith) and actinomycetes share the proteasome core.

No one I know argues that HslUV evolved from the proteasome. But the presence of the core proteasome in "neomura" proves mostly that this complex is ancestral in Archaea + Eukarya. To assert that it is also ancestral in actinomycetes, more argumentation is needed, for which the following passes:

> *If proteasomes have never been lost from free-living bacteria, they evolved only in the immediate common ancestor of Actinomycetales, and thus may be only half as old as actinobacteria. If that is correct and proteasomes have always been vertically inherited, neomura must be more closely related to*

> *Actinomycetales (as several other characters such as cholesterol biosynthesis also suggested ...), making Actinobacteria paraphyletic.*

Note that the cholesterol connection does not hold water, as has been discussed previously in this chapter, whereas the truth of the whole statement is dependent on "if always vertically inherited."

The most obvious response to the latter proposition is that proteasome indeed may not always be vertically inherited. Cavalier-Smith is aware of such an argument, which he dismisses without evidence:

> *The red herring of lateral gene transfer might be raised against the above interpretation. Gille et al. assumed that proteasome genes were laterally transferred from archaebacteria to the common ancestor of actinomycetes. However, they presented no phylogenetic analysis to support this assumption; unpublished trees give no support for lateral transfer, but as the a- and b-subunits and HslV proteins are very divergent and with too long branches for satisfactory phylogenetic analysis, such a possibility cannot be excluded with total confidence (J. Archibald, pers. comm.).*

Here, as in the previous case with specific affinity between archaeal and actinomycete proteins, Cavalier-Smith seems to be selling his own argument short: In fact, preliminary analysis shows that actinomycete catalytic subunits of proteasome form two clades on a tree, one of which is very deep (unpublished).

I presented my arguments. I believe they show that, with all the valuable insights into cell physiology (and in his 2006 paper, Cavalier-Smith seems to be more welcoming to gene-content phylogenies, although he apparently refrains from building any tree), with all the inspired guesses about many other events in the evolution of life, and with the possible genomic evidence of a specific relationship between Archaea and Actinomycetes, the case for recent (850 My ago) timing of Archaea appearance from within Gram-positive bacteria is just not there.

We now discuss the other story about evolution. The nearly instant triumph of the Ecdysozoa hypothesis in the late 1990s is in contrast with the traditional emphasis on the importance of the body cavity—coelom—in organization and evolution of body plan. The Coelomata clade seemed to be better supported by anatomical and developmental evidence, although other theories were also proposed. Anatomical evidence is not plentiful for Ecdysozoa: Rigid exoskeleton that requires molting is very nearly the only morphological trait that unites nematodes and arthropods. Lack of ciliate epithelia is another shared feature, but it is more than likely a consequence of having the exoskeleton not an independent character (see later).

Simultaneously with the ecdysozoan hypothesis, another major clade of invertebrates, Lophotrochozoa, has been proposed. This is an even more puzzling entity, where evolutionarily primitive Plathelminthes with two germ layers are united with such highly evolved three-layered animals as, for example, Mollusca. Lophotrochozoa are also synapomorphy poor. The word itself is Humpty Dumpty's portmanteau, produced by fusion of "lophophore" and "trochophore," which only sound similar but in fact refer to two different things, found in different subsets of taxa. Lophophore is a horseshoe-shaped feeding apparatus surrounded by cirri, found in such groups as *Brachiopoda*, *Bryozoa*, and *Phoronida*, whereas trochophore is a swimming larva found in other taxa, such as *Annelida*. Thus, "Lophotrochozoa" means more or less "animals that have either a particular type of a larva, or a particular type of feeding apparatus in adults." Except for the fact that all these forms have ciliate epithelia, this is not really a basis on which to derive a natural group of organisms.

In an attempt to unify taxa within Lophotrochozoa, spiral cleavage of the embryo has been proposed as a better synapomorphy. This has two problems: first, that many of lophophoran animals lack it, and second, that the pattern of embryo cleavage is known to be prone to homoplasy (repeated gain and loss of a trait in different lineages) in response to several factors, notably the size and position of the yolk sac, which itself is extremely variable in invertebrates, sometimes differing significantly even among closely related species.

So why the surrender to the new taxonomy of invertebrates? I will examine the evidence, focusing on Ecdysozoa and molecular traits that come from completely sequenced genomes and functional genomic studies. No such resources are available for Lophotrochozoa, although the pipeline of genome sequencing includes at least *Schmidtea mediterranea* (*Plathelminthes*; *Turbellaria*) and *Aplysia* sp. (*Gastropoda*; *Mollusca*). Give sequencing machines and people some time, and we can examine molecular evidence for Lophotrochozoa as well.

The Ecdysozoa hypothesis in its modern form was precipitated by two types of observations. One is cladistic analysis of a large collection of morphological and molecular traits [in this work, I focus on the latest and most detailed such study—that by Peterson and Eernisse (2001)]. The other is the work on rRNA-based phylogeny, where many investigators started noticing that nematode sequences tended to give very long branches. This led to a realization that some or all nematodes are fast-evolving species. Comparison of various nematodes to other animals in standard relative rate tests indicated that rRNAs in many nematodes indeed evolved much faster than the others. For example, rhabditid nematodes, including *Caenorhabditis elegans*, are fast evolvers, whereas a basal enoplean nematode *Trichinella* seems to evolve relatively slowly.

Aguinaldo *et al.* (1997) examined a phylogentic tree built from 18S rRNAs representing a large variety of invertebrates, with some effort to sample many species within each type of animals. The main conclusion from these trees was that all molting animals form one clade, which was given the name Ecdysozoa. The ecdysozoan clade, however, would not have been noticed by an average practicing phylogeneticist; in fact, the majority of nematode species had rRNAs that, in most cases, did not cluster with arthropod rRNAs. Only inclusion of certain nematode species, which have been determined to evolve slowly, produced the Ecdysozoan topology. The authors concluded that the Coelomata topology, accepted by many zoologists for many decades on the strength of the underlining concept of gradual increase in body plan complexity, is caused by the long branch attraction artifact.

Several of the rules of invoking the LBA artifact (see Chapter 11) were not observed in that study. In particular, rule 5 (never assume the attraction to the out-group, if there is more than one long-branched in-group) was not given much thought. This is too bad because many arthropods, including well-sequenced dipteran insects, are also fast evolvers, and their presence adds long branches, which, as we have seen, may affect the tree in unknown ways. The comparative anatomy concerns that I discussed previously were not discussed either. The importance of ecdysis (molting), however, seems to have swayed the "evo-devo" community, and the theory spread like wildfire. But let us look at the data more closely.

The ecdysozoan hypothesis was corroborated by phylogeny of 28S rRNA (Mallatt and Winchell, 2002). As already mentioned, cladistic analysis of morphological traits also supported the monophyly of Ecdysozoa, but, interestingly, it failed to support monophyly of Lophotrochozoa (Peterson and Eernisse, 2001). We will return to that analysis later.

When protein coding genes were examined, however, nothing really worked in favor of the Ecdysozoa. The topology of a tree obtained from aligned homeoboxes from Hox genes, frequently cited as supportive of the ecdysozoan hypothesis (Balavoine *et al.*, 2002), is not reliable because at least 20% of orthologs from different species were misassigned in this study (my unpublished observations). This tree was criticized on several other grounds (Zdobnov *et al.*, 2005). Trees built from myosin II sequence and also purported to favor Ecdysozoa (Ruiz-Trillo *et al.*, 2002) in fact gave virtually no statistical support to that clade, except for a single case (one tree with Bayesian posterior probability of 71% among multiple trees without statistical support for Ecdysozoa). Also, putative synapomorphy of internal sequence triplication in beta-thymosin (Manuel *et al.*, 2000) turned out to lack evolutionary signal (Telford, 2004).

In 1998, we attempted the first proteome-based analysis of the ecdysozoan hypothesis. We collected 42 quartets of likely orthologous proteins from humans, flies, worms, and yeast (this number was small because fly and human genome sequences were far from completion) and built neighbor-joining trees for each quartet (Mushegian *et al.*, 1998). The result was interesting: Approximately two-thirds of the trees favored the Coelomata hypothesis [i.e., tree topology (((human; fly)worm)yeast)], and approximately one-third looked Ecdysozoan [((human(fly; worm))yeast)]. There were indications that unequal evolutionary rates may play a role in inconsistent tree topologies: Trees with shorter branches tended to support the ecdysozoan hypothesis, and trees with longer branches favored Coelomata. On the other hand, trees with Coelomata-like topology had much better statistical support than trees that favored Ecdysozoa. Finally, after completion of human and fly genomes, orthologs could be identified more robustly, and some of our 42 quartets turned out to contain paralogs. After they were removed, the distribution of the remaining trees shifted even more toward supporting Coelomata (Xie and Ding, 2000).

In recent years, much larger sets of orthologs from various eukaryotes were compiled. The databases were also scouted, and targeted sequencing was occasionally performed, in order to include more genes and more species. For example, Blair Hedges and colleagues at Pennsylvania State University obtained protein sequences by translating ESTs from *Trichinella*, a nematode that is supposed to be a relatively slow evolver (Aguinaldo *et al.*, 1997). The outcome of these efforts was very similar to that of our work: The majority of protein-based trees favored Coelomata, although there was always a smaller fraction of trees supporting Ecdysozoa. The correlation between branch lengths and observed tree topology, however, all but disappears in these larger, more comprehensive data sets (Blair *et al.*, 2002). In another study, Wolf *et al.* (2002) investigated the problem from a different angle by simulating series of trees with fixed topology (coelomate or ecdysozoan) and controlled variation of branch lengths. They studied this parameter space with a number of standard phylogeny methods and found that some of these methods, particularly those that use the maximum likelihood approach, were robust to LBA artifacts. That is, when the branch length ratios in simulated trees were set to be in the range of what was actually observed in nematodes and arthropods, the tree topology was correctly reconstructed in at least 70% of cases. The majority of actual trees supported Coelomata, and the LBA artifacts appear to have been ruled out in this case.

Consistency of conclusions from different kinds of observations is of high value in those sciences in which direct experiments in controlled environments are difficult to perform. Therefore, a premium should be put on such a topology, be it Coelomata- or Ecdysozoa-favoring, which is supported by trees built on the basis of different types of characters—especially if these characters evolve under reasonably diverse models. In the test of consistency of various types of trees, the Coelomata hypothesis wins hands down: It is supported by individual and concatenated protein sequence alignments, by measures of shared gene content and conserved domain architecture, by chromosomal syntheny of protein coding genes, and by positions of indels in exons. In contrast, support of Ecdysozoa, after all, comes only from rRNA and, under some models of gain and loss, from positions of orthologous introns in protein coding genes (reviewed in Zdobnov *et al.*, 2005).

Dopazo *et al.* (2004) proposed several tests aimed at finding protein sequences with the same relative rate of evolution in arthropods (i.e., fruit fly and mosquito) and nematodes (i.e., two species of *Caenorhabditis*). There are not many of these genes, but those that possess this property of similar evolutionary rate tend to support Ecdysozoan topology, whereas the majority of other genes supports Coelomata. In another work, Herve Philippe and co-authors (2004) made two important contributions. First, they presented the largest set of full and partial alignments of many sequences from 35 diverse species of invertebrates. The set consists of 146 genes (more than 35,000 positions occupied in at least two-thirds of species). Second, they proposed what might be called the "sliding out-group test." The idea here is that if the data is suffering from the LBA, and if this attraction is between a long in-group and an out-group, then by choosing another, more recently evolved and/or slowly evolving out-group, one may reduce the attraction and eliminate some artifactual clades (Fig. 12.2).

The results of this approach were not exactly as expected: When a usual out-group (yeast) was replaced with a more closely related and slower evolving basal metazoan, cnidarian Hydra, the grouping of *Plathelminthes* and *Nematoda* was observed, which, according to the authors, "does not make any biological sense." Yet, in fact, this grouping has been making all the sense in the world to generations of zoologists, even if the initial definition of a group is mostly by negation: These are both primitive worms lacking coelom (although a hypothesis also has been put forward that some worms may be acoelomic secondarily). True body cavity allows a major innovation in the animal kingdom—that is, uncoupling between movements in the digestive tract and the motility of the whole animal—lifting many constraints on morphogenesis and behavior. Thus, worms without genetic program for true coelom development are dead-enders on the road of morphological progress (although, obviously, nematodes enjoy considerable biological progress on Earth in terms of sheer biomass and taxonomic diversity). Hypotheses that *Plathelminthes/Nematoda* is a true clade or, possibly, that flatworms and roundworms are both basal branches are worth serious consideration.

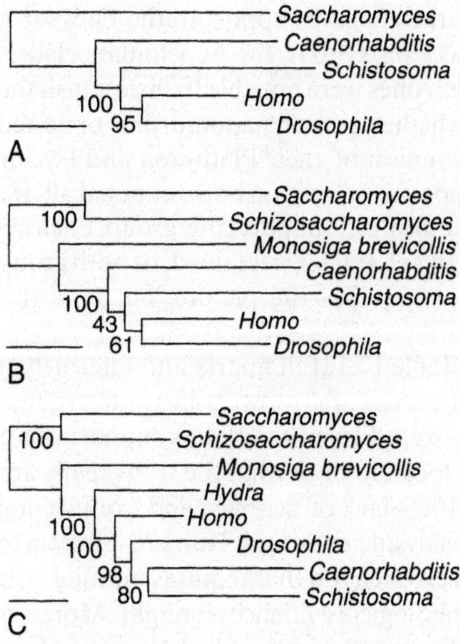

Figure 12.2. Topology of a tree with long branches is sensitive to the length of the out-group: A shorter out-group branch may be a weaker attractor of a long in-group. Reproduced from Philippe, H., Snell, E. A., Bapteste, E., Lopez, P., Holland, P. W., and Casane, D. (2004). Phylogenomics of eukaryotes: Impact of missing data on large alignments. *Mol. Biol. Evol.* **21,** 1740–1752, by permission from Oxford University Press.

Thus, these new results again can be seen as providing evidence for Coelomata (and, in one act of joining flatworms and roundworms, disrupting both Ecdysozoa and Lophotrochozoa clades). Note also that Philippe's group, which did much to promote Felsenstein's discovery of LBA and to quantify its effects, seems to be oblivious to the fifth rule of LBA: They seem to always assume that the long branch of their interest is always to the out-group and not to another in-group.

Thus, ecdysozoan topology is observed in trees that are based on rRNA sequences and subsets of protein sequences, if these data are adjusted to account for unequal evolutionary rates. On the other hand, Coelomata topology is consistently seen in trees based on different genome characters, some of which are much less prone to LBA artifacts than others. It seems that rapid evolution of everyone's favorite nematode, *C. elegans*, notwithstanding, there is simply not enough LBA artifacts out there to explain away all support for Coelomata. It will take much more than shopping single-sequence trees to refute the theory of body plan evolution.

But what about morphology? Peterson and Eernisse (2001) collected the evidence on the state of 138 morphological characters and analyzed the matrix of these characters using maximum parsimony methods, with resampling of taxa and characters. Their results supported monophyly of Ecdysozoa. The authors followed the "total evidence" approach, being agnostic about the evolutionary theories and modes of character evolution and admitting some degree of arbitrariness in how they coded the character states. I was interested in finding out which of their characters gave the most specific contribution to the monophyly of Ecdysozoa. For that purpose, I examined their Table 1 in detail.

Among 138 characters used by Peterson and Eernisse, at least 28 were completely uninformative with regard to Ecdysozoa and Lochotrophozoa, in that they changed states only within groups external to these lineages (i.e., either within the groups basal to Eubilateria or within Deuterostomes). I examined the remaining 110 characters using the compatibility approach, asking which characters gave support to the Ecdysozoa hypothesis, which to the Coelomata hypothesis (or, as a proxy to it, the Acoelomata clade, in which flatworms joined with roundworms), and which ones were not able to distinguish the two hypotheses. To do so, I asked, for each character, whether it was synapomorphic or nested within Ecdysozoa, within Acoelomata (defined by the union of their Plathyzoa and Cycloneuralia; i.e., roughly flatworms and roundworms, respectively), within both, or not at all. If a character state is synapomorphic or nested within one group, it supports this group. Each character may support either one grouping, in which case this character is retained, or both groupings, in which case it is dismissed. This test inevitably simplifies the picture, but nevertheless it provides an initial overview of the problem.

The results are shown in Table 12.1 (full matrix and intermediate steps of the analysis are available on my Web site, http://research.stowers-institute.org/bioinfo). The numeric advantage seems to be with Ecdysozoa: Eleven characters support this clade, and only six support Coelomata. However, if we look closer at what the traits really are, the result is turned on its head. Indeed, traits 44 and 108—lack of neoblast and frontal complex—may not be synapomorphic at all; they are probably ancient states. Traits 78–80 seem to be correlated and go back to the defining trait of rigid exoskeleton that requires molting—that is, the initial hypothesis for which independent morphological evidence is sought. Moreover, traits 10, 13–14, 15, and, probably, 19 essentially say the same thing, namely that nematodes and arthropods do not have cilia. This is most likely a direct consequence of the exoskeleton. For animals with ciliated epithelia, the type of cilia may be an important phylogenetic marker, but for animals without cilia this is essentially one character counted as four or five.

Thus, analysis of morphology, when taken beyond the "bag of characters," fails to recover any traits common to Ecdysozoa and not directly related to molting. The support for this clade

Table 12.1. Morphological and Developmental Characters Informative with Regard to Either Ecdysozoa or Coelomata Hypotheses[a]

Trait	Compatible with Ecdysozoa	Compatible with Coelomata or Acoelomata	Trait description
13	Yes	No	Ciliated epidermis
14	Yes	No	Densely multiciliated epidermis
15	Yes	No	Step in cilia
19	Yes	No	Spermatozoa with accessory centriole
42	No	Yes	Teloblastic segmentation
44	Yes	No	Neoblasts
78	Yes	No	Cuticle with chitin
79	Yes	No	Trilaminate epicuticle
80	Yes	No	Trilayered cuticle
86	No	Yes	Head separated into three segments
91	No	Yes	Food modified with limbs
104	No	Yes	Circumesophageal nerve ring
105	No	Yes	Dorsal and ventral nerve cords
108	Yes	No	Frontal comples
109	No	Yes	Tanycytes

[a]Raw data were taken from Peterson and Eernisse (2001) and filtered as described in the text.

in Peterson and Eernisse's trees may be explained, in part, by counting lack of cilia as five independent traits—now, that's a long branch! Truly, "unweighted parsimony is stupid" (Cavalier-Smith, 2002a).

Finally, what about the molting process? Perhaps this is indeed a morphological and developmental synapomorphy—a derived trait arisen only once in evolution? I have my doubts. What molts in arthropods and what molts in nematodes are biochemically different entities—rigid shell made mostly of polysaccharide chitin in the former, and flexible cuticle on the foundation of protein collagen, encrusted on the outside by thin layers of glycoproteins and lipids/glycolipids, in the latter. Chitin in nematodes has been detected only in pharynx (where, in fairness, it behaves similarly to the arthropod exoskeleton, being shed and replaced with each molt) and in eggshells. Several decades of research did identify only a few shared factors of molting in nematodes and arthropods. For example, ecdysones were found in both *Drosophila* and *C. elegans*, and yet, ortholog of fly ecdysone receptors is missing from the worm. All this begs the question whether molting in invertebrates is monophyletic.

Analysis of molecular characters that make up the molting pathways would be of help in answering this question. This can initially be done without much regard to sequence similarity or evolutionary rate by high-throughput genetic and genomic assays in worm and fly. When many genetic factors are found, this can be supplemented with the power of sequence analysis, looking for fast-evolving and slow-evolving homologs of each factor in other species. A study of exactly this type was published as I was writing this chapter (Frand *et al.*, 2005). A genomewide RNAi screen in *C. elegans* was done, and various molting phenotypes resulting from knock-down of a specific worm gene were recorded. Between 50 and 100 genes had high penetrance and phenotype highly specific to molting. One-fifth of them had no orthologs in fruit fly, and most of those were worm specific (i.e., they also lacked paralogs in *Drosophila* and other metazoa). There are only two candidate synapomorphies: Nuclear receptors noah-1 and noah-2 are orthologous in worm and fly but missing in humans. Time will tell whether these candidates hold against sequencing of other invertebrate genomes (finding them in nonmolting animals would not help their case). Most other molting factors were found in all three species, sometime in complex paralogous or co-orthologous relationships. Interestingly, one gene involved

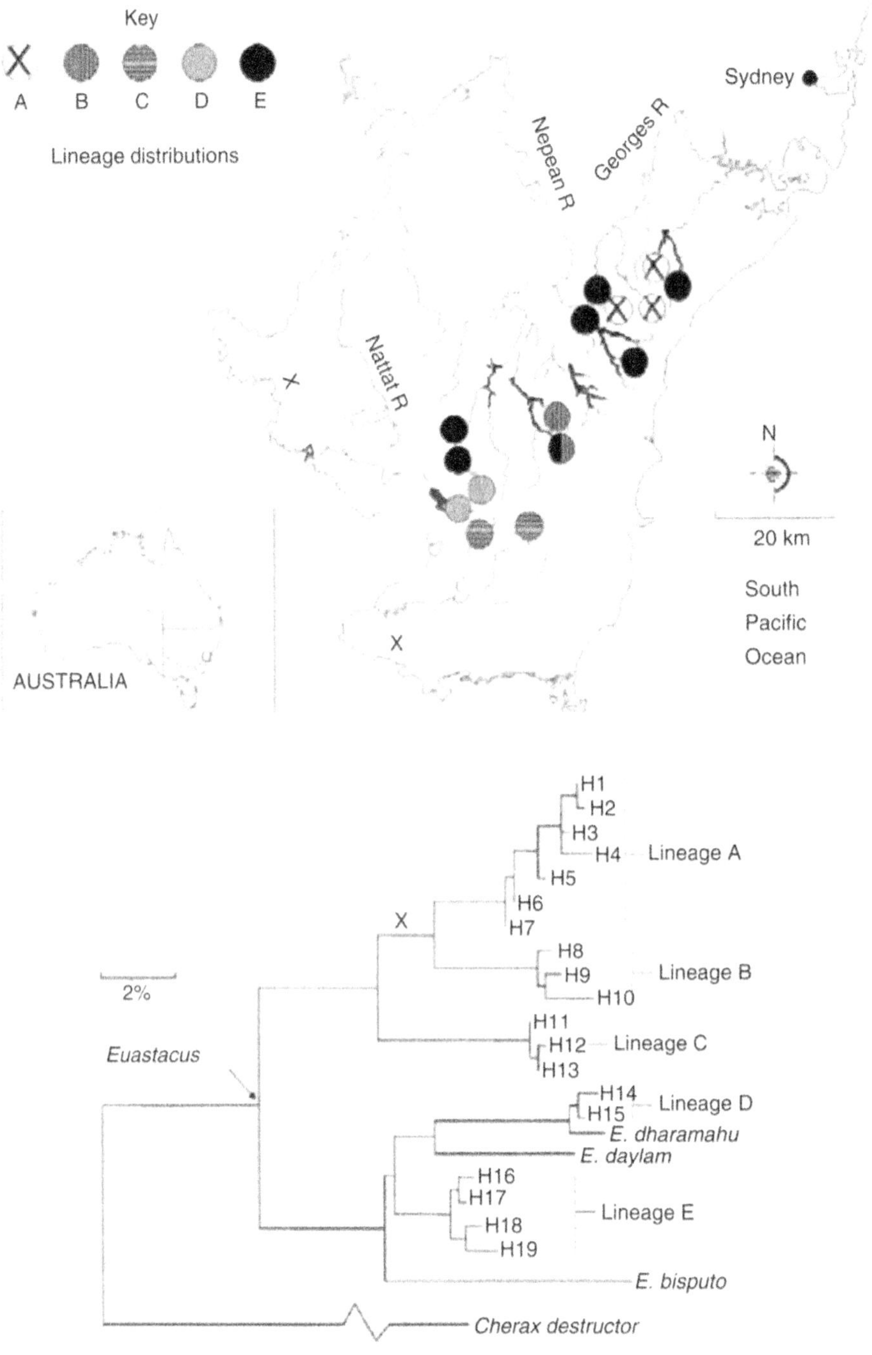

Figure 12.3. Use of phylogenetic trees to assess the strategy of biodiversity preservation. Rapid loss of habitats for lineage A of the spiny crayfish *Euastacus* puts a premium on preservation of the habitats of lineage B, as otherwise a deeper clade (*small x*) will also be lost. Reprinted from Faith and Baker (2006) under Creative Commons License.

in molting (F20G4.1, an uncharacterized protein containing WD40 repeats) is shared by worms and humans, to the exclusion of the fly. There is no physiological basis for such a phyletic pattern, which, I think, is explained by random lineage-specific loss. What is interesting, however, is that the rate of discovery of molting-related worm–fly synapomorphies thus far barely exceeds the occurrence of this random pattern. Thus, the initial outline of genetic blueprint for molting is not lending any support for the ecdysozoan hypothesis.

In this chapter, we discussed phylogenies of deep clades, such as major groups of prokaryotes or invertebrate animals. These trees are genuinely important for fundamental evolutionary biology, but can they, too, be of more immediate practical use? When our 1998 article on 43 quartets of orthologs was under revision, some colleagues and even one reviewer of our manuscript were concerned about social repercussions of Ecdysozoa and Coelomata. They thought that the order of branching in Metazoa may affect the choice of model system for study of human diseases. The argument was approximately as follows: If Coelomata are real, then flies are closer to humans, and fly geneticists may get an advantage in funding over the worm geneticists; however, if Ecdysozoa are the true clade, then flies and worms are equidistant from humans, and postdocs from worm labs have as good of job prospects as those from the fly labs. Under the same logic, researchers studying the biology of Archaea should be fighting tooth and nail for acceptance of Neomura because this would nest the archaean clade into economically much more important actinomycetes. These considerations, of course, are as ridiculous as they sound.

However, similar considerations regarding the topology of the Tree of Life are indeed relevant to some areas of public policy, such as protection of biodiversity on Earth. Many problems in this domain can be cast as the problem of statistical sampling over the evolutionary tree, followed by optimization of costs of species' preservation. I refer the interested reader to the insightful paper of Daniel Faith on this theme (Faith and Baker, 2006; Fig. 12.3).

13

Minimal and Ancestral Genomes

This chapter is about reconstruction of two types of genetic systems that may seem similar, and are often discussed together, but that are in fact quite different. One is ancestral genome, which is the inference of the genetic makeup of a common ancestor of two or more contemporary genomes. For example, we may be interested in reconstructing the genome of the last common ancestor of all known yeasts, all spirochetes, or indeed, of all living forms currently known on Earth. The other construct is minimal genome, which is the smallest gene set capable of sustaining itself.

The first construct is one of the modern applications of the research program set forth by Pauling and Zuckerkandl (see Chapter 1). We compare genes (and gene products) of the existing species and want to infer the set of genes that their ancestor had ("what the ancestor was"), as well as the process that is required to transform this ancestor into the currently living species. Ancestral genomes are relative, in the sense that different groups of the existing species will have different last common ancestors. On the other hand, there was probably only one last common ancestor of all life on Earth. At a meeting in Les Treilles, France, in 1996, Christos Ouzounis (then at Stanford Research Institute and currently at the Centre for Research and Technology Hellas) proposed the name last universal common ancestor (LUCA) for it.

The second construct requires more explanation. Minimal genome is an object of genetic engineering, or "synthetic biology." Most often, the "minimal genome approaches" assume two things. First, we are interested in a minimal genome that sustains a modern-type cell. That is, it must get by with genes that have homologs in some existing species, so we are not considering genes that are completely unheard of, i.e., genes that have no homologs or at least analogs in the sequence databases. Second, we supply the minimal cell with nutrients that we define ourselves. Thus, we do not concern ourselves with any biochemistry that we cannot provide in the laboratory. The minimal genome is also relative, in the sense that not all its genes are required under all growth conditions: Modifications of the growth medium, for example, may add or remove genes. To put together the list of genes that are expected to enable functioning of minimal genome, we use a computer, but the proof of "minimality" has to come from the lab.

Neither of the minimal genome assumptions is set in stone. We can speculate about genetic systems completely different from those that we see in the extant organisms. We can also try to construct genes, proteins, and other molecules that have never been found in nature [for examples of nonnatural nucleotides and amino acids incorporated into cell gene expression program, see Martinot and Benner (2004), Wu *et al.* (2004), Deiters and Schultz (2005), and Wang *et al.* (2006); for more general discussion of synthetic biology and the types of new chemistries it might be able to accommodate, see Benner *et al.* (2004) and Benner and Sismour (2005)].

All these projects rely on building models, and all models can be classified into several broad types, differing in their general temporal direction. In one approach ("forward in time"), we may use the knowledge about the earlier, perhaps prebiotic, stages of Earth existence, and about possibilities of inorganic and organic chemistry. For example, we can ask which conditions on Earth favored the origin of life, what this primitive life might look like, and what functions ancestral genomes may have possessed. This same list of functions is also relevant for defining the list of parts that minimal genome should have (although, again, there is no requirement for a minimal genome to survive in a habitat that existed at the dawn of life on Earth; present-day minimal genomes live on other, better defined media).

The other approach ("backward in time") takes the knowledge of the contemporary genomes and infers earlier life-forms using some sort of evolutionary model, and it also uses genome comparison to identify those components of present-day cells that are indispensable for life and have to be included in a minimal genome. Obviously, completely sequenced genomes are particularly useful for the approaches of this second type. (Synthetic biology, perhaps, should be classified as the "sideways in time" approach.)

Comparative genomics helps to define the gene content of the ancestral genomes, and it is also useful for understanding the gene composition of minimal genomes. Many of the computational approaches to inferring the ancestral genomes and to building minimal gene sets are essentially the same, but this does not mean that the two constructs are interchangeable: It is important to not confuse one with the other.

In fact, it is not too difficult to avoid the confusion. All we have to do is define two conditions, one for ancestral and one for minimal genome. The condition for the ancestral genome is the phylogenetic position of the ancestor that we are trying to reconstruct. In the rest of this chapter, of all ancestors of different phylogenetic clades we will be most interested in the last common ancestor of all living forms, the LUCA. The condition for minimal genome is the defined environment in which we want a cell with minimal genome to survive.

Given these two conditions, every gene belongs to exactly one of the following categories: (1) minimal and ancestral, (2) minimal but not ancestral, (3) ancestral but not minimal, and (4) neither minimal nor ancestral. Let us take a closer look at each of these categories.

Minimal and ancestral genes (Min + Anc+): These genes are required for sustaining a modern-type cell (as noted, on a given growth medium), and their orthologs must have been present in the LUCA. Ribosomal proteins, some aminoacyl-tRNA synthetases, components of DNA-dependent RNA polymerase, and diverse other proteins, including many metabolic enzymes, belong to this class.

Minimal but not ancestral (Min+ Anc−): These genes are required for survival of the modern-type cell, but their orthologs were not found in LUCA. In some cases, the function was missing in LUCA altogether. For example, one modern hypothesis holds that LUCA had an RNA genome and did not require the enzymes of DNA replication. If this is correct, then the main processive DNA polymerase has been recruited twice in the history of life—once in the bacterial lineage and another time in archaeo/eukaryal lineages. In other cases, a gene displacement (DOG; see Chapter 6) removes the ancestral gene and replaces it with an isofunctional gene, as may have happened with ancestral flavin-dependent thymidylate synthase, which was likely present in LUCA but is replaced by folate-dependent thymidylate synthase in most living species (and, for good measure, in many viruses). The irony of placing thymidylate biosynthesis into LUCA while also considering a hypothesis that the same ancestor had no DNA is not lost on me. I will discuss this problem later in this chapter.

Ancestral but not minimal (Min– Anc+): There is no requirement, nor any evidence, that LUCA had minimal genome, which contained the handful of genes necessary and sufficient for survival in LUCA habitat. On the contrary, many authors, particularly Steven Benner (of the Swiss Polytechnic Institute in the 1980s and 1990s, when much of his relevant work was published, and currently of the University of Florida), have developed a well-reasoned argument that LUCA was probably metabolically rich (Benner *et al.*, 1987, 1989, 1993). For example, LUCA most likely contained a considerable number of enzymes for biosynthesis of amino acids *de novo*. But if the growth medium for minimal genome contains some amino acids, then the enzymes for their biosynthesis are not part of minimal cell, and their genes are not part of minimal genome on that medium.

Neither minimal nor ancestral (Min–Anc–): Most of the known genes probably belong to this category. Such are all genes gained in individual lineages since the LUCA. Their functions were either not needed in LUCA or may have been played by different genes there. Minimal genome can get by without them, too.

Thus, minimal genome is a construct that is distinct from LUCA or indeed any genome ancestral to any group of species. Because it is made on the basis of the modern-type genes, it is also distinct from the early forms of life. Furthermore, there was most likely only one LUCA of all life on Earth, but there may be many minimal genomes, each corresponding to a particular growth medium or habitat. Finally, minimal genome is not the same as the smallest of the currently known genomes of the autonomously living microorganisms, even though some of these organisms, notably mycoplasmas, are used in the experimental work on genome-size reduction.

The interest in mycoplasmas, a group of extracellular bacterial parasites of plants and animals, was fueled by sustained argument for their extreme simplicity, made throughout the years by Harold Morowitz at Yale (currently at George Mason University) and some other authors. These bacteria, with their small size, small amount of genomic DNA, and the "unit membrane" (i.e., a single lipid membrane with a thin, if any, peptidoglycan layer), were thought to be the closest approximation of a "minimal cell" among all known living forms (Morowitz and Cleverdon, 1959; Morowitz and Tourtellotte, 1962; Morowitz, 1964, 1984). The possibility that mycoplasmas may also be primitive or ancestral cells, however, was put to rest 25 years ago by analysis of ribosomal RNA, which showed that mycoplasmas are derived Gram-positive bacteria (Woese *et al.*, 1980).

The smallest known genome among the mycoplasmas is the 580-kb chromosome of *Mycoplasma genitalium*. It was the second completely sequenced cellular genome, and the gene set encoded by *M. genitalium* has been called "minimal" (Fraser *et al.*, 1995). In fact, it was not known whether it is minimal in any sense.

There are two ways to proceed from the smallest known genome to the minimal genome. One is to employ molecular genetics and try to reduce genome size, deleting one or more genes at a time. The other strategy relies on the computational methods of comparative genomics. I discuss the results of these computer studies first and then return to the wet-lab experiments.

When *M. genitalium* genome sequence was published in 1995, the only other fully sequenced genome was *Haemophilus influenzae*. There are 468 protein coding genes in *M. genitalium* and 1711 protein coding genes in *H. influenzae* (genes that code for functional RNA other than mRNA also count, of course; some of those are discussed later).

Eugene Koonin and I decided that the set of genes shared between these two species may be a better approximation of the minimal genome than the actual genome of *M. genitalium* (Mushegian and Koonin, 1996a). One reason we thought this to be the case was that Gram-positive bacteria (the clade that includes mycoplasmas) and Proteobacteria (the clade that includes *Haemophilus*) separated a very long time ago, at least 1.2 Gy. Since their divergence, genomes of these species must have had ample time to gain and lose genes, so the orthologs that are still shared by the two species are there because their presence is most likely strongly selected for: Perhaps life without them is not sustainable.

On the other hand, if long absolute times since the split and short generation time in both bacteria resulted in significant sequence divergence, then the orthologous genes might be there, but we may be unable to recognize their relationship. Thus, substantial evolutionary distance between two fully sequenced genomes may suggest the minimal gene set of shared orthologs, if only we were able to overcome the effect of this same distance on sequence similarity.

The other reason to expect that orthologous genes in *H. influenzae* and *M. genitalium* approximate minimal genome was the observation that both bacteria had substantially reduced their repertoire of biosynthetic enzymes, as they came to rely on the host for many classes of nutrients. We thought that the genes that were not deleted in both species are more likely to be strictly required for cell function. (On the other hand, some of the shared gene products might be parallel or convergent adaptations to this lifestyle, for example, virulence factors.)

When all sequence comparisons were done, we found 244 shared orthologous genes. Thus, despite large evolutionary distance (and in agreement with the first fact of comparative genomics, discussed in Chapter 5), more than half of the smaller genome had recognizable orthologs in the larger genome. Ten years later, we know of several orthologs that we missed in that analysis because of low sequence similarity, but I think that at least 90% of all shared orthologs in these two species have been identified already in 1996.

A few features of the shared set of orthologs were obvious: The minimal set built on the basis of *H. influenzae–M. genitalium* comparison would not have a gene if it is missing from any one of the two genomes. For example, *M. genitalium* has no genes coding for the enzymes of citric acid cycle and no genes coding biosynthesis of fatty acids or any amino acids: Accordingly, minimal genome would not have any of those pathways. But we attempted to reduce this shared set even further by eliminating the "parasitism-specific" genes. Function prediction suggested two genes potentially involved in the interaction with human host, and we deleted them from the minimal gene set.

This turned out to be an overcorrection. One gene, putative hemolysin, is most likely indeed a factor involved in adaptation to parasitic lifestyle (iron is a limiting nutrient for parasitic bacteria that live in humans and animals, so most of them have evolved strategies of scavenging iron, for example, by breaking open red blood cells). But the other, an endopeptidase prototyped by *Escherichia coli ygjD* gene product (MG046 in *M. genitalium*, COG00533), should have stayed in the minimal set. We deleted it because it contained a leader peptide, and some of its homologs in other species were annotated as sialoglycoproteases. We concluded, most likely erroneously, that this protein may be involved in the interaction of parasite with the sialylated extracellular proteins of the host. Later genome sequencing, however, indicated that this is one of extremely widespread and well-conserved proteins: in the latest release of the COG database, each genome had an ortholog of this gene. The exact function of this metalloprotease is still not known with certainty, but it is an essential protein in *E. coli* and *Bacillus subtilis* (Arigoni *et al.*, 1998).

The other small handful of proteins that we eliminated from the list of orthologs were three components of a putative phosphotransferase system (PTS)—a specialized system for simultaneous uptake and phosphorylation of sugar molecules. The set of orthologs was not sufficient for making the full, functioning PTS complex, and a minimal gene set already contained a predicted sugar permease and a few predicted sugar kinases, which would achieve the same result as the PTS. Specificity of the transporter may be a problem, of course, but we noted that the transport proteins included in the minimal gene set tended to be of broad specificity.

For example, the only amino acid transport system that survived the filter of shared orthology turned out to import nonspecific oligopeptides, thus elegantly solving the problem of supplying a variety of amino acids for protein biosynthesis and other processes.

There was also another systematic source of missed orthologs. Operational definition of the orthologs started with the symmetrical best matches between *H. influenzae* and *M. genitalium* proteins (the seeds of the algorithm that would become the main way to define COG in later work from Koonin's group; see Chapter 5). However, after provisional lists were assembled, we analyzed the phylogenetic tree of each family, including all homologs that could be found in the database, and compared the topology of the tree with what was thought to be the order of branching in the tree of life [i.e., Woesean topology ((AE)B)]. We wanted to see that the two bacterial proteins were separated from each other by a smaller number of branching events than either of them was from what we thought was an out-group. In particular, if one of the bacterial proteins was too close to an archaeal ortholog (e.g., one archaeal and one bacterial protein formed a clade to the exclusion of the other bacterial protein, or the order of speciation was not well resolved), this gene would not make the list of 244. Several proteins, including three glycolytic enzymes, were in this category. Of course, little did we know at the time that phylogenies operational genes by and large did not follow the Woesean topology and instead tended to be ((AB)E). This was not discovered until later in 1996 (published the next year; Koonin *et al.*, 1997), when the complete sequence of *M. janaschii* was properly analyzed (see Chapter 11).

The reader will notice that, contrary to what I proposed previously, in our 1996 work we did not fix the contents of the growth medium in advance. Instead, we decided to examine whether the properties of the minimal organism, including its nutritional requirements, could be deduced from the minimal gene set that we had thus far. In order to do that, we related the orthologs to the map of biochemical pathways, as they were known at the time. Of course, very little biochemical study had been done for either of the "parental" genomes of the minimal gene set, but there were model organisms with well-studied biochemistry related to each parent: gammaproteobacteria *E. coli* and *Salmonella*, of a low-GC Gram-positive bacterium *B. subtilis*, and partial biochemical information about a few mycoplasmas.

When the orthologs found in *H. influenzae* and *M. genitalium* proteomes were superimposed on the composite metabolic map, we saw that several pathways were present in an incomplete form. For example, not all enzymes of glycolysis were among the products of minimal gene set, and if we brought the "archaeal/bacterial" glycolytic enzymes back into the minimal gene list, then it became almost complete, but phosphoglycerate mutase remained at large. Similarly, all functions needed for salvage of all nucleotides were present, except for nucleoside diphosphate kinase. These "gaps" could be explained in one of three ways: A function could be truly missing in both "parents"; it was present in both species, performed by orthologous proteins, but we were unable to recognize their orthology; or the same function was performed by proteins that were not orthologous at all.

To determine whether we could explain some of these "missing links" in one of these three ways, we did the following. First, we asked whether the two genomes contained nonorthologous proteins with the same predicted function. We found that this was indeed the case: for example, *H. influenzae* and *M. genitalium* each had a gene coding for a protein with phosphoglycerate mutase activity, but to the best of our ability to detect distant sequence relationships, we could not find any evidence of similar sequence or structure in these two isofunctional proteins (later studies indicated that there indeed was no similarity; see Chapter 6). In another example, two predicted lipoate–protein ligases not orthologous to one another were found, one in each genome: *H. influenzae* had an ortholog of *E. coli* LplA, and *M. genitalium* contained the ortholog of the second *E. coli* ligase, LipB. Thus came about the idea of nonorthologous gene displacement (Koonin *et al.*, 1996 and Chapter 6).

Second, we searched for candidate gene displacements (i.e., the predicted proteins that could possess the needed activity) using analysis of sequence homology and early version of genomic context approaches (see Chapters 7 and 8). Not all of our guesses about these missing links were equally lucky. For example, RNAase H was presumed missing in *M. genitalium*; this was the only unaccounted-for enzyme of basic DNA replication machinery. Because it is indispensable in replication, due to the need to remove the RNA primers left behind by primase at the replication initiation sites, we nominated gene MG262, which contains conserved phosphohydrolase motifs, as a displacer of RNAase H. This answer to the problem was wrong: The real, orthologous RNAase H was encoded by *M. genitalium* all along (MG099; COG00164), and we simply missed it because of low sequence similarity, not tractable by the ungapped, noniterative BLAST program available to us at the time. Today, finding members of this COG is easier; this is one of the most pervasive COGs, found in more than 95% of all species included in the COG database.

On the other hand, nucleoside diphosphate kinase (NDK), encoded by the essential, exceptionally well-conserved gene, was missing in *M. genitalium* (today we know that there are also no homologs of this gene in other mycoplasmas as well as in such Gram-positive bacteria as *Fusobacterium nucleatum* and one species of *Clostridia*). The NDK activity, however, is essential in every cell because it supplies triphosphates for DNA and RNA synthesis; deficiency of this enzyme in humans is responsible for metastases of several tumors (Ouatas *et al.*, 2003). We proposed MG268 (COG01428) as a candidate displacement because of its distant similarity to deoxynucleotide kinases. On the other hand, it has been shown that several overexpressed and partially purified small-molecule kinases in mycoplasmas can phosphorylate nucleoside diphosphates (Pollack *et al.*, 2002). Whether they moonlight in this way *in vivo*, however, remains unknown, and MG268 could still turn out to be the principal NDK in *Mycoplasma*. These hits and misses, however, do not change the basic fact that orthologous genes are displaced in evolution, and that any method of minimal genome reconstruction that relies only on counting shared orthologs will underestimate the minimal gene set.

Returning to the picture that emerged after supplementing the set of orthologs by proteins that have been displaced between the two species, we found that the resulting amended set of 256 genes could provide for a coherent enough, if pared down, metabolism. Nutritional requirements of a minimal genome would be quite extensive, including all amino acids, nucleobases, at least some sugars (although transketolase was available to make longer skeleton monosaccharides from the shorter ones), and fatty acids. Coenzymes represented a special and interesting case. Minimal gene set contained enzymes dependent on the following cofactors: NAD(P), FAD, *S*-adenosyl methionine, lipoate, pyridoxal, thiamine, folate, and coenzyme A. None of these coenzymes can be synthesized *de novo* by *M. genitalium* nor, consequently, by minimal genome. But minimal genome included most of the enzymes required for the last step of coenzyme assembly or activation, including FAD synthase that conjugates flavine and adenine nucleotides, SAM synthase that makes *S*-adenosyl methionine from ATP and methionine, lipoate–protein ligase, and dephospho-CoA kinase. There were also a few putative kinases that might activate pyridoxal and thiamine by phosphorylation. The complete set of enzymes enabling C1 turnover by folate was also present (this cycle is the source of functional groups for two important syntheses, those of formylmethionine and of thymidylate).

Thus, the gene set identifies the nutritional requirements of the organism with minimal genome. Informational genes (see Chapter 11) appear to have a much higher retention rate in the minimal genome than operational genes: The translation apparatus of *M. genitalium* is placed into the minimal genome without much gene loss. The main transcription enzyme, multisubunit RNA polymerase, also makes it into the minimal gene set in its entirety (except for the tiny sigma subunit that has not been discovered in any mycoplasmas yet), as do the sigma factor and two RNA elongation factors. In contrast, none of specific transcription factors made it to the minimal genome, underscoring the difference in transcription mechanisms in mycoplasmas and proteobacteria (however, it is quite likely that there are still undiscovered DOGs in this class of proteins). Replication machinery of *Mycoplasma* remains intact in the minimal genome, although the number of functional DNA repair systems is strongly reduced.

In the decade that passed since that first publication on the minimal genome, researchers gained access to many more genome sequences. If we tried to repeat the experiment today, we would see the same trends as 10 years ago, but magnified in interesting ways. The two first fully sequenced genomes comprised 2300 genes, approximately 250 of which (50% of the smaller and 14% of the larger genome) were shared between both genomes, and perhaps at least 20 more genes were isofunctional, mutually displacing genes. The last release of the National Center for Biotechnology Information (NCBI) COG database includes genomes of archaea, bacteria, and unicellular eukaryotes, and they contain approximately 10^7 genes, most of which belong to one of ~14,000 COGs. Drastically, only 50 of these COGs are found in all genomes without exception.

The list of COGs that are found in all completely sequenced genomes is of some interest (Table 13.1). Approximately two-thirds of these proteins are involved in translation, and only

Table 13.1. COGs Found in All Completely Sequenced Unicellular Genomes or in at Least 95% of Them[a]

COG No.	Functional category	Molecular function
	COGs found in all completely sequenced species	
COG00037	Cell cycle	Predicted ATPase of the PP-loop superfamily implicated in cell cycle control
COG00358	DNA replication, recombination, and repair	DNA primase (bacterial type)
COG01109	Monosaccharide metabolism	Phosphomannomutase
COG00528	Nucleotide metabolism	Uridylate kinase
COG00533	Protein folding and repair	Metal-dependent proteases with possible chaperone activity
COG00492	Protein folding and repair	Thioredoxin reductase
COG00201	Protein secretion	Preprotein translocase subunit SecY
COG00541	Protein secretion	Signal recognition particle GTPase
COG00552	Protein secretion	Signal recognition particle GTPase
COG00018	RNA metabolism and translation	Arginyl-tRNA synthetase
COG00030	RNA metabolism and translation	Dimethyladenosine transferase (rRNA methylation)
COG00008	RNA metabolism and translation	Glutamyl- and glutaminyl-tRNA synthetases
COG00124	RNA metabolism and translation	Histidyl-tRNA synthetase
COG00143	RNA metabolism and translation	Methionyl-tRNA synthetase
COG00016	RNA metabolism and translation	Phenylalanyl-tRNA synthetase alpha subunit
COG00012	RNA metabolism and translation	Predicted GTPase; probable translation factor
COG00442	RNA metabolism and translation	Prolyl-tRNA synthetase
COG00081	RNA metabolism and translation	Ribosomal protein L1
COG00244	RNA metabolism and translation	Ribosomal protein L10
COG00080	RNA metabolism and translation	Ribosomal protein L11
COG00102	RNA metabolism and translation	Ribosomal protein L13
COG00093	RNA metabolism and translation	Ribosomal protein L14
COG00200	RNA metabolism and translation	Ribosomal protein L15
COG00197	RNA metabolism and translation	Ribosomal protein L16/L10E
COG00256	RNA metabolism and translation	Ribosomal protein L18
COG00198	RNA metabolism and translation	Ribosomal protein L24
COG00087	RNA metabolism and translation	Ribosomal protein L3

Table 13.1 COGs Found in All Completely Sequenced Unicellular Genomes or in at Least 95% of Them[a]

COG No.	Functional category	Molecular function
COG00088	RNA metabolism and translation	Ribosomal protein L4
COG00094	RNA metabolism and translation	Ribosomal protein L5
COG00097	RNA metabolism and translation	Ribosomal protein L6P/L9E
COG00100	RNA metabolism and translation	Ribosomal protein S11
COG00099	RNA metabolism and translation	Ribosomal protein S13
COG00052	RNA metabolism and translation	Ribosomal protein S2
COG00522	RNA metabolism and translation	Ribosomal protein S4 and related proteins
COG00049	RNA metabolism and translation	Ribosomal protein S7
COG00096	RNA metabolism and translation	Ribosomal protein S8
COG00103	RNA metabolism and translation	Ribosomal protein S9
COG00172	RNA metabolism and translation	Seryl-tRNA synthetase
COG00441	RNA metabolism and translation	Threonyl-tRNA synthetase
COG00231	RNA metabolism and translation	Translation elongation factor P (EF-P)/translation initiation factor 5A (eIF-5A)
COG00480	RNA metabolism and translation	Translation elongation factors (GTPases)
COG00532	RNA metabolism and translation	Translation initiation factor 2 (IF-2; GTPase)
COG00180	RNA metabolism and translation	Tryptophanyl-tRNA synthetase
COG00162	RNA metabolism and translation	Tyrosyl-tRNA synthetase
COG00525	RNA metabolism and translation	Valyl-tRNA synthetase
COG00202	Transcription	DNA-directed RNA polymerase; alpha subunit/40 kDa subunit
COG00085	Transcription	DNA-directed RNA polymerase; beta subunit/140 kDa subunit
COG00086	Transcription	DNA-directed RNA polymerase; beta′ subunit/160 kDa subunit
COG00250	Transcription	Transcription antiterminator
COG00195	Transcription	Transcription elongation factor
	COGs found in 95–99% of the completely sequenced species	
COG00128	Amino acid metabolism	5-Enolpyruvylshikimate-3-phosphate synthase
COG00601	Amino acid metabolism	ABC-type dipeptide/oligopeptide/nickel transport systems; permease components
COG01173	Amino acid metabolism	ABC-type dipeptide/oligopeptide/nickel transport systems; permease components
COG00436	Amino acid metabolism	Aspartate/tyrosine/aromatic aminotransferase
COG00136	Amino acid metabolism	Aspartate-semialdehyde dehydrogenase
COG00527	Amino acid metabolism	Aspartokinases
COG00082	Amino acid metabolism	Chorismate synthase
COG00329	Amino acid metabolism	Dihydrodipicolinate synthase/*N*-acetylneuraminate lyase
COG00112	Amino acid metabolism	Glycine/serine hydroxymethyltransferase
COG00520	Amino acid metabolism	Selenocysteine lyase
COG00169	Amino acid metabolism	Shikimate 5-dehydrogenase
COG00006	Amino acid metabolism	Xaa-Pro aminopeptidase
COG00206	Cell cycle	Cell division GTPase
COG00190	Coenzyme metabolism	5;10-Methylene-tetrahydrofolate dehydrogenase/methenyl tetrahydrofolate cyclohydrolase
COG00237	Coenzyme metabolism	Dephospho-CoA kinase
COG00294	Coenzyme metabolism	Dihydropteroate synthase and related enzymes
COG00171	Coenzyme metabolism	NAD synthase
COG01136	Detoxification	ABC-type antimicrobial peptide transport system; ATPase component
COG01132	Detoxification	ABC-type multidrug transport system; ATPase and permease components

Continued

Table 13.1. COGs Found in All Completely Sequenced Unicellular Genomes or in at Least 95% of Them[a]—Cont'd

COG No.	Functional category	Molecular function
COG00258	DNA replication, recombination, and repair	5′-3′ exonuclease (including N-terminal domain of PolI)
COG00470	DNA replication, recombination, and repair	ATPase involved in DNA replication
COG00178	DNA replication, recombination, and repair	Excinuclease ATPase subunit
COG00556	DNA replication, recombination, and repair	Helicase subunit of the DNA excision repair complex
COG00582	DNA replication, recombination, and repair	Integrase
COG00350	DNA replication, recombination, and repair	Methylated DNA–protein cysteine methyltransferase
COG00084	DNA replication, recombination, and repair	Mg-dependent Dnase
COG00494	DNA replication, recombination, and repair	NTP pyrophosphohydrolases including oxidative damage repair enzymes
COG00322	DNA replication, recombination, and repair	Nuclease subunit of the excinuclease complex
COG00177	DNA replication, recombination, and repair	Predicted *Endo*III-related endonuclease
COG00468	DNA replication, recombination, and repair	RecA/RadA recombinase
COG00164	DNA replication, recombination, and repair	Ribonuclease HII
COG00210	DNA replication, recombination, and repair	Superfamily I DNA and RNA helicases
COG00550	DNA replication, recombination, and repair	Topoisomerase IA
COG00188	DNA replication, recombination, and repair	Type IIA topoisomerase (DNA gyrase/topo II; topoisomerase IV); A subunit
COG00187	DNA replication, recombination, and repair	Type IIA topoisomerase (DNA gyrase/topo II; topoisomerase IV); B subunit
COG00636	Energy supply	F0F1-type ATP synthase; subunit c/Archaeal/vacuolar-type H+-ATPase; subunit K
COG00039	Energy supply	Malate/lactate dehydrogenases
COG01249	Energy supply	Pyruvate/2-oxoglutarate dehydrogenase complex; dihydrolipoamide dehydrogenase (E3) component; and related enzymes
COG01028	Lipid metabolism	Dehydrogenases with different specificities (related to short-chain alcohol dehydrogenases)
COG00142	Lipid metabolism	Geranylgeranyl pyrophosphate synthase
COG00558	Lipid metabolism	Phosphatidylglycerophosphate synthase
COG00020	Lipid metabolism	Undecaprenyl pyrophosphate synthase
COG00575	Lpid metabolism	CDP-diglyceride synthetase
COG00126	Monosaccharide metabolism	3-phosphoglycerate kinase
COG00148	Monosaccharide metabolism	Enolase
COG00697	Monosaccharide metabolism	Permeases of the drug/metabolite transporter (DMT) superfamily
COG00477	Monosaccharide metabolism	Permeases of the major facilitator superfamily
COG00061	Monosaccharide metabolism	Predicted sugar kinase
COG00469	Monosaccharide metabolism	Pyruvate kinase
COG00149	Monosaccharide metabolism	Triosephosphate isomerase
COG00057	Monosaccharide metabolism/ Energy supply	Glyceraldehyde-3-phosphate dehydrogenase/erythrose-4-phosphate dehydrogenase

Table 13.1. COGs Found in All Completely Sequenced Unicellular Genomes or in at Least 95% of Them[a]

COG No.	Functional category	Molecular function
COG00563	Nucleotide metabolism	Adenylate kinase and related kinases
COG00504	Nucleotide metabolism	CTP synthase (UTP-ammonia lyase)
COG00537	Nucleotide metabolism	Diadenosine tetraphosphate (Ap4A) hydrolase and other HIT family hydrolases
COG00167	Nucleotide metabolism	Dihydroorotate dehydrogenase
COG00518	Nucleotide metabolism	GMP synthase; glutamine amidotransferase domain
COG00519	Nucleotide metabolism	GMP synthase; PP-ATPase domain/subunit
COG00105	Nucleotide metabolism	Nucleoside diphosphate kinase
COG00461	Nucleotide metabolism	Orotate phosphoribosyltransferase
COG00284	Nucleotide metabolism	Orotidine-5′-phosphate decarboxylase
COG00462	Nucleotide metabolism	Phosphoribosylpyrophosphate synthetase
COG00209	Nucleotide metabolism	Ribonucleotide reductase; alpha subunit
COG00125	Nucleotide metabolism	Thymidylate kinase
COG00127	Nucleotide metabolism	Xanthosine triphosphate pyrophosphatase
COG00449	Polysaccharide metabolism, including cell wall metabolism in prokaryotes	Glucosamine 6-phosphate synthetase; contains amidotransferase and phosphosugar isomerase domains
COG00438	Polysaccharide metabolism, including cell wall metabolism in prokaryotes	Glycosyltransferase
COG00463	Polysaccharide metabolism, including cell wall metabolism in prokaryotes	Glycosyltransferases involved in cell wall biogenesis
COG00750	Polysaccharide metabolism, including cell wall metabolism in prokaryotes	Predicted membrane-associated Zn-dependent proteases 1
COG00472	Polysaccharide metabolism, including cell wall metabolism in prokaryotes	UDP-*N*-acetylmuramyl pentapeptide phosphotransferase/UDP-*N*-acetylglucosamine-1-phosphate transferase
COG00459	Protein folding and repair	Chaperonin GroEL (HSP60 family)
COG00484	Protein folding and repair	DnaJ class molecular chaperone with C-terminal Zn finger domain
COG01214	Protein folding and repair	Inactive homolog of metal-dependent proteases; putative molecular chaperone
COG00443	Protein folding and repair	Molecular chaperone
COG00576	Protein folding and repair	Molecular chaperone GrpE (heat shock protein)
COG00526	Protein folding and repair	Thiol-disulfide isomerase and thioredoxins
COG00681	Protein secretion	Signal peptidase I
COG00013	RNA metabolism and translation	Alanyl-tRNA synthetase
COG00215	RNA metabolism and translation	Cysteinyl-tRNA synthetase
COG00073	RNA metabolism and translation	EMAP domain
COG00060	RNA metabolism and translation	Isoleucyl-tRNA synthetase
COG00024	RNA metabolism and translation	Methionine aminopeptidase
COG02890	RNA metabolism and translation	Methylase of polypeptide chain release factors
COG00130	RNA metabolism and translation	Pseudouridylate synthase
COG00101	RNA metabolism and translation	Pseudouridylate synthase
COG00009	RNA metabolism and translation	Putative translation factor (SUA5)
COG00343	RNA metabolism and translation	Queuine/archaeosine tRNA-ribosyltransferase
COG00090	RNA metabolism and translation	Ribosomal protein L2
COG00091	RNA metabolism and translation	Ribosomal protein L22
COG00089	RNA metabolism and translation	Ribosomal protein L23
COG00255	RNA metabolism and translation	Ribosomal protein L29
COG00051	RNA metabolism and translation	Ribosomal protein S10
COG00048	RNA metabolism and translation	Ribosomal protein S12
COG00199	RNA metabolism and translation	Ribosomal protein S14
COG00184	RNA metabolism and translation	Ribosomal protein S15P/S13E

Continued

Table 13.1. COGs Found in All Completely Sequenced Unicellular Genomes or in at Least 95% of Them[a] — Cont'd

COG No.	Functional category	Molecular function
COG00186	RNA metabolism and translation	Ribosomal protein S17
COG00185	RNA metabolism and translation	Ribosomal protein S19
COG00092	RNA metabolism and translation	Ribosomal protein S3
COG00098	RNA metabolism and translation	Ribosomal protein S5
COG00361	RNA metabolism and translation	Translation initiation factor 1 (IF-1)
COG00454	Transcription	Histone acetyltransferase HPA2 and related acetyltransferases
COG00517	Unknown	CBS domain
COG00596	Unknown	Predicted hydrolases or acyltransferases (alpha/beta hydrolase superfamily)
COG00500	Unknown	SAM-dependent methyltransferases

[a]Genomes are from the extended version of the unicellular subset of the NCBI COG database (www.ncbi.nlm.nih.gov/COG/grace/uni.html and Yu. Wolf, personal communication), in which Eukarya are represented by a subset of Fungi and Microsporidia.

one biological function, namely basic transcription of DNA, is substantially complete. If we add to the list those genes that are found in more than 95% of all genomes (103 more COGs), the picture becomes more coherent, but not by much. We gain several proteins required for translation, but not all of them; synthesis of membrane lipids is hardly represented at all (distal part of isoprenoid biosynthesis that emerges in the 95% set may have other functions in bacteria); some enzymes of nucleotide and cofactor salvage also come into the picture; and there are very few replication genes, none of them directly concerned with processive synthesis of DNA genome.

The list of omnipresent COGs corresponds to 3.5% of all COGs and to approximately 10% of genes encoded by the smallest free-living genome. Granted, some of the genes may have been missed, either in the process of genome annotation (Nielsen and Krogh, 2005) or even, perhaps, in the process of COG construction. However, more detailed analysis shows that scarcity of the omnipresent genes is not an artifact of gene prediction. When the complete COG database is ranked by the number of genomes that have each COG (Fig. 13.1), there is a near-perfect log-linear dependency across most of the data range. The decay is rapid: Less than 1% of all COGs are found in at least 95% of all genomes, and only approximately 5% are found in more than two-thirds of all genomes. In contrast, the decrease in the COG frequency among the minimal gene set is slow: 93% of the minimal set COGs are found in more than two-thirds of all genomes (Fig. 13.1). Thus, the minimal gene set constructed in 1996 from just two distantly related bacterial genomes was strongly enriched in the universally conserved, omnipresent genes.

It is worth remembering that nearly a half of all species included in the COG database (53 genomes out of 109) are pathogenic bacteria, which have experienced profound reduction of gene content. What happens if we count shared orthologs only in genomes that have more autonomous lifestyles and, on average, larger gene numbers than parasites? I excluded human, animal, and plant parasites from the COG database (as an aside, this removes almost 3000 of the COGs, many of which may be "parasitism-specific" genes and are of great interest for future understanding of the fundamental mechanisms of pathogenesis and identifying potential drug targets). The distribution of COGs that survived the removal of parasitic species was qualitatively the same as with the complete data set (see Fig. 13.1). The properties of the very top percentile, however, were different. If we focus on the proteins present in more than 95%

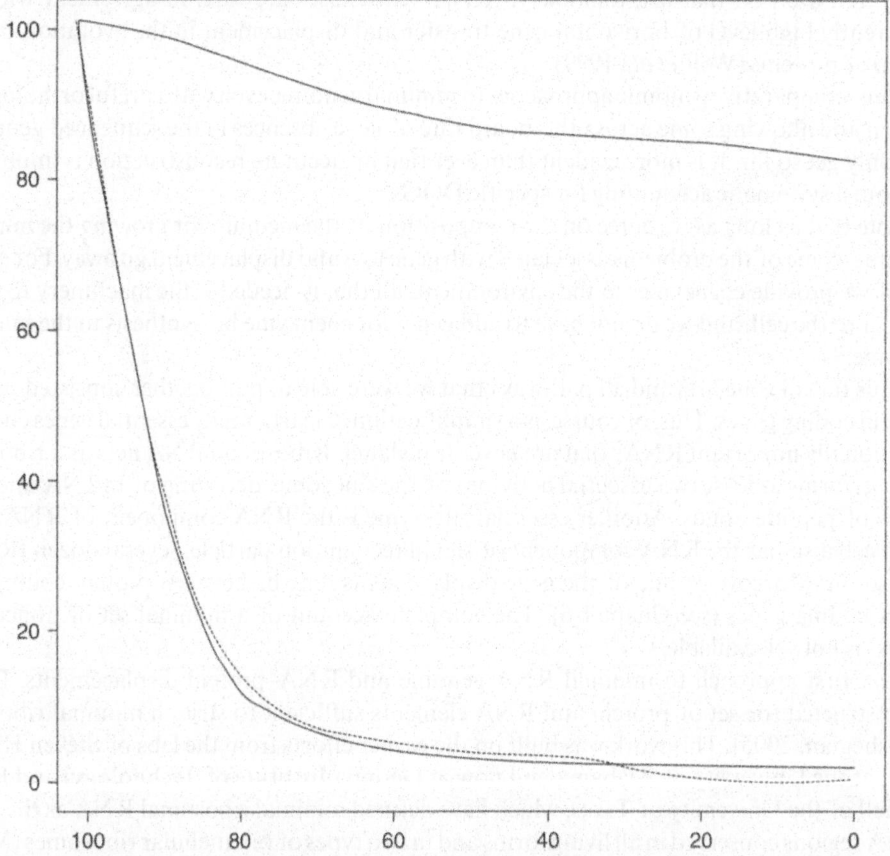

Figure 13.1. Distribution of COGs from the unicellular subset of the NCBI COG database in completely sequenced genomes. (**Horizontal axis**) COGs ranked by their frequency. (**Vertical axis**) Frequency of each COG. Each axis is scaled from 100 to 0%. *Solid and dotted curves* are complete data set (1.4×10^3 COGs in 10 genomes) and COGs that are found in free-living species (1.12×10^3 COGs in 56 genomes), respectively. The *top curve* corresponds to ~260 COGs that belong to the minimal gene set defined by comparison of two completely sequenced bacteria genomes in 1996.

of all free-living species, we find almost twice as many proteins as in the case of the equivalent percentile for the complete data set. Among 98 proteins that are found in almost all free-living species, are all COGs shown in Table 13.1, as well as 50 additional COGs, 40 of which are enzymes for *de novo* biosynthesis or salvage of amino acids, nucleotides, and coenzymes. On the other hand, gain in operational genes is modest: there are no additions among ribosomal proteins, just one new aminoacyl-tRNA synthetase, etc. The reason is quite obvious: the loss of genes in parasites has more impact on operational genes than on informational genes.

We can work our way down the ranked lists of COGs, admitting the most widespread COGs to the minimal set first and then accepting the COGs that are found in increasingly lower proportion of all genomes, on the premise that many of these genes code for essential functions, but in some genomes they are displaced by isofunctional genes. In this way, many classes of functions become better represented, and the corresponding model of minimal metabolism becomes more coherent, with fewer gaps. For example, the top 15th percentile of frequency-ranked COGs consists of 298 genes, 63 of which are involved in translation and include 26 ribosomal proteins, 7 aminoacyl-tRNA synthetases, 5 enzymes of rRNA modification, etc.

Note that even on that list, aminoacyl-tRNA synthetases are rare, in agreement with the apparently high level of horizontal gene transfer and displacement in the evolution of this group of proteins (Wolf *et al.*, 1999).

Thus, comparative genomic approaches to minimal genome, even with careful ortholog definition and allowing some across-the-board rate of gene absences in the sequenced genomes, can only get so far. It is more evident than ever that an accurate reconstruction is impossible without a systematic accounting for specific DOGs.

Note that as long as we agree on the composition of the medium for growing the minimal genome, some of the problems associated with gene loss and displacement go away. For example, if we provide coenzymes in the environment, all that is needed is the machinery to bring them into the cell, and we do not have to add genes for coenzyme biosynthesis to the minimal genome.

Thus far, all gene lists and all pathways that we were able to put together consisted only of protein coding genes. This, or course, is a simplification. Clearly, many essential genes code for functionally important RNAs that are never translated. Ribosomal RNA acts as a ribozyme that performs at least two essential activities of the ribosome, decoding of mRNA and synthesis of peptide bonds. Another essential ribozyme is the RNA component of RNAase P. Each cell also has the RNA component of signal recognition particle, several dozen tRNAs, and so on. Moreover, some of the gene displacements may be between protein coding and RNA coding genes (see Chapter 6). The complete account of a minimal set of noncoding RNAs is not yet available.

As a first approach to minimal RNA genome and RNA–protein displacements, I have reconstructed the set of protein and RNA elements sufficient to sustain minimal ribosome (Mushegian, 2005). This work was built on the earlier efforts from the labs of Steven Harvey (then at the University of Alabama and now at Georgia Institute of Technology) and Robin Guttell of the University of Texas, which have defined minimal ribosomal RNA as the set of rRNA regions conserved in all living forms and in two types of organellular ribosomes (Mears *et al.*, 2002). This approach is obviously very similar to finding shared orthologs that we have used to obtain the minimal protein set. Mears *et al.*, however, dealt with domains of one long rRNA molecule, rather than with individual protein coding genes. This difference is not as great as it might seem: First, the definition of orthologous proteins also involves domain dissection (see Chapter 5), and second, two ribosomal RNAs are known to contain separate functional domains that may have even been separate entities in the past, and some of the currently known RNA genes are even transcribed as several distinct fragments that are subsequently spliced.

There are many lineage-specific indels in all ribosomal RNAs; one can use the alignment of all rRNAs from diverse life-forms and to find all regions that are not deleted in any of these sequences. This intersection of all rRNAs would then be a candidate minimal rRNA. Of course, this would not take account of domain displacements: What if two nonaligned segments of rRNA play the same roles, and one is deleted only if the other is inserted? Fortunately, nature provided a (significant part of) an answer: This minimal, or omnipresent, portion of universally shared rRNA is very close to the pair of extremely deleted rRNAs found in mitochondria of nematodes. As the first approximation, every segment found in every other ribosomal rRNA is also found in nematode mitochondrial rRNAs, and almost every segment missing from an rRNA from any species is also missing from the nematode mitochondrial mRNAs. All told, the small-subunit (SSU) rRNA in nematode mitochondria has lost ~55% of bases compared to a bacterial rRNA, the large-subunit (LSU) rRNA lost three-fourths of all bases compared to bacteria, and mitochondrial 5S RNA is not found in nematodes at all (Mears *et al.*, 2002). The remaining, universally conserved rRNA domains include most of the regions with the known function, particularly the peptidyltransferase center in the large subunit; the decoding sites in the small subunits; the A, P, and E sites of interaction with tRNA; and most of the intersubunit interface, which mediates interaction of the large and small subunits with very little help from ribosomal proteins. Thus, rRNA from any species can be partitioned into three sequence components: regions that can be aligned to nematode mito-rRNA—those may be noncontiguous in the sequence alignment, but pairs of such linear elements are typically brought together to form highly ordered conserved structures in rRNA; those that do not align to, but have a functional equivalent within, mito-rRNA, to serve as connectors between well-conserved elements of sequence and structure; and sequences that are aligned to gaps in the nematode mito-rRNA. The first two components together can be viewed as the minimal rRNA model.

Mears *et al.* (2002) did not have much to say about the protein component of the minimal ribosome other than to note that, according to their calculations, 22 ribosomal proteins had orthologs in "3P2O" (i.e., in three major phylae of life—Bacteria, Archaea, and Eukarya—and in two types of organelles). I made a recount and found that the actual number is not 22 but more than 30. I settled on the exact number of 32, allowing for one or two absences among the 66 genomes that the COG database included at the time. These "extra" components were found by running PSI-BLAST searches to convergence and with more permissive parameters. Some of the "novel" sequences that Mears *et al.* did not include were already annotated in the database as ribosomal proteins, and a few were even studied biochemically. Nonetheless, I thought that not all of these proteins properly belong to the minimal ribosome: An additional requirement must be that they have cognate interaction sites in Min rRNA. The intersection of omnipresent proteins and proteins that bind Min rRNA gives the set of 25 ribosomal proteins, 14 in SSU and 11 in LSU. I called this protein set Min1. Interestingly, attrition of proteins was higher in LSU than in SSU (45% of the SSU proteins were included in Min1, compared to 20% of the LSU proteins), in agreement with the earlier suggestion that proteins are more important for holding together three relatively mobile domains of the SSU RNA, whereas the LSU RNA can pack on its own (Brodersen *et al.*, 2002; Bashan *et al.*, 2003).

Min1, however, did not account for gene displacements. In fact, until very recently, little was known about functions of most ribosomal proteins, so it was not known which ribosomal proteins (nor which rRNA segments) may be isofunctional in different species. By 2004, however, we could already benefit from the high-resolution x-ray structures of complete SSU from the bacterium *Thermus thermophilus* and complete LSUs structures from bacterium *Deinococcus radiodurans* and archaeon *Haloarcula marismortui* (Brodersen *et al.*, 2002; Bashan *et al.*, 2003; Klein *et al.*, 2004). The high-resolution structure of an archaeal SSU, or of any complete ribosome, remains to be determined. But at least with two types of LSU structures in hand, some of the putative displacements can be proposed using the location of each protein with regard to the homologous RNA segments as the proxy of its function.

Ribosomal proteins exist in two main types: Bacterio/organellar and Archaeo/Eukaryal. Specifically, if a ribosomal protein has orthologs in all living species, the phylogenetic tree of such orthologs is (B(AE)), and if a protein is not omnipresent, its phyletic pattern is either B– or -AE (see Chapters 11 and 12 for discussion of this theme and other issues related to the Tree of Life). Therefore, to find genes that replace each other in bacterial/organellar vs. archaeo/eukaryal ribosome, I collected all proteins found either only in Bacteria or only in Archaea/Eukarya and asked whether the two sets included any proteins that share an interaction site on Min-rRNA. This comparison of LSUs from *H. marismortui* and *D. radiodurans* identified several probable displacements in the large subunit, which are listed in Table 13.2. Interestingly, most of them are not one-to-one displacements: There are six distinct locations on the rRNA, with which either nine bacterial or eight archaeal proteins could interact [see the discussion of most of the same displacements by Harms *et al.* (2001) and Klein *et al.* (2004), both of whom, however, examine only functional, not evolutionary, implications].

Table 13.2. Protein–Protein and Protein–RNA Displacements in Minimal and Ancestral Ribosome[a]

Ribosome functional site	Bacterial protein	Archaeo-Eukaryal protein	Comments
Exit tunnel opening	L17	L31E	Both proteins are from the alpha-beta fold class, but the number and arrangement of secondary structure elements are different, and two families do not show sequence similarity.
Exit tunnel interior	L23 L34	L39E L37E	
Helix 25 (common part)	L20 L21	L32E	In the middle of L32E protein, there is a 45-amino-acid stretch forming a twisted loop. It penetrates the ribosome body, interacts with the min-rRNA region of Helix 25, and appears to be taking over the RNA contacts which, in bacteria, are enabled by interacting, extended regions of two proteins, L20 and L21. Both L32E and L21 have many positively charged residues in these regions (as many other ribosomal proteins do), but there is no clear indication of an evolutionary relationship.
P-loop, P-tRNA interaction	L27	L21E	Two proteins belong to the same fold class and share additional specific features, such as the strong twist of the beta sheet. There is, however, no sequence similarity between L27 and L21E.
Helix 75, E-tRNA interaction	L31 L33	L15E L44E	Elongated loopy regions of L15E and L44E contact each other, Helix 75, and tRNA at the exit site. This composite protein structure is structurally and functionally replaced by a single protein, L31, in Bacteria. There remains the globular part of L44E, which also interacts with tRNA at the exit site and is replaced by a smaller globular protein L33 in Bacteria
Intersubunit bridge B6	L19	L24E	L19 and L24E belong to the all-beta fold class but are dissimilar at the sequence level. Both L19 and L24E, however, donate two beta strands in a similar way, to form an interprotein beta sheet with S14.

[a]Modified from Mushegian (2005) by permission of the RNA Society.

Furthermore, some DOGs within ribosome may involve a protein and an RNA; that is, the function of a deleted rRNA element in one clade is taken over by an extra protein in another clade, or an insertion of additional nucleotides in rRNA in one clade obviates a need for a ribosomal protein in another clade. The most relevant observation here concerns the interactions between proteins and a nonminimal region of the LSU RNA. There is a 50-base insert in *Haloarcula* RNA, which extends RNA helix 25. This region interacts with the globular domain of archaeal L32E protein. In bacteria, L32E is displaced by L20 and L21 (see Fig. 3.3), but there is no RNA site in bacteria that would interact with these displacers. Interestingly, however, bacterial L20 makes extensive contacts with the globular portion of L21, which in turn contacts L15. Thus, the Archaea-specific extension of RNA helix 25 and the globular domain of Bacteria-specific L21 may be isofunctional, yet they are obviously not orthologous.

There is less evidence of gene displacement in the SSU because structural corroboration cannot be done for SSU until we have a high-resolution structure from archaea. There are, however, three Bacteria-specific proteins, S6, S16, and S18, which contact Min-rRNA in the SSU, and at least seven proteins that belong to the small ribosomal subunit in Archaea/Eukarya. Some of these may displace each other. Interestingly, SSU may also harbor a possible RNA–protein displacement. It is suggested by the strange case of the S8 protein, which interacts with rRNA mostly via helices 21, 22, and 25, which do not belong to Min rRNA. S8 is found in all major clades (although not in every species) and has extensive interactions with Min1 protein S2 and S17. A bacterial-like S8 homolog is present in nematode and probably works in mitochondrial translation (although, like some other nuclear-encoded mitochondrial ribosomal proteins, it lacks a defined mitochondria-targeting peptide). However, nematode mitochondrial RNA has no S8 interaction site, which therefore has to be replaced by something else—perhaps by extra proteins.

This is the outline of the computational approaches that can be used to construct a minimal genome on the basis of the knowledge of gene sets encoded by completely sequenced genomes of microorganisms. These reconstructions are confined to prokaryotic type of cell organisation because the addition of nucleus and other organelles requires a much larger complement of genes, and the experimental gene knockouts in yeast indicate that there are many more essential genes in eukaryotes than in bacteria (Mushegian, 1999). The main conclusions thus far are the following:

1. The minimal genome is an object of synthetic biology, and its characteristics can be defined differently depending on the goal of the experimentator. The estimates of the size of a minimal genome, therefore, is all relative. In particular, it is dependent on the parameters of the environment, particularly the nutrients that are available to the minimal cell.
2. A minimal cell that shares many features of the smallest known bacterial genomes can probably sustain itself with approximately 300 protein coding genes. Such a cell is simplified even in comparison with the small genomes of mycoplasmas. It would have glycolysis as the sole source of energy; would need to import all amino acids, nucleobases, some sugars, fatty acids, and precursors of most coenzymes; would have only rudimentary systems of DNA repair, gene regulation, and signal transduction; and would lack the cell wall.
3. The minimal genome also must contain several dozen genes that encode untranslated RNAs (tRNAs, rRNAs, the RNA components of SRP and RNAase P, and some others).
4. The completeness of the minimal genome is dependent on our ability to identify gene displacements, including RNA–protein displacements.

A different, and most direct way to learn about minimal genome is to start knocking out genes of your favorite prokaryote one by one—either randomly and without recourse to computational analysis or according to a plan that relies on comparative genomics. The first experiment of this sort was reported by Mitsuhiro Itaya of Mitsubishi Kasei Institute of Life Sciences before the completion of the first genome sequencing projects (Itaya, 1995). Itaya used rare *Not*I restriction sites, quasi-randomly distributed in the genome of *B. subtilis*, and attempted to insert a minitransposon into as many such sites as possible. He was able to identify insertions into 79 loci, only 6 of which had any impaired-growth phenotype on rich medium. Statistical analysis indicated that the indispensable DNA may constitute 318–562 kbp or, given that the size of an average bacterial open reading frame is close to 1 kbp, 300–500 genes (Itaya, 1995). Of course, *B. subtilis* is a large genome with relatively large PHISO (see Chapter 5) and, probably, much functional redundancy between genes. In any case, PHISO in Itaya's projected minimal gene set would have to be lower than in *B. subtilis*, and many genes that are dispensable in *B. subtilis* because of functional redundancy would become unique and possibly indispensable in a smaller genome.

It is thus better to use a smaller genome as a starting point. This is exactly what Craig Venter, Hamilton Smith, and Clyde Hutchison and co-workers have done at the J. Craig Venter Foundation (Hutchison is also professor at the University of North Carolina). They applied a series of increasingly precise protocols to insert tetracycline resistance cassette into as many genes of *M. genitalium* as possible, characterized the insertion mutants at the single-colony level, controlled for possible complementation by mutants in mixed colonies, and verified the phenotypes in multiple passages (Hutchison *et al.*, 1999; Glass *et al.*, 2006). All told, 100 protein coding genes could be disrupted without lethal effects, although some mutants grew slowly (interestingly, mutants in 3 genes grew faster than the wild type). On the other hand, none of the known 43 RNA-encoding genes could be successfully disrupted. Thus, 382 nondisrupted protein coding genes of *M. genitalium* and its 43 RNA coding genes are the closest current experimental approximation of the minimal gene set. Almost all genes identified in our 1996 article, as well as the most likely candidate essential genes discussed in this chapter, belong to this group of 382 genes. One notable exception is *RecA*, the multifunctional DNA strand exchange ATPase. This gene is found in every completely sequenced genome and is included in all lists of essential and minimal genes, and yet Glass *et al.* (2006) were able to recover a viable mutant with insertion in this gene.

Almost 400 essential genes obtained in this experiment are most likely an underestimation: Some genes that are dispensable when singly mutated may actually form synthetic lethals in double-deletion experiments. One reason for such a large difference between this experiment and our theoretical expectation has already been stated: All counts of shared orthologs ignore the isofunctional genes in two distantly related species, and even though our analysis of minimal gene set (Mushegian and Koonin, 1996a) alerted us to the phenomenon of nonorthologous gene displacement (Koonin *et al.*, 1996), we were still unable to identify all displacements correctly.

After all nonessential genes are engineered out of the *M. genitalium* genome (which will become a very small genome already called *Mycoplasma laboratorium*; Glass *et al.*, 2006), what may be the next steps to reduce the genome size even further? One approach may be to explore DOGs in more depth, especially those that are of many-to-one nature (see Chapter 6). If needed, viruses can be brought into comparison: For example, RNA synthesis in bacteria is performed by the holoenzyme that consists of four types of constitutive subunits—alpha, beta, beta′, and omega (although omega subunit has not been found in mycoplasmas thus far)—but phages of the T3/T7 group use a single-subunit RNA polymerase. Replacement of polymerase would result in the net removal of two or three genes; of course, this would require redesigning all promoters and much of the transcription regulation system, which would be a formidable engineering problem.

Polyfunctionality of proteins may also be exploited. As we have previously seen, *M. genitalium* may be doing this already, for example, if the function of its missing nucleoside diphosphate kinase is performed by another small-molecule kinase on a side (Pollack *et al.*, 2002). Perhaps some of the enzymes encoded in minimal genome may be reengineered to a broader specificity. Another, indirect way to enable further reductions is to redesign the metabolism in a more modular form. For example, the pair of genes, folate-dependent thymidylate synthase and dihydrofolate reductase, both of which are essential in *M. genitalium*, could be replaced by flavin-dependent thymidylate synthase (one gene), which is the only enzyme of *de novo* thymidylate synthesis in many bacteria (Mylykallio *et al.*, 2002). This, however, will not result in net elimination of a gene because dihydrofolate reductase plays an additional essential role in C1 turnover by folate, which is also needed for several other cellular processes. Uncoupling of thymidylate synthesis from folate metabolism, however, may facilitate further engineering of the latter pathway. Finally, reduction of genome size, as opposed to gene number, can be undertaken when nonessential portions of all genes (and of intergenic regions) are defined and deleted.

Compendium of orthologs shared by different lineages of living forms may also be used to reconstruct the ancestral genome. Unlike minimal genome, which is the man-made construct, the reconstructed ancestral genome is supposed to be our best guess of the genetic makeup of an organism that really existed. As already mentioned, gene content in the common ancestor is an important source of information about the properties of such ancestor and the conditions in which it might have lived.

Before discussing some recent work, in which comparative genomics was ingenuously used to reconstruct LUCA, let us review a few current assumptions about early evolution of life:

1. The current world operates under the central dogma of molecular biology. DNA is the genome of cellular organisms, and RNA has multiple functional roles, including catalytic functions but not genomic function (except in some viruses; see Chapter 4).
2. There had been an earlier stage in the evolution of life, which is often referred to as the "RNA world." The RNA world has been initially, and perhaps most rigorously, defined as the world in which life consisted of species with RNA genomes and RNA enzymes. There was no DNA to play a genomic role, and there were no encoded proteins. More recently, the RNA world came to have two more meanings. One of them is a stage in evolution when genomic DNA was still not around, but RNA already started to encode proteins that (eventually) took over many enzymatic functions. Yet another meaning of "RNA world" is quite loose and simply refers to the variety of RNA functions observed in the ancestral as well as contemporary cells.
3. Many names have been given to various organisms that are thought to have existed in the RNA world in the first two senses. The transition from the RNA–RNA world to the RNA–protein world was modeled as the evolution from "the last riboorganism" to "the breakthrough organism" (Benner *et al.*, 1987, 1989). An entity similar to the breakthrough organism has been called "progenote" by Carl Woese. There are many speculations, and some detailed reconstructions, of these earlier stages of evolution (Benner *et al.*, 1987, 1989, 1993; Penny and Poole, 1999; Koonin and Martin, 2005). However, as elsewhere in this book, my main attention is on the bread and butter of comparative genomics—that is, the analysis of protein coding genes—and on the "backwards" direction of evolutionary reconstruction, when we use information about the proteins we know today to learn as much as we can about the ancestral species. In this way, we may obtain only limited information about riboorganisms and will have to focus on the genomes that encoded proteins.
4. The entity that is of much interest is LUCA. This is the organism that lived at the root of the three main kingdoms of life—Bacteria, Archaea, and Eukarya.

5. There are several thorny questions about LUCA. One concerns its cellular organization: Was it a prokaryote? A bacterium? An archaeon? An entity that combined some features of bacteria and archaea (and/or eukaryotes, for good measure)? The other concerns its genetic discreteness: Was it an organism in a modern sense, or perhaps a loose association of semiautonomous, self-replicating "sense-carrying units," some of which encoded products that provided benefits to other such units (Woese, 1998a, 1999, 2002; Leipe *et al.*, 1999; Koonin and Martin, 2005; Koonin, 2006; see in particular the open discussion accompanying the latter publication)?

At the same time, there are many questions about the status of individual genes in LUCA. The two molecules that are long-lived and always physically passed between generations of the present-day organisms, DNA genome and plasma membrane. Machinery for synthesis of both these molecules exists in two drastically different forms in modern organisms. The systems of DNA replication contain several homologous components in all species (Leipe *et al.*, 1999), but the main catalytic (nucleotidyltransferase) domain of processive DNA polymerase, the replicative helicase, the origin recognition ATPase, and DNA primase are nonorthologous in Bacteria and Archaea/Eukarya. Thus, what was the status of DNA replication machinery, and of the DNA genome, in LUCA? Likewise, the pathway of lipid side chain biosynthesis is also different in the three domains of life. In this case, however, Archaea uniquely have isoprenoid side chains in their lipids, which they synthesize via a modified mevalonate pathway. Bacteria and Eukarya have fatty acid side chains, and they also make isoprenoids, either by the mevalonate pathway in animals, fungi, a subset of bacteria, and plant cytoplasm, or by the alternative, methylerythritol phosphate pathway in most bacteria and plant chloroplasts. Thus, what were the lipid side chains in LUCA?

Of course, the status of all other cellular systems in LUCA is also of interest. Rampant gene loss and displacement will complicate our efforts to infer the ancestral gene content, same as it interferes with most other inferences, whether they have to do with phylogenetic tree, metabolic pathways, or minimal genome.

When talking about reconstructing the ancestral genes, I will be mostly concerned with gene lists—that is, the statements of presence and absence of the ancestral orthologs in LUCA. A different line of investigation may use the contemporary offspring of each ancestral gene in order to estimate the actual sequence of the ancestor. This is a tremendously interesting endeavor (recall the example of reconstructed ancestral alcohol dehydrogenase from yeast in Chapter 11), but I will not address it here in any detail for several reasons, one of which is that at the very large evolutionary distances that separate LUCA from the modern organisms, such reconstruction is almost impossible, except perhaps for a few very slowly evolving genes. In contrast, the gene content of LUCA may be inferred with reasonable accuracy.

One of the first purposeful reconstructions of the ancestral set of molecular functions was done by Steven Benner and co-workers in the early 1990s (Gonnet *et al.*, 1992; Benner *et al.*, 1993). The central ingredient of their work was the database of highly scoring sequence pairs, obtained by exhaustive matching of the sequences in the SWISSPROT database. By applying single-linkage algorithm and joining pairs that shared a protein into clusters, they determined "connected components." In the circumstances in which sequence sampling of different divisions of life was highly uneven (before the advent of fully sequenced genomes), the authors selected such connected components that include proteins from two or more "superkingdoms" (Bacteria, Archaea, or Eukarya). Those connected components were more likely to represent proteins with universally important molecular functions. Even though archaeal sequences were the limiting resource at the time, there were 36 connected components that contained proteins from archaea and at least one other domain of life.

Twenty of these components were families of proteins with enzymatic activity, representing glycolysis, tricarboxylic acid cycle, amino acid biosynthesis, urea cycle, synthesis of ATP, and DNA transcription. The other 16 were ribosomal proteins. This early work was an encouraging proof of concept: Computer methods were shown to be sufficiently mature to be systematically applied to reconstruct at least fragments of an ancient metabolism, despite the incompleteness of all genome sequences at the time. The main biological conclusion from that study was that the ancestor of all modern life-forms may have been metabolically quite complex.

In 1996, Christos Ouzounis and Nikos Kyrpides, of IMBB in Greece, now at the Joint Genome Institute at the U.S. Department of Energy, produced a similar kind of reconstruction. The protein database available to them was larger and reflected the more advanced stages of several large-scale genome projects. Instead of exhaustive all-against-all comparison, they used the PROSITE patterns, selected all proteins with matches to these diagnostic expressions, and for each family of proteins that matched one and the same PROSITE pattern, examined the phylogenetic position of the genomes in which these proteins were found. There were 944 protein families identified in that way, 77 of them found in all three kingdoms of life. The authors analyzed the composition of the universal set, finding substantial diversity of function, and, having observed two glycolytic enzymes, predicted that if we find a few enzymes from one pathway in the set of universal families, then the genes completing the pathway are likely to be discovered later (they noted that completion of a genome can bring surprises, such as the lack of at least three enzymes of the TCA cycle in fully sequenced *H. influenzae*, but were optimistic that gene loss is nevertheless rare). As we now know, this expectation was not quite correct: Gene losses and DOGs appear to be common enough so as to make it difficult to infer the status of the pathway on the basis of its few components, either in the extant genome or, especially, in the ancestral one. Interestingly, in the same work, Ouzounis and Kyrpides were one of the first to raise the possibility of what they called "mosaicism" of archaeal genomes, on the strength of only a few examples, such as the presence of both bacteria-like (HU) and eukaryotic-like histones. This prescient observation was confirmed soon after their publication (see Chapter 11).

The next major contribution in LUCA reconstruction was the work by Boris Mirkin of Birkbeck College and Eugene Koonin at NCBI (Mirkin *et al.*, 2003). They realized that phyletic patterns (see Chapter 6) contain information about gene gains and losses, and consequently may inform about the status (presence or absence) of many, if not all, genes in the common ancestor. They formulated the problem as follows:

> *Given a species tree and a set of orthologs with a particular phyletic pattern of presence–absence of the species within the analyzed set of species (this set of species should be the same as in the tree), find the most parsimonious mapping of the set of orthologs on the tree. Such a mapping corresponds to the most parsimonious evolutionary scenario for the given set of orthologs, i.e., the scenario with the smallest possible number of events.*

As always with parsimony, it helps to define the reason why we are searching for the "simplest" or "most economical" scenario. Parsimony should perhaps not be its own justification (see more detailed discussion in Sober, 1991, 2004; Felsenstein, 2003), and evolution is most likely not parsimonious in several respects. Cavalier-Smith (2002a; see Chapter 12) noted that it is not parsimonious with regard to gene losses, which can be most easily illustrated by the following kind of argument: Earlier in this chapter, we saw that the pathways of amino acid biosynthesis *de novo* tend to be quickly lost upon switch of free-living bacteria to parasitic lifestyle. If the parsimony principle was unabashedly used for studying evolution of genomes, and if data on amino acid biosynthesis were used as the set of characters, the most parsimonious solution would be to place all parasites into one clade, to the exclusion of all free-living bacteria. This makes no sense.

Evolution may also be nonparsimonious with regard to horizontal gene transfers (HGTs). Thus, any realistic assumptions about evolution have to include some quantitative estimation of the rate of gene loss and gene gain (of which HGT is a special case)—that is, to assign costs, or weights, to the gene gain and gene loss events. This, in fact, is what Mirkin and co-authors did.

There were several further assumptions. First, Mirkin and co-authors did not examine the topology of trees that could be obtained from sequence alignments of each family. Instead, only the presence–absence patterns of genes (phyletic vectors) were studied. This is despite many examples showing that phylogenetic trees of individual sequence families may disagree with any consensus species tree (see Chapters 6, 11, and 12). Second, genomes were treated as "bags of genes," and dependencies between gene losses were not studied (at least one recent investigation, however, indicates that losses of genes within the same pathway are nonrandom, with both rate of loss and the order in which genes are lost being influenced by earlier loss events; Tanaka *et al.*, 2005, 2006; Makino and Gojobori, 2006). Finally, the method is sensitive to the topology of the species tree.

Mirkin and co-workers considered three types of events: gene loss, emergence of a new gene/COG (either by duplication or perhaps, less commonly, from a noncoding sequence), and acquisition of a gene/COG by HGT. Emergence of a new COG, whether by duplication or xenology, is considered a gain in the remainder of their argument. The basic procedure for counting gain and loss events is illustrated in Fig. 13.2.

Suppose that a gene is found in lineages *B, C,* and *D,* whose known phylogeny is shown in Fig. 13.2A. Such phyletic pattern may be explained either by a gene gain by the last common ancestor of *B, C,* and *D* (open circle; one event) or by gene appearance in the last common ancestor of all four species (open circle; first event) followed by gene loss in the lineage leading to *A* (black circle; second event). According to the parsimony principle, the

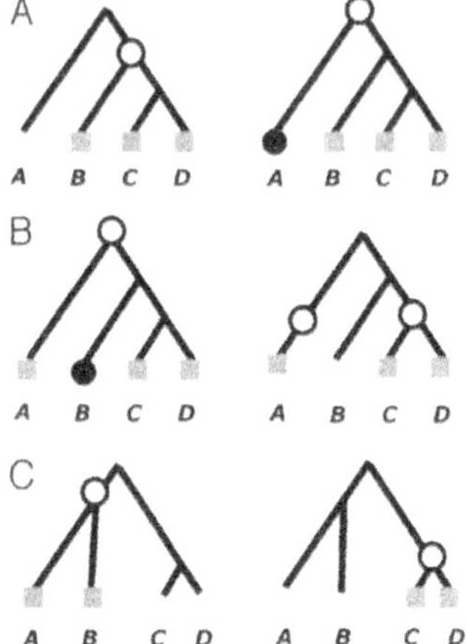

Figure 13.2. Inference of the presence or absence of a COG at the root of the species' phylogenetic tree, given that tree and the status of the COG in the extant species. *Gray boxes* indicate the species that have a COG. Gene gains are indicated by *open circles*, and gene losses are indicated by *solid circles*. Panels A and B are reprinted with modification from Mirkin *et al.* (2003) under BioMed Central Open Access license agreement.

one-event explanation is better than the two-event scenario. In Fig. 13.2B, a more complex situation is shown. It is a toss-up between two scenarios of two events each (either one gain + one loss or two gains).

After the raw counts of possible events explaining each phyletic pattern at the background of the species tree are obtained, there may be ties for the minimal count. That is where biological realism has to come into play. As already said, there is no reason to consider all gains and all losses equally likely: It is generally easier to lose genes than to gain them, so in the case of a tie for event number, it is sensible to choose the scenario with the smallest number of gains. All this may be good for breaking the ties, but in the complex reconstructions with many gains and losses, the relative costs of gains and losses have to be substituted in order to provide a measure of the least costly scenario. General mathematical approaches for character weighting have been developed (Swofford and Madison, 1992; Swofford *et al.*, 1996; Bruno *et al.*, 2000), but the question is what the actual value of the weights should be. If the extreme values of the weights are chosen, there are interesting consequences. For example, if we decided that gene losses are rare, and that their use in the evolutionary scenario has to be strongly penalized, whereas gene gains were low-cost events, the scenarios for patterns with many absences will tend to push losses deeper into the past because one gene loss in a distant ancestor would take care of all absences, and regaining the gene is cheap and thus is allowed to occur many times. Conversely, if gains are to be strongly penalized (which is essentially what is argued by the proponents of the rareness of HGT; see Chapter 11) and losses are thought to have low cost, then losses will tend to be more numerous and more recent.

Mirkin *et al.* (2003) call the crucial parameter the "gain penalty" (although, of course, it is a factor that can be applied to either favor or penalize gene gains). Gain penalty, *g*, was estimated indirectly by varying its values and comparing solutions for each value. Since the value of *g* directly influences the list of genes that are placed into the reconstructed LUCA, we can seek, for example, such a LUCA that is more functionally coherent than the others. The authors studied the range of *g* values from 0.1 (10 gains are equivalent to 1 loss) to 10 (10 losses score the same as 1 gain). In the first case, gains would be prevalent in the history of most COGs, and in the second case phyletic patterns would be explained mostly by genes losses. Expectedly, the total number of gains needed to explain all observed phyletic patterns was becoming increasingly smaller with an increase of *g*, and the number of losses grew with an increase of *g*.

Perhaps less expected was a sharp difference in the ratio of two types of events experienced between two neighboring *g* values: At $g = 0.9$, there were almost four times as many HGT events as there were losses, but at $g = 1.0$, the numbers of the two types of events were almost equal. The LUCA gene sets reconstructed for these two values of the *g* parameter indeed had some special properties.

As *g* increased from 0.1, small fragments of different cellular pathways were slowly gained. LUCA0.9 was the first genome that had significant portions of many metabolic pathways. LUCA1.0, with 572 genes, had the complete translation system, except for glycyl-tRNA synthetase (a well-known displacement); the set of basal RNA polymerase subunits, transcription termination factors, and several helix–turn–helix transcription regulators; the complete set of the bacterial-type H^+-ATPase subunits; and many complete or nearly complete metabolic pathways, including almost complete glycolysis (again, a missing phosphoglyceromutase is a gene displacement well-known to us by now), complete TCA cycle, nucleobase biosynthesis, and nucleotide salvage and substantially full pathways for biosynthesis of amino acids. Coenzyme biosynthesis pathways, however, continued to be incomplete.

In a recurring theme, the replicative DNA polymerase, helicase, and replication initiation ATPase were missing from LUCA1.0. Bacterial DNA polymerase I was present, but this may have been an artifact of it being shared by bacteria and eukaryotes, in which it replicates mitochondrial DNA and must have been gained by HGT from the protomitochondrial alphaproteobacterial endosymbiont. It is not clear whether this protein really belongs to LUCA or what, if any, was the identity of the DNA replication enzyme.

Thus, the full account of HGT and gene displacement cannot be taken even in this second-generation approach. Undoubtedly, phylogenetic trees of individual gene families have to be studied to detect some HGTs and perhaps to improve the cost function. One other parameter that would become explicit in this analysis, and that would improve the reconstruction, is the relative lengths of branches in different parts of the species' tree. Consider two cases of gene gain (single event in each case), mapped onto the same phylogenetic tree (see Fig. 13.2). When deciding whether the gene has been present at the root, the parsimony principle will not distinguish between these cases—it states only that both distributions are explained by the same number of events. If, however, the branch lengths are included in the model, the left-hand case becomes more suggestive of the ancestral presence of the gene because it appears to persist in evolution since more ancient times.

What about the status of DNA genome in LUCA? This is a topic of a vivid discussion in recent years. One idea is that there was none: LUCA might have had an RNA genome, together with a rich repertoire of proteins, and DNA replication could have been invented twice independently, once in a lineage leading to Bacteria and the other time in the clade leading to Archaea/Eukarya. This hypothesis explains the lack of orthology between several major components of DNA replication machinery. However, it does not illuminate the peculiar evolutionary history of several enzymes that work with DNA, such as three subunits of DNA-dependent RNA polymerase and bacterial-type DnaG primase, as well as some very specialized enzymes that are involved in biosynthesis of deoxyribonucleotides, such as flavin-dependent thymidylate synthase and two subunits of ribonucleotide reductase. Most reconstructions confidently place all these enzymes into LUCA, suggesting that deoxyribonucleotides might have been present already (Mirkin *et al.*, 2003; Koonin, 2003a; Ouzounis *et al.*, 2006).

One of the explanations elaborated by Patrick Forterre in recent years hypothesizes that LUCA had an RNA genome, and DNA genomes first emerged in virus-like parasites, as a resistance mechanism against the host surveillance systems that we are searching for and destroying foreign genomes. Under this hypothesis, deoxyribonucleotide precursors and DNA replication systems of different viruses have been independently hijacked by cellular genomes two or maybe even three times (Filee *et al.*, 2003; Forterre, 2005, 2006). Another hypothesis, which seems to be more general than that of Forterre, is that LUCA may have had an RNA genome with a virus-like strategy, similar to what is found in retroid viruses of Baltimore class VI/Agol DDRD class (see Chapter 4). Under this hypothesis (Leipe *et al.*, 1999; Koonin *et al.*, 2006), portions of LUCA genome may have been copied into DNA intermediates that were transcribed into mRNAs. This was supplanted by modern-type DNA genomes twice in two main early lineages.

Finally, what about lipid side chains? They are represented by fatty acids in bacteria and eukaryotes and by isoprenoids in archaea (other chemical details are also different but can be left out for the moment). Isoprenoids are also found in all known divisions of life, and almost all nonparasitic species, but outside the Archaea they play no role as lipid side chains (even though their highly elaborated derivatives, such as sterols, are obviously important for membrane function in eukaryotes). Here, again, we have an ancient split, but with the -BE or A—phyletic pattern, different from the -AE or B—pattern observed in DNA replication. Moreover, isoprenoid biosynthesis appears to have been invented twice, once as the mevalonate pathway and another time as the deoxy-D-xylulose phosphate pathway.

What was the status of lipid side chains in LUCA, or were there any? One proposal is that ancient cells lived in microcompartments with inorganic cell walls, which obviated the need of membranes, and the escape from these compartments was only possible with the invention of lipid side chains (Koonin *et al.*, 2006). Under this scenario, the organism with bacteria-like

DNA replication was able to escape when it invented fatty acids biosynthesis. What kind of side chain enabled the escape of the archaeo-eukaryal lineage is anybody's guess. But even if there was an ancestral lipid side chain, its evolution almost certainly involved multiple gene and whole-pathway displacements.

Our trip into the past is over for now. More genome sequencing will undoubtedly bring us closer to the understanding of the common ancestors of present-day genomes and will provide tools and resources for engineering minimal genomes.

14

Comparative Genomics and Systems Biology

Ours are interesting times for comparative genomics. The words "bioinformatics" and "computational biology" barely existed 15 years ago. They were first used as Medline keywords in the 1990 paper describing the first steps of the National Center for Biotechnology Information (Benson *et al.*, 1990), and the achievements in the area at that time could be summarized in a few monographs. But as I am writing this in mid-2006, the count of articles, and even books, on various aspects of bioinformatics is all but lost, and Google search already finds approximately 600 Web sites that contain the expression "traditional bioinformatics," which is typically understood as analysis of biological sequences, the Pauling and Zuckerkandl sense-carrying units.

As people were still trying to get used to traditional as well as perhaps nontraditional bioinformatics, to "comparative genomics," "functional genomics," and a host of other "omics" (Petsko, 2002; Nicholson, 2006; Joyce and Palsson, 2006), a strange thing happened. More or less suddenly, all these formerly scientific disciplines have been relegated to the level of enabling technologies. The real game in town is now systems biology. That expression occurs more than 1300 times in Medline, 99% of them after 2000, when the namesake institute was founded in Seattle by Leroy Hood. "Systems biology" is found at approximately 5 million websites. These are spectacular results for an area of science that is still trying to define itself.

In a commentary on the state of genomics and systems biology, Maureen O'Malley and John Dupre, social scientists at the Center for Genomics in Society at the University of Exeter, noted that "much of the discussion of the status of genomes has been conducted via evaluations of the evolving metaphors in genomic discourse—from the ineptness of the blueprint metaphor to analogies with jazz scores and Theseus's ship" (O'Malley and Dupre, 2005). In this book, I tried to present ideas and generalizations that demonstrate that comparative genomics is neither a set of enabling technologies nor a postmodern juggling of metaphors but, rather, a coherent discipline that discovers new facts about biology and generates new understanding. I hope that the view of genome as a system was at least implicit—or, more often, explicit—in the preceding chapters. So, what is added by the systems biology "discourse"?

O'Malley and Dupre (2005) say:

> *Under the systems biology rubric are two different (but not mutually exclusive) understandings of "system." The first account is given by scientists who find it useful for various reasons (including access to funding) to refer to the interconnected phenomena that they study as "systems." The second definition comes from scientists who insist that systems principles are imperative to the successful*

development of systems biology. We could call the first group "pragmatic systems biologists" and the second "system-theoretic biologists."... The majority of today's systems biologists fall into the former category, united simply by an agreement that systems biology involves the study of interacting molecular phenomena through the integration of multilevel data and models. For them, "system" is a convenient but vague term that covers a range of detailed interaction with specifiable function.... For hard-line systems-theoretic biologists, however, an ad hoc approach to systems is inadequate. It is crucial, they argue, "to analyze systems as systems, and not as mere collections of parts" in order to understand the emergent properties of component interactions.

In the same essay, the early influences on systems theory are sketched, going back to Norbert Wiener and Ludwig von Bertalanffy. But there is another, more immediate source of inspiration for at least "theoretical systems biologists," namely the study of complex networks.

The history of that discipline will be written in the foreseeable future; its roots are in graph theory and statistical physics and also in empirical examination of social relationships. Of more immediate interest to us, in this final chapter of a book about comparative genomics, is the following question: Are there any facts or at least claims about biology that are emerging from the network-level analyses of the living systems? In other words, it is not too difficult to present genome structure and function in the form of a large number of gene nodes connected by some tangled edges, which look like a dandelion or, to a less romantic soul, like a dust bunny. Most of us have seen such figures, but what do these images tell us about biological systems? This final chapter is a brief overview of some biologically significant observations emerging from the examination of genomewide dust bunnies.

Systems biology constructs are built, as a rule, from Pauling and Zuckerkandl's "sense-carrying units"—most often, from some combination of genes, transcripts, proteins, and so on. Often, relationships between these sense-carrying units are represented in the form of graphs. Sometimes, genes/proteins are vertices of the graph, such as in diagrams of gene–gene or protein–protein interactions, and the edges represent either experimentally determined or inferred relationships between genes and proteins. In other representations, genes may be edges, and nodes may represent something else as in the charts of metabolic pathways. In this case, nodes correspond to metabolic intermediates, usually known from the experiment. In the rest of this chapter, we will mostly discuss the examples in which nodes correspond to sense-carrying units. As always in graphs, edges are defined as pairs of nodes: In our case, the edge is a pair of genes that has the right type of relationship. In any given graph, some pairs of nodes may be connected, but usually not all of them.

Second, there are many biologically interesting ways to connect pairs of genes or proteins. For example, genes can be connected because they share sequence similarity, because they are neighbors on a chromosome, because their products interact, or because they have similar phyletic patterns. Some of these relationships are binary (yes or no, all or none, present or absent). Other types of relationships are measured quantitatively on some interval: For example, proteins can interact strongly or weakly; transiently or over the whole lifetime; in all samples that have been examined or in only some of them. Relationships may also be conditional: Two genes/gene products may interact only in some contexts, such as the presence of the third component or factor in appropriate amounts; only at high concentrations of both components; or only at night. These conditions also help to define the edges between pairs of genes/proteins. One interesting set of problems in systems biology has to do with the ways to measure various kinds of (possibly conditional) relationships between genes and to find the conditions or thresholds that define the edges between genes. Obviously, we are most interested in such measures and thresholds that allow us to detect biologically important signals in the data.

The graphs that we were discussing are undirected: Two nodes that constitute an edge are equivalent (i.e., edges do not have starting or ending points). However, edges can be given direction, sometimes using quite obvious rules. For example, an arrow at one end of an edge can be drawn to show that a gene regulates expression of another gene, or that the product of reaction catalyzed by one protein is the substrate for another protein. In many other cases, the ways to direct the edges are not obvious. For example, two genes may have very similar patterns of expression across the range of conditions, and perhaps they should be connected by the edge on this basis; but is there any way to polarize this edge, if genes are not directly acting on each other? Another interesting problem in systems biology is to find general approaches to assigning direction to the edges—again, in a way that would reveal something interesting about the biological systems that we are studying.

Previously discussed graphs are also unweighted: Two nodes either have an edge between them or they do not, and there is just one kind of edge. But edges can be given weights, which can be visually represented, for example, as lines of different thickness. Weights may directly reflect some biological property, such as the strength of interaction between genes. Weights may also reflect some technical aspect of the experiment, such as reproducibility with which a certain type of relationship is observed. Yet another interesting problem in systems biology is to find biologically meaningful ways to assign weights to edges in the genomewide networks. Different edges in the same graph may also be of fundamentally different nature—for example, representing either genetic interactions between genes or physical interactions between their products—which may be represented as different "colors." Issues related to coloring biological networks are also of great interest.

From these general considerations, it is quite clear that the same set of genes—for example, all genes in a genome—can be connected into a network in many different ways, which reflect different types of relationships between genes. A critical question, however, is what to do with the networks once they are constructed. Perhaps the only way to answer this question is to decide which properties of biological networks are worth studying. Going out an a limb, I propose the following as "the first expectation of systems biology":

> *Interesting biology will tend to manifest itself mostly at the level of relatively small, local subgraphs in the gene networks, and not so much at the level of the global properties of the network.*

In other words, even if we define the network of all genes in a genome, it is the pattern of configurations of a small number of nodes (genes/proteins) that will be of main interest. The study of these local configurations should allow us to make statements about modes of gene regulation, metabolic flux through the pathways, stability of the system under different kinds of perturbations, and evolution of all of these. At the end of the chapter, I will mention some directions of this local network analysis. First, however, I briefly discuss the global properties of complex networks, some of which receive much attention these days—a fad that I expect to pass.

A formal description of any network is most properly made in the language of graph theory or, perhaps also satisfactorily, of statistical physics. This gives a way to describe the quantitative properties of the network, such as "network clustering coefficient," "largest node degree," "betweenness centrality," "modularity," and so on (in a recent survey, at least 30 measures of global properties of complex networks were recognized; Costa *et al.*, 2006). Often, however, discussion of networks is overloaded with the metaphors.

Perhaps one of the most famous semimetaphoric constructs produced by this line of inquiry is a "small world" network. Informally, it can be defined as a network in which (almost) any node can be reached from (almost) any other node in a small number of steps. At this point, the "six degrees of separation" meme is usually invoked, which appears to have been introduced in 1967 by the social psychologist Stanley Milgram of Yale University, who sent passport-size packages to a few hundred randomly selected individuals in Nebraska and Kansas and asked them to make a connection with target individuals in the Boston area, with a restriction to only send the packets to someone whom the originator knew on a first-name basis, the same for the next sender, and so on, until the target is reached. His famous result, first publicized in a popular magazine (Milgram, 1967), was that residents of the Midwest and Boston area (and, by implication, the entire United States) are connected by not more than six first-name acquaintances. As later inquiry showed, however, Milgram's evidence was rather limited, with only a few dozen chains completed (Travers and Milgram, 1969; Kleinfeld, 2001). More important, however, in 1998 Duncan Watts and Steven Strogatz of Cornell University published an article on the theory of small world networks, in which they did two main things. First, they showed how a simple set of rules can be used to generate a small world network either from a randomly connected graph or from a highly regular ring-like graph while preserving the numbers of nodes and edges in it (Fig. 14.1). Second, they found that several kinds of real networks, including some social networks but also, interestingly, the network of connections between neurons in nematode *Caenorhabditis elegans*, are small world networks, with the characteristic "degree of separation" between neurons in the nematode much less than the proverbial six (in fact, it is 2.65). Watts is at Columbia University now, where he has initiated the study of e-mail chains, which seems to be consistent with Milgram's prediction—with the qualification that Internet users in the 21st century may be a different social network than U.S. mail users in 1967.

What has not been shown, of course, is that the process described by Watts and Strogatz (or other models that can be set up to produce small world networks) has relevance to biological "small worlds." Moreover, some initial "discoveries" of small worlds in biological systems did not stand up to more detailed examination. For example, the small world nature of a network that connects cellular metabolites has been reported, with characteristic separation degree between 3 and 4 (Jeong *et al.*, 2000; Wagner and Fell, 2001). These observations, however, were refuted when the irreversibility of reactions and additional considerations of the actual fate of individual carbon atoms were taken into account (i.e., when the connections were retained only between those compounds that actually share at least one carbon atom), which resulted in characteristic separation degree of more than 8 (Ma and Zeng, 2003; Arita, 2004). Thus, the very extent to which different biological systems have small world properties is an open question, and the reasons for such properties, when they are indeed observed, remain unclear.

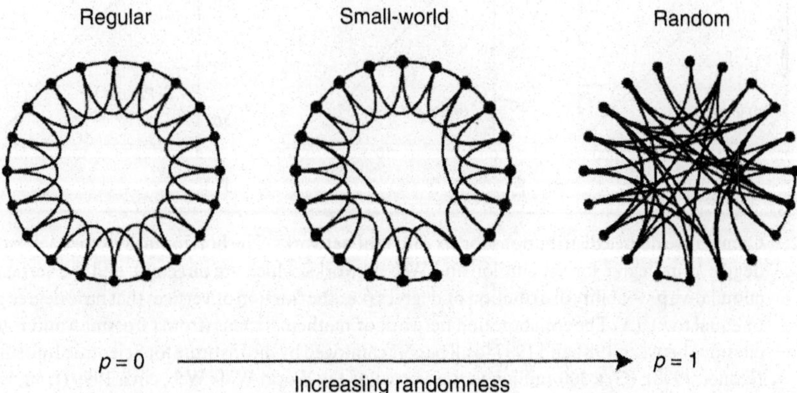

Figure 14.1. Random rewiring procedure that can produce a "small world" network from a regular ring lattice, without altering the number of vertices or edges in the graph. A vertex and the edge that connects it to its nearest clockwise neighbor are chosen, and with a constant probability p, this edge is reconnected to a vertex chosen at random over the entire ring, with duplicate edges forbidden. This is repeated clockwise for each vertex in turn, then the edges that connect vertices to their second-nearest neighbors are randomly rewired with the same probability, and the process is repeated with increasingly more distant neighbors after each full turn. For $p = 0$, the original ring is unchanged; for $p = 1$, all edges are rewired randomly; and for certain intermediate values of p, the graph is a small world network. Reproduced from Watts and Strogatz (1998) by permission from Nature Publishing Group.

The other extremely popular concept, also mixing metaphor with quantitative observation, is "scale-free" networks. The formal definition of "scale-free" with regard to networks is still under debate. In fact, the only type of distribution that satisfies all attempted definitions of scale-free is the power law distribution, well-known in mathematics. Therefore, scale-free probably has no independent meaning and is subsumed by the power laws (for details, see Newman, 2005). The intuition here is the famous 80/20 rule and other similar rules, which state that there are many poor people but few very rich people, that there are many small towns but few very large towns, or that some words are very frequent in most texts but the majority of words are rarely encountered. More formally, the power law distribution is described as $f_i = Ci^{-\gamma}$, where f_i is the number of nodes with degree i, C is a term that is generally not of interest (it is introduced to ensure that frequencies sum to 1), and γ is a parameter called the power law exponent, which in many power law networks takes values from 1 to 3 (Fig. 14.2). Not every power law distribution comes from analysis of a network (e.g., I know of no natural or social network that directly induces the distribution of cities by size), but some networks exhibit the power law distribution of the edge degrees (i.e., the number of edges connecting to each node). Unlike the case of small world networks, at least one good reason why power laws may be frequently observed in biological data sets is known.

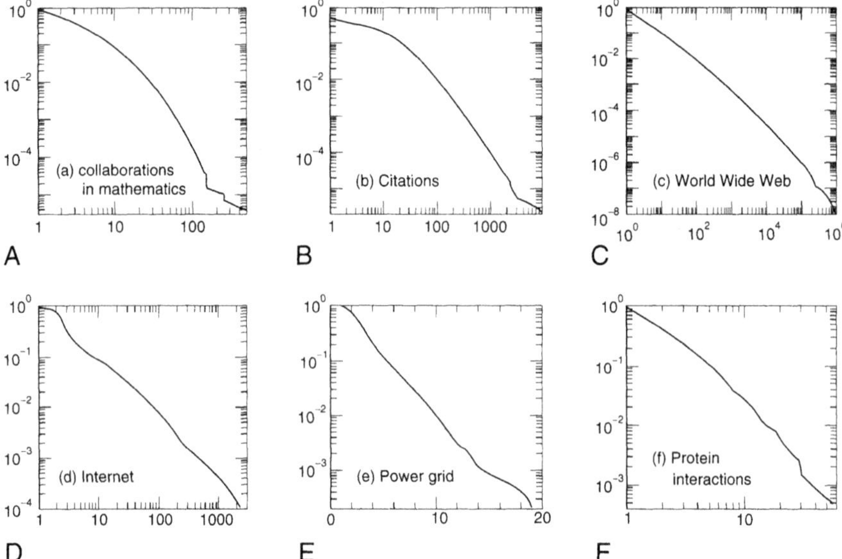

Figure 14.2. Cumulative degree distributions for six different networks. The horizontal axis for each panel is vertex degree k (in-degree for the citation and Web networks, which are directed), and the vertical axis is the cumulative probability distribution of degrees (i.e., the fraction of vertices that have degree greater than or equal to k). (**A**) The collaboration network of mathematicians (from Grossman and Ion, 1995); (**B**) citations between 1981 and 1997 to all papers cataloged by the Institute for Scientific Information (from Redner, 1998); (**C**) a 300-million vertex subset of the World Wide Web, circa 1999 (from Broder *et al.*, 2000); (**D**) the Internet at the level of autonomous systems, April 1999 (from Chen *et al.*, 2002); (**E**) the power grid of the western United States (from Watts and Strogatz, 1998); and (**F**) the protein interaction network of proteins in the metabolism of the yeast *Saccharomyces cerevisiae* (from Jeong *et al.*, 2001). When not carefully analyzed, each of these distributions can be fitted to a power law curve of some sort, and yet only networks c, d, and f appear to have power law degree distributions. Network b is approximately power law-like in the tail, but it has a different behavior when the edge degree is small. Network e has an exponential degree distribution (note a different scale on the horizontal axis). Network a may have a truncated power law degree distribution or possibly two separate power law regimes with different exponents. See Newman (2003) for more detailed description of each network. Reproduced with permission from Newman, M. E. J. (2003). The structure and function of complex networks. *SIAM Rev.* **45,** 167–256. Copyright © Society for Industrial and Applied Mathematics.

The analysis of degree distribution in biological networks has resulted in the reporting power laws everywhere, from ecosystems to virus epidemics, cellular populations, and gene interactions (Harrison and Gerstein, 2002; Jose and Bishop, 2003; Gatenby and Frieden, 2004; Marquet *et al.*, 2005). I expect, however, that this system-theoretic binge will very soon be moderated by the following cautionary observations. First, often the power law-like behavior is reported on the basis of fitting a single family of curves: Basically, what is asked is which value of the γ parameter gives the best fit to the data. However, what should be asked instead is whether other distributions, such as logarithmic or stretched exponential, would give a better fit. Second, it is known that many real-life data sets exhibit power law behavior only on an interval of i (Newman, 2005; Adamic, 2006). Third, power law properties of the observed sample may not be representative of population, and some real-life networks with unique properties are not easily tractable by sampling (Stumpf *et al.*, 2005). Finally, it has been proposed that many networks, when analyzed using a popular but misleading technique of frequency degree plots, can give the illusion of the power law distribution, or they can incorrectly estimate the γ parameter when the power law distribution is indeed present (Tanaka *et al.*, 2005; Doyle *et al.*, 2005).

All these reservations about the role of power laws in biological networks are serious, and yet there are biological data sets that, although not necessarily taking the network form, are described by power laws in a natural way. This is because biological data sets are made of sense-carrying units or of operational taxonomic units. And the standard way of making new genes and proteins is by duplication of already existing genes, much like the standard way of making new OTUs is speciation. Enter the so-called Yule process, which can be summarized as follows [the line of explanation is from Newman (2005), which offers more quantitative detail].

Suppose we have a set of taxa or protein families (similar rules may apply to social objects, such as cities, scientific articles, or Web pages; we discuss them in parallel for now, but it should be quite clear that all these examples deal with different types of entities that may have unique features in addition to general themes). New entities are produced occasionally: Genes undergo duplications, and species split (or new cities appear on the map, and scientists publish new papers). Each entity is characterized by some quantitative property, such as the number of proteins in a family, species in a genus, or citations that each paper receives. Each new entity is associated with some initial value of that property, often but not always equal to one: When a new genus emerges, it may have just one species in it (towns, however, usually are designated as towns when they have at least a few dozen residents, and new scientific papers are usually cited zero times at the moment of their publication).

Some families (genera/cities/papers) continue to acquire new genes (species/residents/citations), but some do not. As a first approximation, all these systems are characterized by a property that the probability of gaining new entities is proportional to the number of already existing entities. Indeed, the more genes in a family (species in a genus), the higher the chance that at least some of them will undergo duplication (speciation). Note that in the social entities, the mechanism of gaining new members may be different (a research paper that is widely cited becomes better known, which makes it more likely that someone else will cite it; large cities may attract more migrants for social reasons). This process, studied by Yule (1925), is also called "rich get richer," "cumulative advantage," or, in the context of networks, "preferential attachment." It can be shown that this generates data sets that have power law distribution, as initially proposed by the Nobel prize winner in economics Herbert Simon (Simon, 1955), applied to citation networks by de Solla (1976), and popularized for biological networks by several researchers, especially Albert-Lazslo Barabasi of the University of Notre Dame, starting at least from Barabasi and Albert (1999).

Elsewhere in this book, especially in Chapters 5 and 10, we examined the attempts to make sense of the distribution of gene families and protein folds in completely sequenced genomes and in sequence databases. New appreciation of the power laws has resulted, in the past several years, in new models for this process. It appears now that the distribution of protein folds by the number of families, of families by the number of members in each complete genome, and perhaps of genes by the number of transcripts at any given time in a cell is best represented by a modified power law formula or, more precisely, by the so-called generalized Pareto function $f_i = C(i+a)^{-\gamma}$, where a is another parameter that is not well-understood, and the values of a and γ are different for different data sets (Kuznetsov, 2001). This function is asymptotically close to power law when i is large.

Thus, power law-like behavior, or something close to it, is seen in distributions of sense-carrying units and OTUs. This is not the same as stating that power laws are observed in all biological networks—not all of them are necessarily produced by a "rich get richer" type of process (on the other hand, there are other types of processes that may produce the power law distribution of node degrees, although none of these processes have been shown to operate in biological systems). Thus, we return to the set of problems listed at the beginning of this chapter, particularly to the ways of deriving networks in the first place. Indeed, without a proper way for building a network, we cannot hope to discover the laws that describe it.

Most of this book (and, I argue, most of comparative genomics) is built on the foundation of detecting and evaluating sequence similarities between sense-carrying units. As discussed in Chapter 2, "nothing in genomics makes sense except in light of sequence comparison." Indeed, sequence similarity is instrumental in establishing homology (see Chapter 3) and was essential in trying whole-genome approaches on viruses (see Chapter 4). "The first fact of comparative genomics" (see Chapter 5) is a direct result of applying sensitive methods of sequence similarity searches to complete genomes of cellular organisms. "The second fact" (see Chapter 6) is about the reverse situation, when the top-of-the-line methods of sequence and structure comparison reject the common ancestry hypothesis for isofunctional proteins. Whole-genome metabolic reconstruction (see Chapters 7 and 8) is about using homology and posthomology methods for understanding functions of gene products. Phylogenetic questions in the era of complete genomes are based on comparison of molecular characters (see Chapters 11 and 12). So how about "nontraditional bioinformatics" and systems biology—can it also be built on some kind of similarity that would be of fundamental importance? I believe that the answer to this question is "Yes!" (which is why I am asking it, of course).

The genomewide measurements associated with every gene in a completely sequenced genome have been introduced before. Such sets of numbers connected with a gene are (or should be) called gene vectors. In different experiments, the same gene or gene product can be associated with a phyletic gene vector, an expression gene vector, a protein–protein interaction gene vector, a phenotypic vector, a subcellular localization vector, and so on. I suggest that, much like sequence similarity has become the main organizational principle of comparative genomics, nothing in systems biology will make sense except in light of similarity between gene vectors. Let us have a closer look at them.

A gene vector space, or vector database, is a set of vectors $X_{ij} = (x_{i1}, x_{i2}, ..., x_{iN})$, where $I = 1, ..., M$, and $j = 1, ..., N$. M and N indicate, respectively, the number of genes and the number of data points/experimental conditions produced by a genomewide experiment and associated with each gene. Measurements may be relative, as with data obtained in two-color printed gene expression arrays, or absolute, as in gene expression measured on the Affymetrix high-density array platform. Some types of measurements may be numerically encoded discrete states, such as gene presence vs. absence, or change vs. no change of gene expression. Furthermore, vectors can represent not only direct experimental measurements, but also models derived from groups of related vectors. The following are but a few examples of gene vectors:

Phyletic vectors: These were first introduced in Chapter 5, when we discussed the COG database. Each of the 14,000 COGs in the current release of the COG resource is associated with a phyletic vector, where the *j*th coordinate ($j = 1, ..., 110$) is set at 1 if it is represented in the *j*th genome and 0 if it is not. This is a simplification; some COGs contain in-paralogs, so they could also be viewed as vectors in which some coordinates are neither 1 nor 0 but represent either actual or normalized counts of in-paralogs. If we compare COGs in the 110-dimensional genome space, we can find groups of coinherited COGs, which may correspond to biochemical pathways (see Chapter 8). But we can also compare genomes in 14,000-dimensional vector space to see groups of genomes that share more genes with one another and to discern phylogenetic signal in these vectors (see Chapters 11 and 12).

Protein interaction vectors: Screening of protein–protein interaction (PPI) at a large scale can be done with the yeast two-hybrid system and other similar technologies that register only pairwise PPI, as well as with various affinity purification schemes, which record the protein content of a complex but does not directly discern individual interacting pairs. The PPI-related vectors can be binary (e.g., protein is present/absent), or they may include information about relative or absolute abundance of each protein in each sample. The PPI vector space can be analyzed in several ways. For example, purification vectors can be compared in the space of protein coordinates, protein vectors can be compared in the purifications' space, or proteins can be compared in the protein space. In the first case, the search result would be the set of similar purifications; in the second and third cases, the results are the sets of proteins copurifying with each other. Along with physical interaction between gene products, one can also study genetic interactions in model organisms using any number of clever genomewide schemes (Tong *et al.*, 2001; Giaever *et al.*, 2002; Schuldiner *et al.*, 2005)

Gene expression vectors: These are familiar from numerous publications on array-based expression profiling. In this case, the coordinates of a vector correspond to different samples, treatments, or conditions under which the measurements were made. For instance, they may represent tissue samples from different patients or observations made at different time points.

Protein localization vectors. In this case, the coordinates of a vector represent all cellular locations in which at least one protein has ever been observed. The complete set of possible locations is not known, and analysis of localization vectors indicates that the current vocabulary used to describe these locations is drastically inadequate (Robert Murphy, Carnegie Mellon University, personal communication).

In each of these cases, finding groups of gene vectors related by some sort of similarity are of great interest. But all too often, the discussion of ways to find groups in the data starts and ends with the discussion of the clustering algorithms. The terms of the art, such as "hierarchical clustering," "K-means," "partitioning around medoids," "self-organizing maps," "support vector machines," and many others, are all around us. However, in my opinion, two things are often overlooked in this discussion, one of which comes before and another after the clustering process.

In Chapter 2, we examined an analogous situation with sequence similarities. Much attention has been given by researchers and by textbook authors to algorithms for finding highly scoring sequence pairs. But these algorithms, important as they are for efficient database searches, are not capable of finding biologically important signals unless two other ingredients are also present, namely a good measure of distance/similarity between sequences and a statistical theory that allows us to evaluate the significance of the similarities that we observe. The same is true in comparing gene vectors. We should be concerned not only about fast ways to partition vectors into groups but also (or perhaps mostly) about finding a good way to measure distance/similarity, without which we would not be able to sensibly define groups of related vectors, and about evaluating the significance of clusters.

In the rest of this chapter, I focus on binary vectors (i.e., such vectors that have only ones or zeroes as coordinates). Most of my examples will deal with phyletic vectors, for which binary form (presence or absence of a given gene/COG in a given genome) may be quite natural. One problem with distances between phyletic vectors was discussed in Chapter 8: If $\mathbf{x}_1 = (1011110)$, $\mathbf{y}_1 = (0111110)$, $\mathbf{x}_2 = (1000000)$, $\mathbf{y}_2 = (0000001)$, and we are interested in whether there is a special relationship between genes x_1 and y_1, and between genes x_2 and y_2, we need a distance/similarity measure that distinguishes pairs of genes such as x_1 and y_1, which are indeed found together, from pairs such as x_2 and y_2, which are not. But some of the measures frequently advocated in the literature for comparing phyletic and other vectors, such as Hamming or Euclidean distances, actually give the same distance value for both pairs of vectors.

There are other properties of distance measures that are not desirable. For example, the popular Jaccard similarity coefficient is written as

$$J = \frac{M_{11}}{M_{11} + M_{01} + M_{10}},$$

where M_{11} is the number of coordinates set to "one" in both vectors, and M_{01}, M_{10} are, respectively, the number of coordinates set to "one" in the first but not the second, and in the second but not the first vector. The distances derived from it may give counterintuitive results when binary vectors are concerned. Suppose that we want to compare two genomes using the number of shared orthologs, as is commonly done in phylogenetic analysis (see Chapters 12 and 13). If the two genomes have a similar number of genes, and approximately half of the genes in each genome are also found in another genome, we would expect the similarity to be approximately one-half. However, the Jaccard coefficient gives one-third, and for this reason there has been a suggestion to normalize it differently (Mirkin and Koonin, 2003). Yet another awkward property of some distances is that they behave differently when the number of "ones" is large and when it is small (Fig. 14.3).

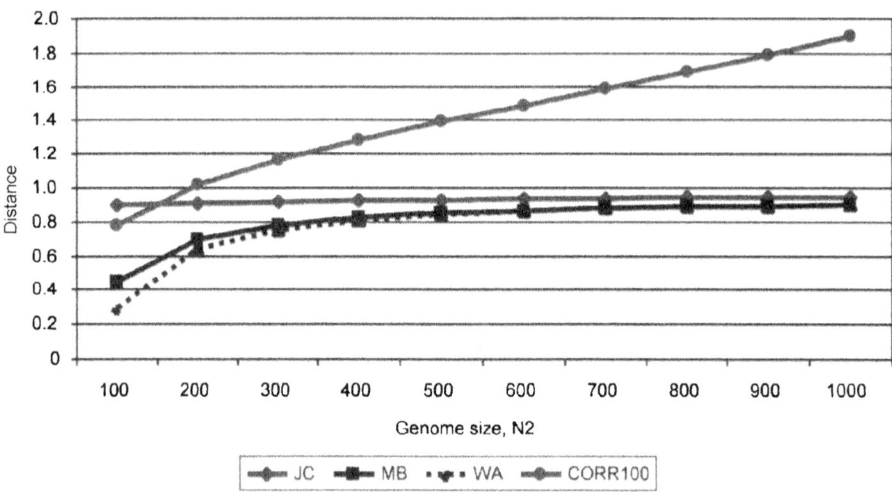

Figure 14.3. Measuring distances between genomewide binary vectors: differential sensitivity of different distance measures to vector spaces dominated by zeroes. The genome content distances between two simulated genomes are shown. One genome, N_1, has a constant size of 1000 genes, and the size of the other genome, N_2, is shown on the horizontal axis. The number of genes N_{12} shared by N_1 and N_2 is 100 in all cases. Three types of weighted-average distances (Glazko *et al.*, 2005), namely Jaccard coefficient (JC), Maryland Bridge distance (MB), and Dutilh weighted average (WA), tend to have low resolution either on the whole interval or on large parts of it.

In fact, all these problems are related: They have to do with the behavior of the distance measure when vectors are dominated by zeroes. Therefore, in many cases, distance/similarity measures that are based on correlation coefficient or on mutual information, where zeroes do not contribute to the result, work better than the others. In fact, it has been shown that for the case of binary vectors, these two measures have an exact relation (for the proof and details, see Li, 1990). Recently, there has been much empirical testing of different distance measures and their ability to find groups of related genomewide vectors, and there appear to be two main conclusions from these studies. First, even though we can measure pairwise distances between every two vectors in a given space, in fact only relatively short distances are indicative of a biological signal—the rest is a gray zone or pure noise. Second, high pairwise similarity between vectors defines (usually small) clusters of vectors, representing genes/proteins that have some sort of functional or evolutionary connection. For example, cluster of expression vectors defines genes likely to be involved in the same transcriptional program, cluster of phyletic vectors defines genes likely to belong to the same biochemical pathway, and so on. However, the relationship between such clusters and other types of information about the same genes is usually only approximate. In other words, clusters of "linked" genes obtained using different types of genomewide data—the units of transcriptional coregulation, the pathways from the biochemistry textbook, the patterns of gene coinheritance, the groups of genes adjacent on the chromosome, and so forth—map only imperfectly onto each other.

This can be illustrated by a study mentioned in Chapter 8 (Glazko and Mushegian, 2004), in which we clustered all COGs on the basis of correlation distance between their phyletic vectors (see Fig. 8.4). There, we wanted to determine whether genes/COGs that belong to the same pathway or functional system can be efficiently assigned to the same cluster of phyletic patterns (i.e., a group of coinherited genes). But whatever we tried, we rarely recovered clusters that would correspond to an entire biochemical pathway and nothing but that pathway. Only a few pathways or complexes, such as the deoxy-D-xylulose phosphate pathway of terpenoid biosynthesis, lipid A biosynthesis, aerobic branch of cobalamine biosynthesis, and the NADH–ubiquinone oxidoreductase complex, were recovered in their entirety by at least one clustering method. The majority of clusters of phyletic vectors that were found mostly belonged to one of two types. On one end of the spectrum, there were three very large clusters with mostly phylogenetic, rather than functional, signal: One set of COGs was found in all species, another only in bacteria, and the third only in archaea/eukarya. The former type of pattern is a subset of the minimal gene set that takes no account of gene displacement (see Chapter 13), and the latter two patterns indicate divergence of some pathways and independent origin of other pathways in bacterial clade and in archaeal/eukaryal clade of life (see Chapter 12). Each of these three classes includes COGs from many different functional systems. For example, COGs found in all bacteria and nothing but bacteria include ribosomal proteins, factors of transcription and translation, several enzymes involved in DNA replication (including catalytic subunit of replicative DNA polymerase III and NAD-dependent DNA ligase), components of secretion apparatus, and several enzymes with predicted molecular function but unknown biological role. Even though this large cluster is formally well-defined, it represents a mix of many functions and systems, not one discrete module.

Most of the cluster space, however, consisted of small clusters that included proteins from the same pathway but, as a rule, excluded a significant portion of the same pathway. Indeed, 48 of the 52 metabolic pathways that we examined in detail were distributed among two, three, or four clusters. For example, the path of riboflavin biosynthesis was split between two clusters, one of which also included the components of two pathways for biosynthesis of several different amino acids. In this case, "pathway fragmentation" may represent an artifact of our clustering method when applied to the evolutionary noisy (i.e., prone to gene gain and loss) data.

In other cases, however, the split of a seemingly wholesome pathway into fragments may represent a genuine functional signal. For example, the bacterial type IV secretion apparatus came out as four distinct clusters, one of which was made of genes *virB8*, *virB9*, *virB10*, and *virB4*, which is exactly the subset of the *Agrobacterium tumefaciens VirB* operon that has been shown to constitute a functionally and structurally discrete module that acts on the side of the recipient bacterium and is sufficient for DNA uptake by the recipient (Liu and Binns, 2003). Here, "pathway fragmentation" may reveal the existence of closely associated groups of genes and gene products that work together in different circumstances and remain tightly associated in some way in the evolution. A similar idea has been described by Hartwell *et al.* (1999):

> *A functional module is, by definition, a discrete entity whose function is separable from those of other modules. This separation depends on chemical isolation, which can originate from spatial localization or from chemical specificity. A ribosome, the module that synthesizes proteins, concentrates the reactions involved in making a polypeptide into a single particle, thus spatially isolating its function. A signal transduction system, on the other hand, such as those that govern chemotaxis in bacteria or mating in yeast, is an extended module that achieves its isolation through the specificity of the initial binding of the chemical signal (for example, chemoattractant or pheromone) to receptor proteins, and of the interactions between signaling proteins within the cell. Modules can be insulated from or connected to each other. Insulation allows the cell to carry out many diverse reactions without cross-talk that would harm the cell, whereas connectivity allows one function to influence another. The higher-level properties of cells, such as their ability to integrate information from multiple sources, will be described by the pattern of connections among their functional modules.*

This sounds eminently reasonable, perhaps to the point of being quite obvious: In fact, all this could be a definition of any biological pathway. What is less obvious is that high-throughput and comparative genomic approaches tend to find modules usually composed of a smaller number of genes/proteins than the biochemical pathways taken off the wall map (or than the protein complexes purified in a biochemistry lab). Thus, each new space of genome vectors may split genes into modules in a novel way, not always known from the investigation of other vector spaces.

This has been observed on many types of genomewide vectors, such as protein–protein interaction vectors, where most macromolecular complexes, even those that have been studied for a long time and thought to be well-defined, contain a "nucleus" and "periphery" (Gavin *et al.*, 2006; Fig. 14.4); gene expression vectors in eukaryotes, where usually only a subset of genes in the same pathway or complex is tightly controlled at the transcription level (de Lichtenberg *et al.*, 2005); phyletic vectors, and so on (reviewed in Campillos *et al.*, 2006). Instability of most operons in bacterial evolution (see Chapter 8) may also be seen as manifestation of the same phenomenon.

One way to view these developments is to admit that our knowledge of pathways and complexes had been limited by the relatively small scale of investigation, where we could only study a small number of components at a time and only a handful of model organisms. The era of complete genomes and high-throughput approaches, however, increases dimensionality of the data and introduces, in earnest, evolution into our study of biological function. In the same article by Hartwell *et al.* (1999) we also read,

> *Modular structures may facilitate evolutionary change. Embedding particular functions in discrete modules allows the core function of a module to be robust to change, but allows for changes in the properties and functions of a cell (its phenotype) by altering the connections between different modules. If the function of a protein were to directly affect all properties of the cell, it would be hard to change that protein, because an improvement in one function would probably be offset by impairments in others. But if the function of a protein is restricted to one module, and the connections of that module to other modules are through individual proteins, it will be much easier to modify, make and prune connections to other modules.*

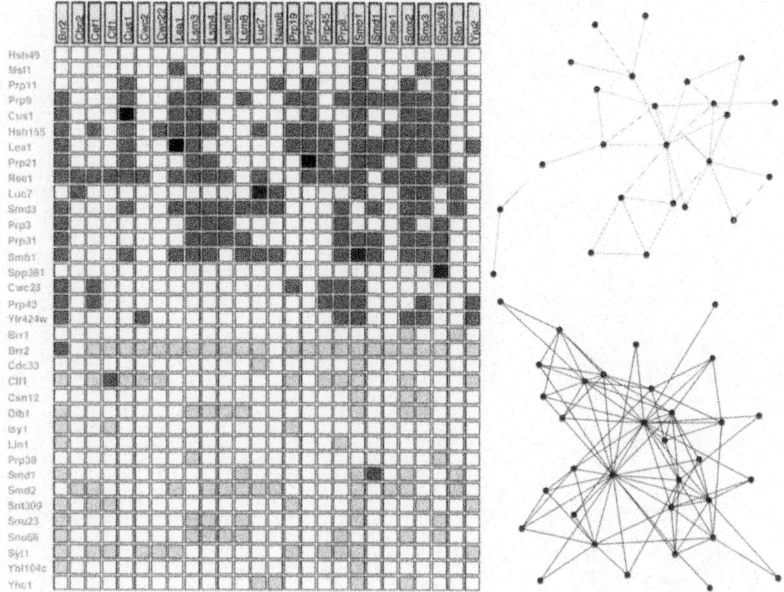

Figure 14.4. Protein–protein interaction matrix and modular network induced by it. Visualization of the yeast U2 snRNP complex is from the on-line resource, "Protein Complexes in Yeast" (available at http://yeast-complexes.embl.de, accessed August 18, 2006). The matrix of protein–protein interactions is shown on the left. *Columns*, proteins used as bait; *rows*, protein–protein interaction vector for each bait. Results of heuristic clustering of these vectors with two different distance thresholds are shown on the right. The smaller network (**top**) corresponds to the core module of the U2 snRNP complex, corresponding to a subset of proteins with stronger degree of clustering (indicated by dark gray shading on the left). The larger network also includes attachment modules, corresponding to more loosely connected proteins (indicated by light gray shading on the left). Data are from Gavin *et al.* (2006).

Of course, modularity is not only a prerequisite, but also the result of evolution—mostly of differential gain, loss, functional takeover, and redistribution of molecular functions between genes that are available in each species. An appropriately selected measure of distance between the genomewide vectors is the clue to defining these modules or to "establishing" and "pruning" connections.

Another level of understanding of biological modules may be afforded by looking at their internal structure. Suppose that the edges in the network have been defined using a sensitive distance measure and a sound way of selecting statistically significant edges and removing the spurious ones. What can be said about the properties of the resulting graphs?

One of the best approaches to this type of analysis comes from the work of Uri Alon's group at Weizman Institute of Science. Alon and co-authors studied several types of networks, most famously the graphs of transcriptional regulation in several species. This is an example of polarized network, in which edges are naturally directed from the nodes representing transcriptional regulators to nodes representing their target genes, some of which may themselves be transcriptional regulators. Instead of examining the global properties of transcriptional regulation network, Alon and co-workers focused on local configurations of small sets of nodes in these networks, which they called "network motifs" (Shen-Orr *et al.*, 2002). In that work and many publications that followed, the idea was generalized for different types of graphs consisting of different but usually small (from three to five) number of nodes. For example, for three nodes and directed edges connecting each of them to at least one other node, there are 13 possible network motifs (see Fig. 14.5).

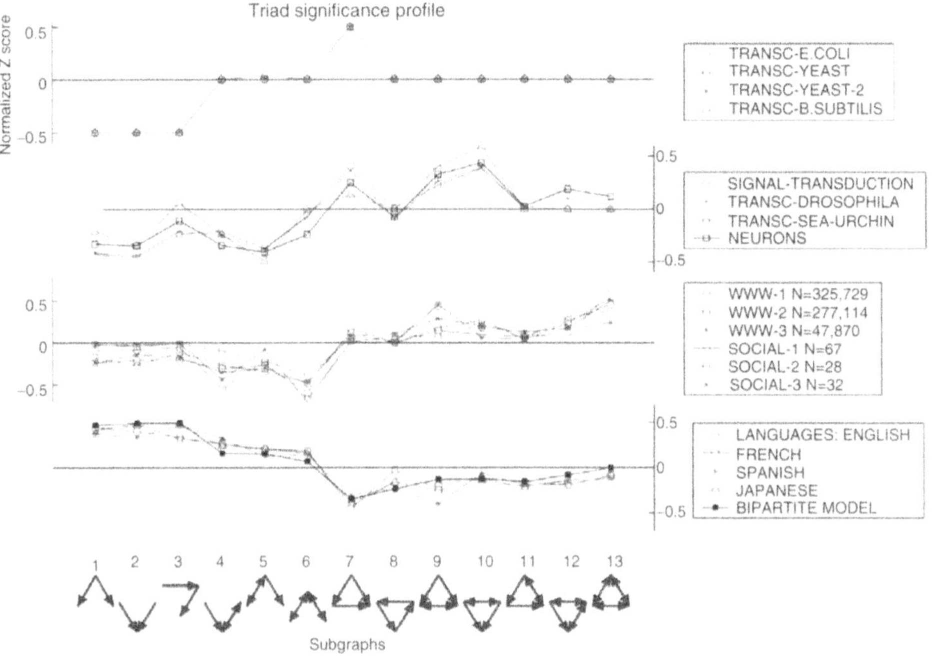

Figure 14.5. Three-node network motifs and their distribution in various networks ("triad significance profiles"). Networks with similar characteristic profiles are grouped into superfamilies. The networks are, from top to bottom, direct transcription interactions in *Escherichia coli* (TRANSC-E.COLI), *Bacillus subtilis* (TRANSC-B.SUBTILIS), and the yeast *S. cerevisiae* (two different sources: TRANC-YEAST and TRANSC-YEAST-2); signal transduction interactions in mammalian cells (SIGNAL-TRANSDUCTION), transcription networks that guide development in fruit fly (TRANSC-DROSOPHILA), endomesoderm development in sea urchin (TRANSC-SEA-URCHIN), and synaptic connections between neurons in *C. elegans* (NEURONS); World Wide Web hyperlinks as a whole and pertaining to different knowledge domains (WWW-1, WWW-2, and WWW-3); and social networks, including inmates in prison (SOCIAL-1), sociology freshmen (SOCIAL-2), and college students in a course on leadership (SOCIAL-3); word-adjacency networks of texts in for languages; and a simulated network with controlled parameters (BIPARTITE). See Milo *et al.* (2004) for more detailed description of each network. Reprinted with permission from Milo *et al.* (2004), copyright 2004, American Association for the Advancement of Science.

These motifs are not equally likely to be found in the well-studied transcriptional circuits of bacteria and yeasts: One type of motif, the so-called feed-forward loop, is strongly overrepresented, and several types of motifs are underrepresented.

Transcription maps of fruit fly development and of sea urchin early embryo show over- and underrepresentation of additional motifs, although the feed-forward loop is also highly represented there. Interestingly, several social networks with naturally defined edge direction, such as links between Web pages or the order of words in texts from several languages, show completely different distribution of the same motifs. Another directed biological network studied by Alon's group is the "wiring diagram" of neurons in nematode. There, direction is also given naturally by axons that make synaptic contacts with other cells. Interestingly, the "triad significance profile" of this network is very similar to the transcriptional network of multicellular eukaryotes.

The statistics used by Alon and co-workers to compute the significance of network motifs may have to be improved (Artzy-Randrup *et al.*, 2004), but this does not change the fact that biological and social networks have their own specific profiles of frequencies of each three-node motif. The feed-forward loop most commonly overrepresented in biological networks, appears to be optimally suited, and probably selected, for a particular biological function. Indeed, both mathematical modeling and experimentation *in vivo* indicate that this motif is optimal for ensuring gene expression in response to a continuing stimulus or inducer, as opposed to a stochastic fluctuation of the inducing agent.

The picture is further elaborated, even for the simplest, three-node motifs, by different polarity of transcriptional regulation and by different gate logic at the signal-receiving end. This offers a large variety of gene regulatory behaviors, many of which are adaptive (for the most recent and most detailed treatise, see Alon, 2006). Thus, biologically significant signals can emerge from analysis of patterns of local connectivity in transcription and signal transduction networks, as well as in the networks of physical connections between cells. Notably, edges in network motifs are directed; without directionality, the diversity of motifs would be low, and most of the interesting biology would be untractable.

What about other types of biological networks—can we add complexity to their local structure by giving direction to at least some edges? Is there any way to direct an edge between two phyletic vectors or between two gene expression vectors when none of them directly regulates the other? It is obvious that we can add some external information, such as the transcription regulation or the order of reactions in a biochemical pathway. But can we direct the edges using information derived from the same vector space? This question has not been examined in earnest thus far, and yet it seems to be quite central for the systems biology framework. I finish this chapter (and this book) with a brief preview of the possibilities.

One idea is to use ranks. A vector may be the nearest neighbor of another vector, the second nearest, or the ith nearest. Likewise, a probabilistic similarity search in vector space, analogous to PSI-BLAST search of sequence similarity (Zhou *et al.*, 2002; Glazko *et al.*, 2006) may find related vectors in one or more steps, also giving a rank order to matching vectors. We may define edges by directing them from a given query gene to all genes whose gene vectors are within certain similarity rank from the query. Analogously to the graphs of sequence similarity (see Fig. 2.6), such relationships are neither commutative nor transitive, providing complex, asymmetrical properties of the gene network.

The other approach, recently proposed for gene expression vectors, relies on teasing out a specific type of stochastic dependency between vectors. For example, we can ask which pairs of genes satisfy the condition $Y = XZ$, where X and Y are random variables representing expression of genes x and y, and Z is another random variable, stochastically independent of X (Klebanov *et al.*, 2005). When this condition is satisfied, one can draw an arrow from X to Y. This does not automatically give us an arrow in the opposite direction because in the expression $X = Y/Z$, the variables Y and $1/Z$ are not necessarily stochastically independent.

When all is said and done, will there be "feed-forward loops," "bi-fans," and other clusters with interesting structure in different types of gene networks? The answer, undoubtedly, is yes, but we do not know which "facts of systems biology" will follow from analysis of these motifs. What we do know, however, is that comparative genomics continues to expand, from pairwise sequence matching to finding patterns in gene networks across many, and more to come, genomes.

References

Aas, P. A., Otterlei, M., Falnes, P. O., *et al.* (2003). Human and bacterial oxidative demethylases repair alkylation damage in both RNA and DNA. *Nature* **421,** 859–863.

Abascal, F., and Valencia, A. (2003). Automatic annotation of protein function based on family identification. *Proteins* **53,** 683–692.

Adamic, L.A. (2006). *Zipf, Power-Laws, and Pareto—A Ranking Tutorial.* Available at www.hpl.hp.com/research/idl/papers/ranking/ranking.html. Accessed October 30, 2006.

Agol, V. I. (1974). Towards the system of viruses. *BioSystems* **6,** 113–132.

Aguinaldo, A. M., Turbeville, J. M., Linford, L. S., *et al.* (1997). Evidence for a clade of nematodes, arthropods and other moulting animals. *Nature* **387,** 489–493.

Aharoni, A., Gaidukov, L., Khersonsky, O., McQ Gould, S., Roodveldt, C., and Tawfik, D. S. (2005). The "evolvability" of promiscuous protein functions. *Nat. Genet.* **37,** 73–76.

Ahlquist, P. (2002). RNA-dependent RNA polymerases, viruses, and RNA silencing. *Science* **296,** 1270–1273.

Ahlquist, P., Schwartz, M., Chen, J., Kushner, D., Hao, L., and Dye, B. T. (2005). Viral and host determinants of RNA virus vector replication and expression. *Vaccine* **23,** 1784–1787.

Albrecht, M., Hoffmann, D., Evert, B. O., Schmitt, I., Wullner, U., and Lengauer, T. (2003). Structural modeling of ataxin-3 reveals distant homology to adaptins. *Proteins* **50,** 355–370.

Almagor, H. (1983). A Markov analysis of DNA sequences. *J. Theor. Biol.* **104,** 633–645.

Alon, U. (2006). *An Introduction to Systems Biology. Design Principles of Biological Circuits.* Chapman & Hall/CRC, New York/Boca Raton, FL.

Altschul, S. F. (1991). Amino acid substitution matrices from an information theoretic perspective. *J. Mol. Biol.* **219,** 555–565.

Altschul, S.F. (2000). *The Statistics of Sequence Similarity Scores.* Online tutorial at NCBI: www.ncbi.nlm.nih.gov/blast/tutorial. Accessed February 13, 2006.

Altschul, S. F. (2006). *The Statistics of Sequence Similarity Scores.* www.ncbi.nlm.nih. gov/BLAST/tutorial/Altschul-1.html. Accessed March 28, 2006.

Altschul, S. F., and Gish, W. (1996). Local alignment statistics. *Methods Enzymol.* **266,** 460–480.

Altschul, S. F., Gish, W., Miller, W., Myers, E. W., and Lipman, D. J. (1990). Basic local alignments search tool. *J. Mol. Biol.* **215,** 403–410.

Altschul, S. F., Madden, T. I., Schaffer, A. A., *et al.* (1997). Gapped BLAST and PSI-BLAST: A new generation of protein database search programs. *Nucleic Acids Res.* **25,** 3389–3402.

Altschul, S. F., Bundschuh, R., Olsen, R., and Hwa, T. (2001). The estimation of statistical parameters for local alignment score distributions. *Nucleic Acids Res.* **29,** 351–361.

Altschul, S. F., Wootton, J. C., Gertz, E. M., *et al.* (2005). Protein database searches using compositionally adjusted substitution matrices. *FEBS J.* **272,** 5101–5109.

Anantharaman, V., Koonin, E. V., and Aravind, L. (2002). Comparative genomics and evolution of proteins involved in RNA metabolism. *Nucleic Acids Res.* **30,** 1427–1464.

Andersson, J. O., Doolittle, W. F., and Nesbø, C. L. (2001). Are there bugs in our genome? *Science* **292,** 1848–1850.

Andreeva, A., Howorth, D., Brenner, S. E., Hubbard, T. J., Chothia, C., and Murzin, A. G. (2004). SCOP database in 2004: Refinements integrate structure and sequence family data. *Nucleic Acids Res.* **32**(Database issue), D226–D229.

Aravind, L., and Koonin, E. V. (1999). Gleaning non-trivial structural, functional and evolutionary information about proteins by iterative database searches. *J. Mol. Biol.* **287,** 1023–1040.

Aravind, L., and Koonin, E. V. (2001). The DNA-repair protein AlkB, EGL-9, and leprecan define new families of 2-oxoglutarate- and iron-dependent dioxygenases. *Genome Biol.* **2,** Research0007.

Aravind, L., Tatusov, R. L., Wolf, Y. I., Walker, D. R., and Koonin, E. V. (1998). Evidence for massive gene exchange between archaeal and bacterial hyperthermophiles. *Trends Genet.* **14,** 442–444.

Aravind, L., Tatusov, R. L, Wolf, Y. I., Walker, D. R., and Koonin, E. V. (1999). Reply. *Trends Genet.* **15,** 299–300.

Aravind, L., Watanabe, H., Lipman, D. J., and Koonin, E. V. (2000). Lineage-specific loss and divergence of functionally linked genes in eukaryotes. *Proc. Natl. Acad. Sci. USA* **97,** 11319–11324.

Arendt, D. (2003). Evolution of eyes and photoreceptor cell types. *Int. J. Dev. Biol.* **47,** 563–571.

Arendt, D., and Wittbrodt, J. (2001). Reconstructing the eyes of *Urbilateria*. *Philos. Trans. R. Soc. London B. Biol. Sci.* **356,** 1545–1563.

Arendt, D., Tessmar, K., de Campos-Baptista, M. I., Dorresteijn, A., and Wittbrodt, J. (2002). Development of pigment-cup eyes in the polychaete *Platynereis dumerilii* and evolutionary conservation of larval eyes in *Bilateria*. *Development* **129,** 1143–1154.

Argos, P., Kamer, G., Nicklin, M. J. H., and Wimmer, E. (1984). Similarity in gene organisation and homology between proteins of animal picornaviruses and a plant comovirus suggest common ancestry of these virus families. *Nucleic Acids Res.* **12,** 7251–7267.

Arigoni, F., Talabot, F., Peitsch, M., *et al.* (1998). A genome-based approach for the identification of essential bacterial genes. *Nat. Biotechnol.* **16,** 851–856.

Arita, M. (2004). The metabolic world of *Escherichia coli* is not small. *Proc. Natl. Acad. Sci. USA* **101,** 1543–1547.

Arratia, R., Gordon, L., and Waterman, M. S. (1986). An extreme value theory for sequence matching. *Ann. Stat.* **14,** 971–993.

Artzy-Randrup, Y., Fleishman, S. J., Ben-Tal, N., and Stone, L. (2004). Comment on "Network Motifs: Simple Building Blocks of Complex Networks" and "Superfamilies of Evolved and Designed Networks." *Science* **305,** 1107.

Ashburner, M., Ball, C. A., Blake, J. A., *et al.* (2000). Gene ontology: Tool for the unification of biology. The Gene Ontology Consortium. *Nat. Genet.* **25,** 25–29.

Aurora, R., and Rose, G. D. (1998). Seeking an ancient enzyme in *Methanococcus jannaschii* using ORF, a program based on predicted secondary structure comparisons. *Proc. Natl. Acad. Sci. USA* **95,** 2818–2823.

Ayala, F. J. (1997). Vagaries of the molecular clock. *Proc. Natl. Acad. Sci. USA* **94,** 7776–7783.

Ayala, F. J. (1999). Molecular clock mirages. *Bioessays* **21,** 71–75.

Babajide, A., Hofacker, I. L., Sippl, M. J., and Stadler, P. F. (1997). Neutral networks in protein space: A computational study based on knowledge-based potentials of mean force. *Fold Des.* **2,** 261–269.

Bairoch, A. (1992). PROSITE: A dictionary of sites and patterns in proteins. *Nucl. Acids Res.* **20**(Suppl.), 2013–2018.

Balavoine, G., de Rosa, R., and Adoutte, A. (2002). Hox clusters and bilaterian phylogeny. *Mol. Phylogenet. Evol.* **24,** 366–373.

Balch, W. E., Fox, G. E., Magrum, L. J., Woese, C. R., and Wolfe, R. S. (1979). Methanogens: Reevaluation of a unique biological group. *Microbiol. Rev.* **43,** 260–296.

Baltimore, D. (1971). Expression of animal virus genomes. *Bacteriol. Rev.* **35,** 235–241.

Ban, C., and Yang, W. (1998). Crystal structure and ATPase activity of MutL: Implications for DNA repair and mutagenesis. *Cell* **95,** 541–552.

Barabasi, A. L., and Albert, R. (1999). Emergence of scaling in random networks. *Science* **286,** 509–512.

Barker, W. C., and Dayhoff, M. O. (1982). Viral src gene products are related to the catalytic chain of mammalian cAMP-dependent protein kinase. *Proc. Natl. Acad. Sci. USA* **79,** 2836–2839.

Bashan, A., Agmon, I., Zarivach, R., *et al.* (2003). Structural basis of the ribosomal machinery for peptide bond formation, translocation, and nascent chain progression. *Mol. Cell.* **11,** 91–102.

Baumstark, T., and Ahlquist, P. (2001). The brome mosaic virus RNA3 intergenic replication enhancer folds to mimic a tRNA TpsiC-stem loop and is modified *in vivo*. *RNA* **7,** 1652–1670.

Bazan, J. F., and Fletterick, R. J. (1989). Detection of a trypsin-like serine protease domain in flaviviruses and pestiviruses. *Virology* **171,** 637–639.

Bellamine, A., Mangla, A. T., Nes, W. D., and Waterman, M. R. (1999). Characterization and catalytic properties of the sterol 14alpha-demethylase from *Mycobacterium tuberculosis*. *Proc. Natl. Acad. Sci. USA* **96,** 8937–8942.

Bellgard, M. I., and Gojobori, T. (1999). Identification of a ribonuclease H gene in both *Mycoplasma genitalium* and *Mycoplasma pneumoniae* by a new method for exhaustive identification of ORFs in the complete genome sequences. *FEBS Lett.* **445,** 6–8.

Bellman, R. (1952). On the theory of dynamic programming. *Proc Natl Acad Sci U S A.* **38,** 716–719.

Benner, S. A., and Sismour, A. M. (2005). Synthetic biology. *Nat. Rev. Genet.* **6,** 533–543.

Benner, S. A., Allemann, R. K., Ellington, A. D., *et al.* (1987). Natural selection, protein engineering, and the last riboorganism: Rational model building in biochemistry. *Cold Spring Harbor Symp. Quant. Biol.* **52,** 53–63.

Benner, S. A., Ellington, A. D., and Tauer, A. (1989). Modern metabolism as a palimpsest of the RNA world. *Proc. Natl. Acad. Sci. USA* **86,** 7054–7058.

Benner, S. A., Cohen, M. A., Gonnet, G. H., Berkowitz, D. B., and Johnsson, K. P. (1993). Reading the Palimpest: Contemporary biochemical data and the RNA world. In *The RNA World* (R. F. Gesteland and J. F. Atkins, eds.), pp. 27–70. Cold Spring Harbor Laboratory Press, Cold Spring Harbor, NY.

Benner, S. A., Cohen, M. A., and Gonnet, G. H. (1994). Amino acid substitution during functionally constrained divergent evolution of protein sequences. *Protein Eng.* **7,** 1323–1332.

Benner, S. A., Ricardo, A., and Carrigan, M. A. (2004). Is there a common chemical model for life in the universe? *Curr. Opin. Chem. Biol.* **8,** 672–689.

Benson, D., Boguski, M., Lipman, D., and Ostell, J. (1990) The National Center for Biotechnology Information. *Genomics*. **6,** 389–391.

Berget, S. M., Moore, C., and Sharp, P. A. (1977). Spliced segments at the 5′ terminus of adenovirus 2 late mRNA. *Proc. Natl. Acad. Sci. USA* **74,** 3171–3175.

Bergsten, J. (2005) A review of long-branch attraction. *Cladistics*. **21,** 163–193.

Bilwes, A. M., Alex, L. A., Crane, B. R., and Simon, M. I. (1999). Structure of CheA, a signal-transducing histidine kinase. *Cell* **96,** 131–141.

Bininda-Emonds, O.R. (2004). Trees versus characters and the supertree/supermatrix "paradox." *Syst. Biol.* **53,** 356–359.

Blair, J. E., Ikeo, K., Gojobori, T., and Hedges, S. B. (2002). The evolutionary position of nematodes. *BMC Evol. Biol.* **2,** 7.

Blanchette, M., Kent, W. J., Riemer, C., *et al.* (2004). Aligning multiple genomic sequences with the threaded blockset aligner. *Genome Res.* **14,** 708–715.

Blocki, F. A., Schlievert, P. M., and Wackett, L. P. (1992). Rat liver protein linking chemical and immunological detoxification systems. *Nature* **360,** 269–270.

Blocki, F. A., Ellis, L. B., and Wackett, L. P. (1993). MIF proteins are theta-class glutathione S-transferase homologs. *Protein Sci.* **2,** 2095–2102.

Blundell, T. L., and Johnson, M. S. (1993). Catching a common fold. *Protein Sci.* **2,** 877–883.

Bonneau, R., Strauss, C. E., and Baker, D. (2001a). Improving the performance of Rosetta using multiple sequence alignment information and global measures of hydrophobic core formation. *Proteins* **43,** 1–11.

Bonneau, R., Tsai, J., Ruczinski, I., *et al.* (2001b). Rosetta in CASP4: Progress in ab initio protein structure prediction. *Proteins* **45,** (Suppl. 5), 119–126.

Bonneau, R., Strauss, C. E., Rohl, C. A., *et al.* (2002). *De novo* prediction of three-dimensional structures for major protein families. *J. Mol. Biol.* **322,** 65–78.

Bork, P., and Koonin, E. V. (1998). Predicting functions from protein sequences—Where are the bottlenecks? *Nat. Genet.* **18,** 313–318.

Bork, P., Sander, C., and Valencia, A. (1993). Convergent evolution of similar enzymatic function on different protein folds: The hexokinase, ribokinase, and galactokinase families of sugar kinases. *Protein Sci.* **2,** 31–40.

Bork, P., Ouzounis, C., Casari, G., *et al.* (1995). Exploring the *Mycoplasma capricolum* genome: A minimal cell reveals its physiology. *Mol. Microbiol.* **16,** 955–967.

Botstein, D. (1980). A theory of modular evolution for bacteriophages. *Ann. N. Y. Acad. Sci.* **354,** 484–491.

Boutanaev, A. M., Kalmykova, A. I., Shevelyov, Y. Y., and Nurminsky, D. I. (2002). Large clusters of co-expressed genes in the *Drosophila* genome. *Nature* **420,** 666–669.

Bradley, P., Chivian, D., Meiler, J., *et al.* (2003). Rosetta predictions in CASP5: Successes, failures, and prospects for complete automation. *Proteins* **53**(Suppl. 6), 457–468.

Bradley, P., Malmstrom, L., Qian, B., *et al.* (2005). Free modeling with Rosetta in CASP6. *Proteins* **61**(Suppl. 7), 128–134.

Braunitzer, G., Braun, V., Hilse, K., *et al.* (1965). Constancy and variability of protein structure in respiratory and viral proteins. In *Evolving Genes and Proteins* (V. Bryson and H. J. Vogel, eds.), pp. 183–195. Academic Press, New York.

Brochier, C., Philippe, H., and Moreira, D. (2000). The evolutionary history of ribosomal protein RpS14: Horizontal gene transfer at the heart of the ribosome. *Trends Genet.* **16,** 529–533.

Broder, A., Kumar, R., Maghoul, F., *et al.* (2000). Graph structure in the web. *Computer Networks* **33,** 309–320.

Brodersen, D. E., Clemons, W. M., Jr., Carter, A. P., Wimberly, B. T., and Ramakrishnan, V. (2002). Crystal structure of the 30 S ribosomal subunit from *Thermus thermophilus*: Structure of the proteins and their interactions with 16 S RNA. *J. Mol. Biol.* **316,** 725–768.

Brown, M., Hughey, R., Krogh, A., Mian, I. S., Sjolander, K., and Haussler, D. (1993). Using Dirichlet mixture priors to derive hidden Markov models for protein families. *Proc. Int. Conf. Intell. Syst. Mol. Biol.* **1,** 47–55.

Bruno, W. J., Socci, N. D., and Halpern, A. L. (2000). Weighted neighbor joining: A likelihood-based approach to distance-based phylogeny reconstruction. *Mol. Biol. Evol.* **17,** 189–197.

Bult, C. J., White, O., Olsen, G. J., *et al.* (1996). Complete genome sequence of the methanogenic archaeon, *Methanococcus jannaschii. Science* **273,** 1058–1073.

Buneman, P. (1974). A note on the metric properties of trees. *J. Combinatorial Theory B* **17,** 48–50.

Burnett, B., Li, F., and Pittman, R. N. (2003). The polyglutamine neurodegenerative protein ataxin-3 binds polyubiquitylated proteins and has ubiquitin protease activity. *Hum. Mol. Genet.* **12,** 3195–3205.

Bushman, F. D. (2001). *Lateral DNA Transfer: Mechanisms and Consequences*. Cold Spring Harbor, NY: Cold Spring Harbor Laboratory Press.

Campillos, M., von Mering, C., Jensen, L. J., and Bork, P. (2006). Identification and analysis of evolutionarily cohesive functional modules in protein networks. *Genome Res.* **16,** 374–382.

Carlson, B. A., Xu, X. M., Kryukov, G. V., *et al.* (2004). Identification and characterization of phosphoseryl-tRNA[Ser]Sec kinase. *Proc. Natl. Acad. Sci. USA* **101,** 12848–12853.

Carmean, D., Crespi, B.J. (1995). Do long branches attract flies? *Nature*. **373,** 666.

Carrell, R. W., and Huntington, J. A. (2003). How serpins change their fold for better and for worse. *Biochem. Soc. Symp*. **70,** 163–178.

Caruthers, J. M., and McKay, D. B. (2001). Helicase structure and mechanism. *Curr. Opin. Struct. Biol.* **12,** 123–133.

Cavalier-Smith, T. (2002a). The neomuran origin of archaebacteria, the negibacterial root of the universal tree and bacterial megaclassification. *Int. J. Syst. Evol. Microbiol.* **52,** 7–76.

Cavalier-Smith, T. (2002b). The phagotrophic origin of eukaryotes and phylogenetic classification of Protozoa. *Int. J. Syst. Evol. Microbiol.* **52,** 297–354.

Cavalier-Smith, T. (2006). Rooting the tree of life by transition analyses. *Biol. Direct*. **1,** 19.

Cello, J., Paul, A. V., and Wimmer, E. (2002). Chemical synthesis of poliovirus cDNA: Generation of infectious virus in the absence of natural template. *Science* **297,** 1016–1018.

Chandonia, J. M., and Brenner, S. E. (2006). The impact of structural genomics: Expectations and outcomes. *Science* **311,** 347–351.

Cheek S., Ginalski K., Zhang H., and Grishin N.V. (2005). A comprehensive update of the sequence and structure classification of kinases. *BMC Struct Biol.* **16,** 5–6.

Chen, Q., Chang, H., Govindan, R., Jamin, S., Shenker, S. J., and Willinger, W. (2002). The origin of power laws in Internet topologies revisited. In *Proceedings of the 21st Annual Joint Conference of the IEEE Computer and Communications Societies*, pp. 608–617. IEEE, New York.

Cheng, H., Shen, N., Pei, J., and Grishin, N. V. (2004). Double-stranded DNA bacteriophage prohead protease is homologous to herpesvirus protease. *Protein Sci.* **13,** 2260–2269.

Chiang, T. M., Reizer, J., and Beachey, E. H. (1989). Serine and tyrosine protein kinase activities in *Streptococcus pyogenes*. Phosphorylation of native and synthetic peptides of streptococcal M proteins. *J. Biol. Chem.* **264,** 2957–2962.

Chivian, D., Kim, D. E., Malmstrom, L., *et al.* (2003). Automated prediction of CASP-5 structures using the Robetta server. *Proteins* **53**(Suppl. 6), 524–533.

Chothia, C. (1992). Proteins. One thousand families for the molecular biologist. *Nature* **357,** 543–544.

Chow, L. T., Gelinas, R. E., Broker, T. R., and Roberts, R. J. (1977). An amazing sequence arrangement at the 5′ ends of adenovirus 2 messenger RNA. *Cell* **12,** 1–8.

Collett, M. S., and Ericson, R. L. (1978). Protein kinase activity associated with the avian sarcoma virus src gene product. *Proc. Natl. Acad. Sci. USA* **75,** 2021–2024.

Copley, R. R., Schultz, J., Ponting, C. P., and Bork, P. (1999). Protein families in multicellular organisms. *Curr. Opin. Struct. Biol.* **9,** 408–415.

Copley, R. R., Ponting, C. P., Schultz, J., and Bork, P. (2002). Sequence analysis of multidomain proteins: Past perspectives and future directions. *Adv. Protein Chem.* **61,** 75–98.

Copley, S. D. (2003). Enzymes with extra talents: Moonlighting functions and catalytic promiscuity. *Curr. Opin. Chem. Biol.* **7,** 265–272.

Costa, L. da F., Rodrigues, F. A., Travieso, G., and Villas Boas, P. R. (2006). *Characterization of Complex Networks: A Survey of Measurements*. Available at http://arxiv.org/abs/cond-mat/0505185. Accessed August 17, 2006.

Coulson, A. F., and Moult, J. (2002). A unifold, mesofold, and superfold model of protein fold use. *Proteins* **46,** 61–71.

Creevey, C. J., and McInerney, J. O. (2005). Clann: Investigating phylogenetic information through supertree analyses. *Bioinformatics* **21,** 390–392.

Crick, F. (1970). Central dogma of molecular biology. *Nature* **227,** 561–563.

Crick, F. H. C. (1958). The biological replication of macromolecules. *Symp. Soc. Exp. Biol.* **12,** 138–163.

Curnow, A. W., Hong, K., Yuan, R., *et al.* (1997). Glu-tRNAGln amidotransferase: A novel heterotrimeric enzyme required for correct decoding of glutamine codons during translation. *Proc. Natl. Acad. Sci. USA* **94,** 11819–11826.

Czernilofsky, A. P., Levinson, A. D., Varmus, H. E., Bishop, J. M., Tischer, E., and Goodman, H. M. (1980). Nucleotide sequence of an avian sarcoma virus oncogene (src) and proposed amino acid sequence for gene product. *Nature* **287,** 198–203.

Dandekar, T., Snel, B., Huynen, M., and Bork, P. (1998). Conservation of gene order: A fingerprint of proteins that physically interact. *Trends Biochem. Sci.* **23,** 324–328.

Daraselia, N., Dernovoy, D., Tian, Y., Borodovsky, M., Tatusov, R., and Tatusova, T. (2003). Reannotation of *Shewanella oneidensis* genome. *OMICS* **7,** 171–175.

Dayhoff, M. O., and Eck, R. V. (1968). *Atlas of Protein Sequence and Structure*, Vol. 3. National Biochemical Research Foundation, Silver Spring, MD.

Dayhoff, M. O., Eck, R. V., Chang, M. A., and Sochard, M. R. (1965). *Atlas of Protein Sequence and Structure*, Vol. 1. National Biochemical Research Foundation, Silver Spring, MD.

de la Torre, J. R., Christianson, L. M., Beja, O., *et al.* (2003). Proteorhodopsin genes are distributed among divergent marine bacterial taxa. *Proc. Natl. Acad. Sci. USA* **100,** 12830–12835.

de Lichtenberg, U., Jensen, L. J., Brunak, S., and Bork, P. (2005). Dynamic complex formation during the yeast cell cycle. *Science* **307,** 724–727.

De Solla, P. (1976). A general theory of bibliometric and other cumulative advantage processes. *J. Am. Soc. Information Sci.* **27,** 292–306.

Deiters, A., and Schultz, P. G. (2005). *In vivo* incorporation of an alkyne into proteins in *Escherichia coli. Bioorg. Med. Chem. Lett.* **15,** 1521–1524.

Del Campo, M., Recinos, C., Yanez, G., *et al.* (2005). Number, position, and significance of the pseudouridines in the large subunit ribosomal RNA of *Haloarcula marismortui* and *Deinococcus radiodurans. RNA* **11,** 210–219.

DeLano, W. L. (2002). *The PyMOL User's Manual.* DeLano Scientific, San Carlos, CA.

Devos, D., and Valencia, A. (2001). Intrinsic errors in genome annotation. *Trends Genet.* **17,** 429–431.

Ding, Q., Lewis, J. J., Strum, K. M., *et al.* (2002). Polyglutamine expansion, protein aggregation, proteasome activity, and neural survival. *J. Biol. Chem. 277,* 13935–13942.

Dobzhansky, T. (1973). Nothing in biology makes sense except in the light of evolution. *Am. Biol. Teacher* **35,** 125–129.

Dong, C., Huang, F., Deng, H., *et al.* (2004). Crystal structure and mechanism of a bacterial fluorinating enzyme. *Nature* **427,** 561–565.

Doolittle, R. F. (1986). *On URFs and ORFs. A Primer on How to Analyze Derived Amino Acid Sequences.* University Science Books, Mill Valley, CA.

Doolittle, R. F. (1992). Reconstructing history with amino acid sequences. *Protein Sci.* **1,** 191–200.

Doolittle, R. F. (1994). Convergent evolution: The need to be explicit. *Trends Biochem. Sci.* **19,** 15–18.

Doolittle, R. F., Hunkapiller, M. W., Hood, L. E., *et al.* (1983). Simian sarcoma virus onc gene, v-sis, is derived from the gene (or genes) encoding a platelet-derived growth factor. *Science* **221,** 275–277.

Doolittle, W.F. (1999). Phylogenetic classification and the universal tree. *Science.* **284,** 2124–2129.

Dopazo, H., Santoyo, J., and Dopazo, J. (2004). Phylogenomics and the number of characters required for obtaining an accurate phylogeny of eukaryote model species. *Bioinformatics* **20**(Suppl. 1), I116–I121.

Douzery, E. J., Snell, E. A., Bapteste, E., Delsuc, F., and Philippe, H. (2004). The timing of eukaryotic evolution: Does a relaxed molecular clock reconcile proteins and fossils? *Proc. Natl. Acad. Sci. USA* **101,** 15386–15391.

Dowell, R. D., and Eddy, S. R. (2004). Evaluation of several lightweight stochastic context-free grammars for RNA secondary structure prediction. *BMC Bioinformatics* **5,** 71.

Doyle, J. C., Alderson, D. L., Li, L., *et al.* (2005). The "robust yet fragile" nature of the Internet. *Proc. Natl. Acad. Sci. USA* **102,** 14497–14502.

Duncan, T., Trewick, S. C., Koivisto, P., Bates, P. A., Lindahl, T., and Sedgwick, B. (2002). Reversal of DNA alkylation damage by two human dioxygenases. *Proc. Natl. Acad. Sci. USA* **99,** 16660–16665.

Dunwell, J. M., Culham, A., Carter, C. E., Sosa-Aguirre, C. R., and Goodenough, P. W. (2001). Evolution of functional diversity in the cupin superfamily. *Trends Biochem. Sci.* **26,** 740–746.

Dunwell, J. M., Purvis, A., and Khuri, S. (2004). Cupins: The most functionally diverse protein superfamily? *Phytochemistry* **65,** 7–17.

Durbin, R., Eddy, S., Krogh, A., and Mitchison, G. (1998). *Biological Sequence Analysis. Probabilistic Models of Proteins and Nucleic Acids.* Cambridge University Press, Cambridge, UK.

Dutilh, B. E., Huynen, M. A., Bruno, W. J., and Snel, B. (2004). The consistent phylogenetic signal in genome trees revealed by reducing the impact of noise. *J. Mol. Evol.* **58,** 527–539.

Dynes, J. L., and Firtel, R. A. (1989). Molecular complementation of a genetic marker in *Dictyostelium* using a genomic DNA library. *Proc. Natl. Acad. Sci. USA* **86,** 7966–7970.

Eddy, S. R. (1998). Profile hidden Markov models. *Bioinformatics.* **14,** 755–763.

Eddy, S. R. (2004a). What is dynamic programming? *Nat. Biotechnol.* **22,** 909–910.

Eddy, S. R. (2004b). Where did the BLOSUM62 alignment score matrix come from? *Nat. Biotechnol.* **22,** 1035–1036.

Eddy, S. R. (2004c). What is Bayesian statistics? *Nat. Biotechnol.* **22,** 1177–1178.

Eddy, S. R. (2004d). What is a hidden Markov model? *Nat. Biotechnol.* **22,** 1315–1316.

Edwards, A. W., and Cavalli-Sforza, L. L. (1965). A method for cluster analysis. *Biometrics* **21,** 362–375.

Edwards, A. W. F., and Cavalli-Sforza, L. L. (1963). The reconstruction of evolution. *Ann. Hum. Genet.* **27,** 105–106.

Edwards, A. W. F., and Cavalli-Sforza, L. L. (1964). Reconstruction of evolutionary trees. In *Phenetic and Phylogenetic Classification* (V. H. Heywood and J. McNeill, eds.), Publication No.6, pp. 67–76. Systematics Association, London.

Eichler, E. E., and Sankoff, D. (2003). Structural dynamics of eukaryotic chromosome evolution. *Science* **301,** 793–797.

Eigen, M. (1971). Self-organization of matter and the evolution of biological macromolecules. *Naturwissenschaften* **58,** 465–523.

Eisen, J. A. (1998). Phylogenomics: Improving functional predictions for uncharacterized genes for evolutionary analysis. *Genome Res.* **8,** 163–167.

Eisen, M. B., Spellman, P. T., Brown, P. O., and Botstein, D. (1998). Cluster analysis and display of genome-wide expression patterns. *Proc. Natl. Acad. Sci. USA* **95,** 14863–14868.

Elofsson, A., and Sonnhammer, E. L. (1999). A comparison of sequence and structure protein domain families as a basis for structural genomics. *Bioinformatics* **15,** 480–500.

Engelhardt, B. E., Jordan, M. I., Muratore, K. E., and Brenner, S. E. (2005). Protein molecular function prediction by bayesian phylogenomics. *PLoS Comput. Biol.* **1,** e45.

Enright, A. J., Kunin, V., and Ouzounis, C. A. (2003). Protein families and TRIBES in genome sequence space. *Nucleic Acids Res.* **31,** 4632–4638.

Ensinger, M. J., Martin, S. A., Paoletti, E., and Moss, B. (1975). Modification of the 5′-terminus of mRNA by soluble guanylyl and methyl transferases from vaccinia virus. *Proc. Natl. Acad. Sci. USA* **72,** 2525–2529.

Fabrega, C., Farrow, M. A., Mukhopadhyay, B., de Crecy-Lagard, V., Ortiz, A. R., and Schimmel, P. (2001). An aminoacyl tRNA synthetase whose sequence fits into neither of the two known classes. *Nature* **411,** 110–114.

Faith, D. P., and Baker, A. M. (2006). Phylogenetic diversity (PD) and biodiversity conservation: Some bioinformatics challenges. *Evol. Bioinformatics Online* at www.la-press.com/EBO-2-Faith(Sc).pdf. Accessed August 15, 2006.

Fang, J., Acheampong, E., Dave, R., Wang, F., Mukhtar, M., and Pomerantz, R. J. (2005). The RNA helicase DDX1 is involved in restricted HIV-1 Rev function in human astrocytes. *Virology* **336,** 299–307.

Felsenstein, J. (1978). Cases in which parsimony or compatibility methods will be positively misleading. *Systematic Zool.* **27,** 401–410.

Felsenstein, J. (2003). *Inferring Phylogenies.* Sinauer, Sunderland, MA.

Felsenstein, J. (2005). PHYLIP (Phylogeny Inference Package) version 3.6. Distributed by the author. Department of Genome Sciences, University of Washington, Seattle.

Fernald, R. D. (2000). Evolution of eyes. *Curr. Opin. Neurobiol.* **10,** 444–450.

Fernald, R. D. (2004). Eyes: Variety, development and evolution. *Brain Behav. Evol.* **64,** 141–147.

Fields, C., and Adams, M. D. (1994). Expressed sequence tags identify a human isolog of the Sui1 translation initiation factor. *Biochem. Biophys. Res. Commun.* **198,** 288–291.

Filee, J., Forterre, P., and Laurent, J. (2003). The role played by viruses in the evolution of their hosts: A view based on informational protein phylogenies. *Res. Microbiol.* **154,** 237–243.

Finkelstein, A. V. (1997). Protein structure: What is it possible to predict now? *Curr. Opin. Struct. Biol.* **7,** 60–71.

Finkelstein, A. V., and Ptitsyn, O. B. (1987). Why do globular proteins fit the limited set of folding patterns? *Prog. Biophys. Mol. Biol.* **50,** 171–190.

Finkelstein, A. V., Gutin, A. M., and Badretdinov, A. Ya. (1993). Why are the same protein folds used to perform different functions? *FEBS Lett.* **325,** 23–28.

Fitch, W. M. (1966a). The relation between frequencies of amino acids and ordered trinucleotides. *J. Mol. Biol.* **16,** 1–8.

Fitch, W. M. (1966b). An improved method of testing for evolutionary homology. *J. Mol. Biol.* **16,** 9–16.

Fitch, W. M. (1967). Evidence suggesting a non-random character to nucleotide replacements in naturally occurring mutations. *J. Mol. Biol.* **26,** 499–507.

Fitch, W. M. (1969). Locating gaps in amino acid sequences to optimize the homology between two proteins. *Biochem. Genet.* **3,** 99–108.

Fitch, W. M. (1970a). Further improvements in the method of testing for evolutionary homology among proteins. *J. Mol. Biol.* **49,** 1–14.

Fitch, W. M. (1970b). Distinguishing homologous from analogous proteins. *Systematic Zool.* **19,** 99–113.

Fitch, W. M. (1995). Uses for evolutionary trees. *Philos. Trans. R. Soc. London B Biol. Sci.* **349,** 93–102.

Fitch, W. M. (2000). Homology: Personal view on some of the problems. *Trends Genet.* **16,** 227–231.

Fleischmann, R. D., Adams, M. D., White, O., *et al.* (1995). Whole-genome random sequencing and assembly of *Haemophilus influenzae* Rd. *Science* **269,** 496–512.

Fomenko, D. E., and Gladyshev, V. N. (2002). CxxS: Fold-independent redox motif revealed by genome-wide searches for thiol/disulfide oxidoreductase function. *Protein Sci.* **11,** 2285–2296.

Forterre, P. (2002). A hot story from comparative genomics: Reverse gyrase is the only hyperthermophile-specific protein. *Trends Genet.* **18,** 236–237.

Forterre, P. (2005). The two ages of the RNA world, and the transition to the DNA world: A story of viruses and cells. *Biochimie* **87,** 793–803.

Forterre, P. (2006). Three RNA cells for ribosomal lineages and three DNA viruses to replicate their genomes: A hypothesis for the origin of cellular domain. *Proc. Natl. Acad. Sci. USA* **103,** 3669–3674.

Fraenkel-Conrat, H. (1990). Virus reconstitution and the proof of the existence of genomic RNA. *Bioessays* **12,** 351–352.

Fraenkel-Conrat, H., and Singer, B. (1980). Effect of introduction of small alkyl groups on mRNA function. *Proc. Natl. Acad. Sci. USA* **77,** 1983–1985.

Fraenkel-Conrat, H., Singer, B., and Williams, R. C. (1957). Infectivity of viral nucleic acid. *Biochim. Biophys. Acta* **25,** 87–96.

Fraenkel-Conrat, H., Staehelin, M., and Crawford, L. V. (1959). Tobacco mosaic virus reconstitution using inactivated nucleic acid. *Proc. Soc. Exp. Biol. Med.* **102,** 118–121.

Fraenkel-Conrat, H., Veldee, S., and Woo, J. (1964). The infectivity of tobacco mosaic virus. *Virology* **22,** 432–434.

Frand, A. R., Russel, S., and Ruvkun, G. (2005). Functional genomic analysis of *C. elegans* molting. *PLoS Biol.* **3,** e312.

Fraser, C. M., Gocayne, J. D., White, O., *et al.* (1995). The minimal gene complement of *Mycoplasma genitalium. Science* **270,** 397–403.

Fry, B. G. (2005). From genome to "venome": Molecular origin and evolution of the snake venom proteome inferred from phylogenetic analysis of toxin sequences and related body proteins. *Genome Res.* **15,** 403–420.

Galagan, J. E., Nusbaum, C., Roy, A., *et al.* (2002). The genome of *M. acetivorans* reveals extensive metabolic and physiological diversity. *Genome Res.* **12,** 532–542.

Galperin, M. Y. (2005). A census of membrane-bound and intracellular signal transduction proteins in bacteria: Bacterial IQ, extroverts and introverts. *BMC Microbiol.* **5,** 35.

Galperin, M. Y. (2006). The Molecular Biology Database Collection: 2006 update. *Nucleic Acids Res.* **34,** D3–D5.

Galperin, M. Y., and Koonin, E. V. (1999). Functional genomics and enzyme evolution. Homologous and analogous enzymes encoded in microbial genomes. *Genetica* **106,** 159–170.

Galperin, M. Y., Walker, D. R., and Koonin, E. V. (1998). Analogous enzymes: Independent inventions in enzyme evolution. *Genome Res.* **8,** 779–790.

Gatenby, R. A., and Frieden, B. R. (2004). Information dynamics in carcinogenesis and tumor growth. *Mutat. Res.* **568,** 259–273.

Gatfield, J., and Pieters, J. (2000). Essential role for cholesterol in entry of mycobacteria into macrophages. *Science* **288,** 1647–1650.

Gavin, A. C., Aloy, P., Grandi, P., *et al.* (2006). Proteome survey reveals modularity of the yeast cell machinery. *Nature* **440,** 631–636.

Geigenmuller-Gnirke, U., Weiss, B., Wright, R., and Schlesinger, S. (1991). Complementation between Sindbis viral RNAs produces infectious particles with a bipartite genome. *Proc. Natl. Acad. Sci. USA* **88,** 3253–3257.

Giaever, G., Chu, A. M., Ni, L., *et al.* (2002). Functional profiling of the *Saccharomyces cerevisiae* genome. *Nature* **418,** 387–391.

Gibbs, A. J., and McIntyre, G. A. (1970). The diagram, a method for comparing sequences. Its use with amino acid and nucleotide sequences. *Eur. J. Biochem.* **16,** 1–11.

Gilks, W. R., Audit, B., De Angelis, D., Tsoka, S., and Ouzounis, C. A. (2002). Modeling the percolation of annotation errors in a database of protein sequences. *Bioinformatics* **18,** 1641–1649.

Ginalski, K., Grishin, N. V., Godzik, A., and Rychlewski, L. (2005). Practical lessons from protein structure prediction. *Nucleic Acids Res.* **33,** 1874–1891.

Glass, J. I., Assad-Garcia, N., Alperovich, N., *et al.* (2006). Essential genes of a minimal bacterium. *Proc. Natl. Acad. Sci. USA* **103,** 425–430.

Glazko, G., Coleman, M., and Mushegian, A. (2006). Similarity searches in genome-wide numerical data sets. *Biol. Direct* **1,** 13.

Glazko, G., Gordon, A., and Mushegian, A. (2005). The choice of optimal distance measure in genome-wide datasets. *Bioinformatics* **21**(Suppl. 3), iii3–iii11.

Glazko, G. V., and Mushegian, A. R. (2004). Detection of evolutionarily stable fragments of cellular pathways by hierarchical clustering of phyletic patterns. *Genome Biol.* **5,** R32.

Goad, W. B., and Kanehisa, M. I. (1982). Pattern recognition in nucleic acid sequences: I. A general method for finding local homologies and symmetries. *Nucleic Acids Res.* **10,** 247–263.

Gogarten, J. P. (1994). Which is the most conserved group of proteins? Homology–orthology, paralogy, xenology, and the fusion of independent lineages. *J. Mol. Evol.* **39,** 541–543.

Gonnet, G. H., Cohen, M. A., and Benner, S. A. (1992). Exhaustive matching of the entire protein sequence database. *Science* **256,** 1443–1445.

Gonnet, G. H., Cohen, M. A, and Benner, S. A. (1994). Analysis of amino acid substitution during divergent evolution: The 400 by 400 dipeptide substitution matrix. *Biochem. Biophys. Res. Commun.* **199,** 489–496.

Gorbalenya, A. E., and Koonin, E. V. (1993). Helicases: Amino acid sequence comparisons and structure–function relationship. *Curr. Opin. Struct. Biol.* **3,** 419–429.

Gorbalenya, A. E., Blinov, V. M., and Donchenko, A. P. (1986). Poliovirus-encoded proteinase 3CL: A possible evolutionary link between serine and cysteine proteinase families. *FEBS Lett.* **194,** 253–257.

Gorbalenya, A. E., Koonin, E. V., Donchenko, A. P., and Blinov, V. M. (1988). A conserved NTP-motif in putative helicases. *Nature* **333,** 22.

Gorbalenya, A. E., Donchenko, A. P., Blinov, V. M., and Koonin, E. V. (1989). Cysteine proteases of positive strand RNA viruses and chymotrypsin-like serine proteases. A distinct protein superfamily with a common structural fold. *FEBS Lett.* **243,** 103–114.

Gorbalenya, A. E., Koonin, E. V., and Lai, M. M. (1991). Putative papain-related thiol proteases of positive-strand RNA viruses. Identification of rubi- and aphthovirus proteases and delineation of a novel conserved domain associated with proteases of rubi-, alpha- and coronaviruses. *FEBS Lett.* **288,** 201–205.

Gorbalenya, A. E., Pringle, F. M., Zeddam, J. L., *et al.* (2002). The palm subdomain-based active site is internally permuted in viral RNA-dependent RNA polymerases of an ancient lineage. *J. Mol. Biol.* **324,** 47–62.

Govindarajan, S., Recabarren, R., and Goldstein, R. A. (1999). Estimating the total number of protein folds. *Proteins* **35,** 408–414.

Grate, L., Herbster, M., Hughey, R., Haussler, D., Mian, I. S., and Noller, H. (1994). RNA modeling using Gibbs sampling and stochastic context free grammars. *Proc. Int. Conf. Intell. Syst. Mol. Biol.* **2,** 138–146.

Green, M. L., and Karp, P. D. (2005). Genome annotation errors in pathway databases due to semantic ambiguity in partial EC numbers. *Nucleic Acids Res.* **33,** 4035–4039.

Green, P., Lipman, D., Hillier, L., Waterston, R., States, D., and Claverie, J. M. (1993). Ancient conserved regions in new gene sequences and the protein databases. *Science* **259,** 1711–1716.

Grishin, N. V. (2001a). Fold change in evolution of protein structures. *J. Struct. Biol.* **134,** 167–185.

Grishin, N. V. (2001b). KH domain: One motif, two folds. *Nucleic Acids Res.* **29,** 638–643.

Grossman, J. W., and Ion, P. D. F. (1995). On a portion of the well-known collaboration graph. *Congressus Numerantium* **108,** 129–131.

Grzymski, J. J., Carter, B. J., Delong, E. F., Feldman, R. A., Ghadiri, A., and Murray, A. E. (2006). Comparative genomics of DNA fragments from six antarctic marine planktonic bacteria. *Appl. Environ. Microbiol.* **72,** 1532–1541.

Gusfield, D. (1997). *Algorithms on Strings, Trees, and Sequences: Computer Science and Computational Biology.* Cambridge University Press, Cambridge, UK.

Hallikas, O., Palin, K., Sinjushina, N., *et al.* (2006). Genome-wide prediction of mammalian enhancers based on analysis of transcription-factor binding affinity. *Cell* **124,** 47–59.

Hannenhalli, S., Chappey, C., Koonin, E. V., and Pevzner, P. A. (1995). Genome sequence comparison and scenarios for gene rearrangements: A test case. *Genomics* **30,** 299–311.

Harel, D. (1992). *Algorithmics: The Spirit of Computing.* Addison-Wesley, Reading, MA.

Harms, J., Schluenzen, F., Zarivach, R., *et al.* (2001). High resolution structure of the large ribosomal subunit from a mesophilic eubacterium. *Cell* **107,** 679–688.

Harrison, P. M., and Gerstein, M. (2002). Studying genomes through the aeons: Protein families, pseudogenes and proteome evolution. *J. Mol. Biol.* **318,** 1155–1174.

Hartmann, E., and Hartmann, R. K. (2003). The enigma of ribonuclease P evolution. *Trends Genet.* **19,** 561–569.

Hartwell, L. H., Hopfield, J. J., Leibler, S., and Murray, A. W. (1999). From molecular to modular cell biology. *Nature* **402,** C47–C52.

Haseloff, J., Goelet, P., Zimmern, D., Ahlquist, P., Dasgupta, R., and Kaesberg, P. (1984). Striking similarities in amino acid sequence among nonstructural proteins encoded by RNA viruses that have dissimilar genomic organisation. *Proc. Natl. Acad. Sc.i USA* **81,** 4358–4362.

Hemmingsen, J. M., Gernert, K. M., Richardson, J. S., and Richardson, D. C. (1994). The tyrosine corner: A feature of most Greek key beta-barrel proteins. *Protein Sci.* **3,** 1927–1937.

Henikoff, S., and Henikoff, J. G. (1992). Amino acid substitution matrices from protein blocks. *Proc. Natl. Acad. Sci. USA* **89,** 10915–10919.

Henikoff, S., and Henikoff, J. G. (1993). Performance evaluation of amino acid substitution matrices. *Proteins* **17,** 49–61.

Hershey, A. D., and Chase, M. (1952). Independent functions of viral protein and nucleic acid in growth of bacteriophage. *J. Gen. Physiol.* **36,** 39–56.

Hodgman, T. C. (1988). A new superfamily of replicative proteins. *Nature* **333,** 22–23.

Holland, P. W. H. (1999). The effect of gene duplications on homology. In *Homology*, Novartis Foundation Symposium 222, pp. 226–242. Wiley, Chichester, UK.

Huelsenbeck, J. P. (1997). Is the Felsenstein zone a fly trap? *Syst. Biol.* **46,** 69–74.

Hutchison, C. A., Peterson, S. N., Gill, S. R., *et al.* (1999). Global transposon mutagenesis and a minimal *Mycoplasma* genome. *Science* **286,** 2165–2169.

Huynen, M., Dandekar, T., and Bork, P. (1999). Variation and evolution of the citric-acid cycle: A genomic perspective. *Trends Microbiol.* **7,** 281–291.

Huynen, M., Snel, B., Lathe, W., 3rd, and Bork, P. (2000). Predicting protein function by genomic context: Quantitative evaluation and qualitative inferences. *Genome Res.* **10,** 1204–1210.

Itaya, M. (1995). An estimation of minimal genome size required for life. *FEBS Lett.* **362,** 257–260.

Ivanov, K. A., Thiel, V., Dobbe, J. C., van der Meer, Y., Snijder, E. J., and Ziebuhr, J. (2004). Multiple enzymatic activities associated with severe acute respiratory syndrome coronavirus helicase. *J. Virol.* **78,** 5619–5632.

Iwabe, N., Kuma, K., Hasegawa, M., Osawa, S., and Miyata, T. (1989). Evolutionary relationship of archaebacteria, eubacteria, and eukaryotes inferred from phylogenetic trees of duplicated genes. *Proc. Natl. Acad. Sci. USA* **86,** 9355–9359.

Iyer, L. M., Aravind, L., Bork, P., *et al.* (2001). *Quod erat demonstrandum*? The mystery of experimental validation of apparently erroneous computational analyses of protein sequences. *Genome Biol.* **2,** Research0051.

Iyer, L. M., Koonin, E. V., and Aravind, L. (2003). Evolutionary connection between the catalytic subunits of DNA-dependent RNA polymerases and eukaryotic RNA-dependent RNA polymerases and the origin of RNA polymerases. *BMC Struct. Biol.* **3,** 1.

Iyer, L. M., Aravind, L., Coon, S. L., Klein, D. C., and Koonin, E. V. (2004). Evolution of cell–cell signaling in animals: Did late horizontal gene transfer from bacteria have a role? *Trends Genet.* **20,** 292–299.

Jacob, F., Perrin, D., Sanchez, C., and Monod, J. (1960). Operon: A group of genes with the expression coordinated by an operator. *C. R. Hebd. Seances Acad. Sci.* **250,** 1727–1729.

Jain R., Rivera, M.C., and Lake, J.A. (1999). Horizontal gene transfer among genomes: the complexity hypothesis. *Proc Natl Acad Sci U S A.* **96,** 3801–3806.

Jaroszewski, L., Rychlewski, L., Li, Z., Li, W., and Godzik, A. (2005). FFAS03: A server for profile–profile sequence alignments. *Nucleic Acids Res.* **33,** W284–W288.

Jeffery, C. J. (2005). Mass spectrometry and the search for moonlighting proteins. *Mass Spectrom Rev.* **24,** 772–782.

Jensen, R. A. (1976). Enzyme recruitment in the evolution of new function. *Annu. Rev. Microbiol.* **30,** 409–425.

Jensen, R. A. (2001). Orthologs and paralogs: We need to get it right. *Genome Biol.* **2,** I002.1–I002.3.

Jeong, H., Tombor, B., Albert, R., Oltvai, Z. N., and Barabasi, A. L. (2000). The large-scale organization of metabolic networks. *Nature* **407,** 651–654.

Jeong, H., Mason, S., Barab'asi, A.-L., and Oltvai, Z. N. (2001). Lethality and centrality in protein networks. *Nature* **411,** 41–42.

Jiang, W., Middleton, K., Yoon, H. J., Fouquet, C., and Carbon, J. (1993). An essential yeast protein, CBF5p, binds *in vitro* to centromeres and microtubules. *Mol. Cell. Biol.* **13,** 4884–4893.

Jones, D. T. (1999). Protein secondary structure prediction based on position-specific scoring matrices. *J. Mol. Biol.* **292,** 195–202.

Jones, N. C., and Pevzner, P. A. (2004). *An Introduction to Bioinformatics Algorithms.* MIT Press, Cambridge, MA.

Jordan, I. K., Makarova, K. S., Spouge, J. L., Wolf, Y. I., and Koonin, E. V. (2001). Lineage-specific gene expansions in bacterial and archaeal genomes. *Genome Res.* **11,** 55–65.

Jordan, I. K., Rogozin, I. B., Wolf, Y. I., and Koonin, E. V. (2002). Essential genes are more evolutionarily conserved than are nonessential genes in bacteria. *Genome Res.* **12,** 962–968.

Jordan, I. K., Wolf, Y. I., and Koonin, E. V. (2004). Duplicated genes evolve slower than singletons despite the initial rate increase. *BMC Evol. Biol.* **4,** 22.

Jose, M. V., and Bishop, R. F. (2003). Scaling properties and symmetrical patterns in the epidemiology of rotavirus infection. *Philos. Trans. R. Soc. London B Biol. Sci.* **358,** 1625–1641.

Joyce, A. R., and Palsson, B. O. (2006). The model organism as a system: Integrating "omics" data sets. *Nat. Rev. Mol. Cell. Biol.* **7,** 198–210.

Jukes, T. H. (1997). Oparin and Lysenko. *J. Mol. Evol.* **45,** 339–340.

Kalmykova, A. I., Nurminsky, D. I., Ryzhov, D. V., and Shevelyov, Y. Y. (2005). Regulated chromatin domain comprising cluster of co-expressed genes in *Drosophila melanogaster. Nucleic Acids Res.* **33,** 1435–1444.

Kamath-Loeb, A. S., Shen, J. C., Loeb, L. A., and Fry, M. (1998). Werner syndrome protein: II. Characterization of the integral $3' \rightarrow 5'$ DNA exonuclease. *J. Biol. Chem.* **273,** 34145–34150.

Kamer, G., and Argos, P. (1984). Primary structural comparison of RNA-dependent polymerases from plant, animal and bacterial viruses. *Nucleic Acids Res.* **12,** 7269–7282.

Kaneda, K., Kuzuyama, T., Takagi, M., Hayakawa, Y., and Seto, H. (2001). An unusual isopentenyl diphosphate isomerase found in the mevalonate pathway gene cluster from *Streptomyces* sp. strain CL190. *Proc. Natl. Acad. Sci. USA* **98,** 932–937.

Kaneko, T., Sato, S., Kotani, H., *et al.* (1996). Sequence analysis of the genome of the unicellular cyanobacterium *Synechocystis* sp. strain PCC6803: II. Sequence determination of the entire genome and assignment of potential protein-coding regions. *DNA Res.* **3,** 109–136.

Karev, G. P., Wolf, Y. I., Rzhetsky, A. Y., Berezovskaya, F. S., and Koonin, E. V. (2002). Birth and death of protein domains: A simple model of evolution explains power law behavior. *BMC Evol. Biol.* **2,** 18.

Karev, G. P., Wolf, Y. I., Berezovskaya, F. S., and Koonin, E. V. (2004). Gene family evolution: An in-depth theoretical and simulation analysis of non-linear birth–death-innovation models. *BMC Evol. Biol.* **4,** 32.

Karev, G. P., Berezovskaya, F. S., and Koonin, E. V. (2005). Modeling genome evolution with a diffusion approximation of a birth-and-death process. *Bioinformatics* **21**(Suppl. 3), iii12–iii19.

Karlin, S., and Altschul, S. F. (1990). Methods for assessing the statistical significance of molecular sequence features by using general scoring schemes. *Proc. Natl. Acad. Sci. USA* **87,** 2264–2268.

Kawashima, S., and Kanehisa, M. (2000). AAIndex: Amino acid index database. *Nucleic Acids Res.* **28,** 374.

Keeling, P. J., and Inagaki, Y. (2004). A class of eukaryotic GTPase with a punctate distribution suggesting multiple functional replacements of translation elongation factor 1alpha. *Proc. Natl. Acad. Sci. USA* **101,** 15380–15385.

Keith, J., and Fraenkel-Conrat, H. (1975). Tobacco mosaic virus RNA carries 5′-terminal triphosphorylated guanosine blocked by 5′-linked 7-methylguanosine. *FEBS Lett.* **57,** 31–34.

Kent, W. J., and Haussler, D. (2001). Assembly of the working draft of the human genome with GigAssembler. *Genome Res.* **11,** 1541–1548.

Khan, Z. A., Hiriyanna, K. T., Chavez, F., and Fraenkel-Conrat, H. (1986). RNA-directed RNA polymerases from healthy and from virus-infected cucumber. *Proc. Natl. Acad. Sci. USA* **83,** 2383–2386.

Kim, H., Klein, R., Majewski, J., and Ott, J. (2004). Estimating rates of alternative splicing in mammals and invertebrates. *Nat. Genet.* **36,** 915–916.

Kimura, K., Wakamatsu, A., Suzuki, Y., *et al.* (2006). Diversification of transcriptional modulation: Large-scale identification and characterization of putative alternative promoters of human genes. *Genome Res.* **16,** 55–65.

Kimura, M. (1983). *The Neutral Theory of Molecular Evolution.* Cambridge University Press, Cambridge, UK.

Kinch, L. N., and Grishin, N. V. (2002). Expanding the nitrogen regulatory protein superfamily: Homology detection at below random sequence identity. *Proteins* **48,** 75–84.

Klebanov, L., Jordan, C., and Yakovlev, A. (2005). A new type of stochastic dependence revealed in gene expression data. *Stat. Appl. Genet. Mol. Biol.* **5,** Article 7.

Klein, D. J., Moore, P. B., and Steitz, T. A. (2004). The roles of ribosomal proteins in the structure assembly and evolution of the large ribosomal subunit. *J. Mol. Biol.* **340,** 141–177.

Kleinfeld, J. (2001). The small world problem. *Society* **39,** 61–66.

Klenk, H. P., Clayton, R. A., Tomb, J. F., *et al.* (1997). The complete genome sequence of the hyperthermophilic, sulphate-reducing archaeon *Archaeoglobus fulgidus. Nature* **390,** 364–370.

Knight, R. D., Freeland, S. J., and Landweber, L. F. (2001). Rewiring the keyboard: Evolvability of the genetic code. *Nat. Rev. Genet.* **2,** 49–58.

Knox, R. A. (1928). *Introduction to the Best Detective Stories of the Year 1928* (with Henry Harrington; U.S. title, *The Best English Detective Stories of 1928*). Horace Liveright, 1929.

Kondrashov, F. A., Rogozin, I. B., Wolf, Y. I., and Koonin, E. V. (2002). Selection in the evolution of gene duplication. *Genome Biol.* **3,** Research0008.1–0008.9.

Koonin, E. (1991). Genome replication/expression strategies of positive-strand RNA viruses: A simple version of a combinatorial classification and prediction of new strategies. *Virus Genes* **5,** 273–282.

Koonin, E. V. (2000). How many genes can make a cell: The minimal-gene-set concept. *Annu. Rev. Genomics Hum. Genet.* **1,** 99–116.

Koonin, E. V. (2001). An apology for orthologs—Or brave new memes. *Genome Biol.* **2,** I005.1–I005.2.

Koonin, E. V. (2003a). Comparative genomics, minimal gene-sets and the last universal common ancestor. *Nat. Rev. Microbiol.* **1,** 127–136.

Koonin, E. V. (2003b). Horizontal gene transfer: The path to maturity. *Mol. Microbiol.* **50,** 725–727.

Koonin, E.V. (2005). Orthologs, paralogs, and evolutionary genomics. *Annu Rev Genet.* **39,** 309–338.

Koonin, E. V. (2006). The origin of introns and their role in eukaryogenesis: A compromise solution to the introns-early versus introns-late debate? *Biol. Direct* **1,** 22.

Koonin, E. V., and Aravind, L. (1998). Genomics: Re-evaluation of translation machinery evolution. *Curr. Biol.* **8,** R266-R269.

Koonin, E. V., and Dolja, V. V. (1993). Evolution and taxonomy of positive-strand RNA viruses: Implications of comparative analysis of amino acid sequences. *Crit. Rev. Biochem. Mol. Biol.* **28,** 375–430.

Koonin, E. V., and Gorbalenya, A. E. (1989). Evolution of RNA genomes: Does the high mutation rate necessitate high rate of evolution of viral proteins? *J. Mol. Evol.* **28,** 524–527.

Koonin, E. V., and Martin, W. (2005). On the origin of genomes and cells within inorganic compartments. *Trends Genet.* **21,** 647–654.

Koonin, E. V., Tatusov, R. L., and Rudd, K. E. (1995). Sequence similarity analysis of *Escherichia coli* proteins: Functional and evolutionary implication. *Proc. Natl. Acad. Sci. USA* **92,** 11921–11925.

Koonin, E. V., Mushegian, A. R., and Bork, P. (1996). Non-orthologous gene displacement. *Trends Genet.* **12,** 334–336.

Koonin, E. V., Mushegian, A. R., Galperin, M. Y., and Walker, D. R. (1997). Comparison of archaeal and bacterial genomes: Computer analysis of protein sequences predicts novel functions and suggests a chimeric origin for the archaea. *Mol. Microbiol.* **25,** 619–637.

Koonin, E. V., Senkevich, T. G., and Dolja, V. V. (2006). The ancient virus world and evolution of cells. *Biol. Direct* **1,** 29.

Korbel, J. O., Jensen, L. J., von Mering, C., and Bork, P. (2004). Analysis of genomic context: Prediction of functional associations from conserved bidirectionally transcribed gene pairs. *Nat. Biotechnol.* **22,** 911–917.

Koski, L. B., and Golding, G. B. (2001). The closest BLAST hit is often not the nearest neighbor. *J. Mol. Evol.* **52,** 540–542.

Kozak, M. (2001). New ways of initiating translation in eukaryotes? *Mol. Cell. Biol.* **21,** 1899–1907.

Kozak, M. (2005). A second look at cellular mRNA sequences said to function as internal ribosome entry sites. *Nucleic Acids Res.* **33,** 6593–6602.

Kozbial, P. Z., and Mushegian, A. R. (2005). Natural history of *S*-adenosylmethionine-binding proteins. *BMC Struct. Biol.* **5,** 19.

Krishna, S. S., and Grishin, N. V. (2005). Structural drift: A possible path to protein fold change. *Bioinformatics* **21,** 1308–1310.

Kriventseva, E. V., Koch, I., Apweiler, R., *et al.* (2003). Increase of functional diversity by alternative splicing. *Trends Genet.* **19,** 124–128.

Krogh, A., Brown, M., Mian, I. S., Sjolander, K., and Haussler, D. (1994). Hidden Markov models in computational biology. Applications to protein modeling. *J. Mol. Biol.* **235,** 1501–1531.

Kumar, S. (2005). Molecular clocks: Four decades of evolution. *Nat. Rev. Genet.* **6,** 654–662.

Kurland, C. G. (2005). What tangled web: Barriers to rampant horizontal gene transfer. *Bioessays* **27,** 741–747.

Kurland, C. G., Canback, B., and Berg, O. G. (2003). Horizontal gene transfer: A critical view. *Proc. Natl. Acad. Sci. USA* **100,** 9658–9662.

Kuznetsov, V. A. (2001). Distribution associated with stochastic processes of gene expression in a single eukaryotic cell. *EUROSIP J. Appl. Signal Processing* **4,** 285–296.

Kyrpides, N. C., and Olsen, G. J. (1999). Archaeal and bacterial hyperthermophiles: Horizontal gene exchange or common ancestry? *Trends Genet.* **15,** 298–299.

Kyrpides, N. C., and Ouzounis, C. A. (1999). Whole-genome sequence annotation: "Going wrong with confidence." *Mol. Microbiol.* **32,** 886–887.

Lamb, D. C., Kelly, D. E., Manning, N. J., and Kelly, S. L. (1998). A sterol biosynthetic pathway in *Mycobacterium*. *FEBS Lett.* **437,** 142–144.

Lamers, M. H., Georgescu, R. E., Lee, S. G., O'Donnell, M., and Kuriyan, J. (2006). Crystal structure of the catalytic alpha subunit of *E. coli* replicative DNA polymerase III. *Cell* **126,** 881–892.

Lander, E. S., Linton, L. M., Birren, B., *et al.* (2001). International Human Genome Sequencing Consortium. Initial sequencing and analysis of the human genome. *Nature* **409,** 860–921.

Lane, T. W., Saito, M. A., George, G. N., Pickering, I. J., Prince, R. C., and Morel, F. M. (2005). Biochemistry: A cadmium enzyme from a marine diatom. *Nature* **435,** 42.

Lasters, I., Wodak, S. J., Alard, P., and van Cutsem, E. (1988). Structural principles of parallel beta-barrels in proteins. *Proc. Natl. Acad. Sci. USA* **85,** 3338–3342.

Lathe, W. C., 3rd, and Bork, P. (2001). Evolution of *tuf* genes: Ancient duplication, differential loss and gene conversion. *FEBS Lett.* **502,** 113–116.

Lawrence, J.G., Hendrickson, H. (2003) Lateral gene transfer: when will adolescence end? *Mol Microbiol.* **50,** 739–749.

Lawrence, J. G., and Ochman, H. (1997). Amelioration of bacterial genomes: Rates of change and exchange. *J. Mol. Evol.* **44,** 383–397.

Lawrence, J. G., and Roth, J. R. (1996). Selfish operons: Horizontal transfer may drive the evolution of gene clusters. *Genetics* **143,** 1843–1860.

Leipe, D. D., Aravind, L., and Koonin, E. V. (1999). Did DNA replication evolve twice independently? *Nucleic Acids Res.* **27,** 3389–3401.

Leipe, D. D., Aravind, L., Grishin, N. V., and Koonin, E. V. (2000). The bacterial replicative helicase DnaB evolved from a RecA duplication. *Genome Res.* **10,** 5–16.

Leipe, D. D., Koonin, E. V., and Aravind, L. (2003). Evolution and classification of P-loop kinases and related proteins. *J. Mol. Biol.* **333,** 781–815.

Leonard, C. J., Aravind, L., and Koonin, E. V. (1998). Novel families of putative protein kinases in bacteria and archaea: Evolution of the "eukaryotic" protein kinase superfamily. *Genome Res.* **8,** 1038–1047.

Lerat, E., Daubin, V., and Moran, N. A. (2003). From gene trees to organismal phylogeny in prokaryotes: The case of the gamma-proteobacteria. *PLoS Biol.* **1,** e19.

Lerat, E., Daubin, V., Ochman, H., and Moran, N. A. (2005). Evolutionary origins of genomic repertoires in bacteria. *PLoS Biol.* **3,** e130.

Lespinet, O., Wolf, Y. I., Koonin, E. V., and Aravind, L. (2002). The role of lineage-specific gene family expansion in the evolution of eukaryotes. *Genome Res.* **12,** 1048–1059.

Letunic, I., Copley, R. R., Schmidt, S., *et al.* (2004). SMART 4.0: Towards genomic data integration. *Nucleic Acids Res.* **32,** D142–D144.

Levinson, A. D., Oppermann, H., Levintow, L., Varmus, H. E., and Bishop, J. M. (1978). Evidence that the transforming gene of avian sarcoma virus encodes a protein kinase associated with a phosphoprotein. *Cell* **15,** 561–572.

Levitt, M., and Chothia, C. (1976). Structural patterns in globular proteins. *Nature* **261,** 552–558.

Lewontin, R., and Levins, R. (1976). The problem of lysenkoism. In *The Radicalisation of Science* (H. Rose and S. Rose, eds.), pp. 32–64. Macmillan, London.

Li, Q., Lee, B. T., and Zhang, L. (2005). Genome-scale analysis of positional clustering of mouse testis-specific genes. *BMC Genomics* **6,** 7.

Li, W. (1990). Mutual information functions versus correlation functions. *J. Stat. Phys.* **60,** 823–837.

Lindbo, J. A., and Dougherty, W. G. (1992a). Pathogen-derived resistance to a potyvirus: Immune and resistant phenotypes in transgenic tobacco expressing altered forms of a potyvirus coat protein nucleotide sequence. *Mol. Plant Microbe Interactions* **5,** 144–153.

Lindbo, J. A, and Dougherty, W. G. (1992b). Untranslatable transcripts of the tobacco etch virus cout protein gene sequence can interfere with tobacco etch virus replication in transgenic plants and protoplasts. *Virology* **189,** 725–733.

Lindbo, J. A., and Dougherty, W. G. (2005). Plant pathology and RNAi: A brief history. *Annu. Rev. Phytopathol.* **43,** 191–204.

Lipman, D. J., and Pearson, W. R. (1985). Rapid and sensitive protein similarity searches. *Science* **227,** 1435–1441.

Liu, J., and Mushegian, A. (2003). Three monophyletic superfamilies account for the majority of the known glycosyltransferases. *Protein Sci.* **12,** 1418–1431.

Liu, J., and Mushegian, A. (2004). Displacements of prohead protease genes in the late operons of double-stranded DNA bacteriophages. *J. Bacteriol.* **186,** 4369–4375.

Liu, J., and Rost, B. (2001). Comparing function and structure between entire proteomes. *Protein Sci.* **10,** 1970–1979.

Liu, J., Glazko, G., and Mushegian, A. R. (2006). Protein repertoire of double-stranded DNA bacteriophages. *Virus Res.* **117,** 68–80.

Liu, Z., and Binns, A. N. (2003). Functional subsets of the virB type IV transport complex proteins involved in the capacity of *Agrobacterium tumefaciens* to serve as a recipient in virB-mediated conjugal transfer of plasmid RSF1010. *J. Bacteriol.* **185,** 3259–3269.

Lowe, J., Stock, D., Jap, B., Zwickl, P., Baumeister, W., and Huber, R. (1995). Crystal structure of the 20S proteasome from the archaeon *T. acidophilum* at 3.4 Å resolution. *Science* **268,** 533–539.

Luc, N., Risler, J. L., Bergeron, A., and Raffinot, M. (2003). Gene teams: A new formalization of gene clusters for comparative genomics. *Comput. Biol. Chem.* **27,** 59–67.

Ludwig, W., Strunk, O., Klugbauer, S., *et al.* (1998). Bacterial phylogeny based on comparative sequence analysis. *Electrophoresis* **19,** 554–568.

Lundstrom, J., Rychlewski, L., Bujnicki, J., and Elofsson, A. (2001). Pcons: A neural-network-based consensus predictor that improves fold recognition. *Protein Sci.* **10,** 2354–2362.

Luttgen, H., Rohdich, F., Herz, S., *et al.* (2000). Biosynthesis of terpenoids: YchB protein of *Escherichia coli* phosphorylates the 2-hydroxy group of 4-diphosphocytidyl-2C-methyl-D-erythritol. *Proc. Natl. Acad. Sci. USA* **97,** 1062–1067.

Lynch, M., and Conery, J. S. (2000). The evolutionary fate and consequences of duplicate genes. *Science* **290,** 1151–1155.

Ma, H., and Zeng, A. P. (2003). Reconstruction of metabolic networks from genome data and analysis of their global structure for various organisms. *Bioinformatics* **19,** 270–277.

Magrum, L. J., Luehrsen, K. R., and Woese, C. R. (1978). Are extreme halophiles actually "bacteria"? *J. Mol. Evol.* **11,** 1–8.

Maheswari, U., Montsant, A., Goll, J., *et al.* (2005). The diatom EST database. *Nucleic Acids Res.* **33,** D344–D347.

Makarova, K. S., and Koonin, E. V. (2005). Evolutionary and functional genomics of the *Archaea. Curr. Opin. Microbiol.* **8,** 586–594.

Makarova, K. S., Wolf, Y. I., and Koonin, E. V. (2003). Potential genomic determinants of hyperthermophily. *Trends Genet.* **19,** 172–176.

Makino, T., and Gojobori, T. (2006). The evolutionary rate of a protein is influenced by features of the interacting partners. *Mol. Biol. Evol.* **23,** 784–789.

Manuel, M., Kruse, M., Muller, W. E., and Le Parco, Y. (2000). The comparison of beta-thymosin homologues among metazoa supports an arthropod–nematode clade. *J. Mol. Evol.* **51,** 378–381.

Mao, Y., Senic-Matuglia, F., Di Fiore, P. P., Polo, S., Hodsdon, M. E., and De Camilli, P. (2005). Deubiquitinating function of ataxin-3: Insights from the solution structure of the Josephin domain. *Proc. Natl. Acad. Sci. USA* **102,** 12700–12705.

Marquet, P. A., Quinones, R. A., Abades, S., *et al.* (2005). Scaling and power-laws in ecological systems. *J. Exp. Biol.* **208,** 1749–1769.

Martinez, M. A., Pezo, V., Marliere, P., and Wain-Hobson, S. (1996). Exploring the functional robustness of an enzyme by *in vitro* evolution. *EMBO J.* **15,** 1203–1210.

Martinot, T. A., and Benner, S. A. (2004). Artificial genetic systems: Exploiting the "aromaticity" formalism to improve the tautomeric ratio for isoguanosine derivatives. *J. Org. Chem.* **69,** 3972–3975.

May, A. C. (2004). Percent sequence identity; The need to be explicit. *Structure* **12,** 737–738.

McCullough, R. M., Cantor, C. R., and Ding, C. (2005). High-throughput alternative splicing quantification by primer extension and matrix-assisted laser desorption/ionization time-of-flight mass spectrometry. *Nucleic Acids Res.* **33,** e99.

Mears, J. A., Cannone, J. J., Stagg, S. M., Gutell, R. R., Agrawal, R. K., and Harvey, S. C. (2002). Modeling a minimal ribosome based on comparative sequence analysis. *J. Mol. Biol.* **321,** 215–234.

Meiners, S., Xu, A., and Schindler, M. (1991). Gap junction protein homologue from *Arabidopsis thaliana*: Evidence for connexins in plants. *Proc. Natl. Acad. Sci. USA* **88**, 4119–4122.

Mendelejeff, D. (1869). Ueber die Beziehumgem der Eigenschaften zu den Atomgewichten der Elemente [On the relationship of the properties of the elements to their atomic weights]. *Z. Chem.* **12**, 405–406.

Meyer, A. (1999). Homology and homoplasy: The retention of genetic programmes. In *Homology* (G. Bock, ed.), pp. 141–157. Wiley, New York.

Mi, H., Vandergriff, J., Campbell, M., *et al.* (2003). Assessment of genome-wide protein function classification for *Drosophila melanogaster. Genome Res.* **13**, 2118–2128.

Milanesi, L., Petrillo, M., Sepe, L., *et al.* (2005). Systematic analysis of human kinase genes: A large number of genes and alternative splicing events result in functional and structural diversity. *BMC Bioinformatics* **6**, S20.

Milgram, S. (1967). The small world problem. *Psychol. Today* **2**, 60–67.

Miller, W. (2001). Comparison of genomic DNA sequences: Solved and unsolved problems. *Bioinformatics* **17**, 391–397.

Milo, R., Itzkovitz, S., Kashtan, N., *et al.* (2004). Superfamilies of evolved and designed networks. *Science* **303**, 1538–1542.

Mira, A., Ochman, H., and Moran, N. A. (2001). Deletional bias and the evolution of bacterial genomes. *Trends Genet.* **17**, 589–596.

Mirkin, B., and Koonin, E. V. (2003). A top-down method for building genome classification trees with linear binary hierarchies. In *Bioconsensus* (M. Janowitz, F. McMorris, B. Mirkin, and F. Roberts, eds.), Vol. 61, pp. 97–112. American Mathematical Society, Providence, RI.

Mirkin, B. G., Fenner, T. I., Galperin, M. Y., and Koonin, E. V. (2003). Algorithms for computing parsimonious evolutionary scenarios for genome evolution, the last universal common ancestor and dominance of horizontal gene transfer in the evolution of prokaryotes. *BMC Evol. Biol.* **3**, 2.

Mongodin, E. F., Nelson, K. E., Daugherty, S., *et al.* (2005). The genome of *Salinibacter ruber*: convergence and gene exchange among hyperhalophilic bacteria and archaea. *Proc. Natl. Acad. Sci. USA* **102**, 18147–18152.

Morowitz, H. J. (1964). Requirements of a minimum free living replicating system. Seventhth COSPAR meeting; 5th International Space Science Symposium, Florence, Italy; United States: May 8–20, 1964.

Morowitz, H. J. (1984). The completeness of molecular biology. *Isr. J. Med. Sci.* **20**, 750–753.

Morowitz, H. J., and Cleverdon, R. C. (1959). An extreme example of the coding problem, avian PPLO 5969. *Biochim. Biophys. Acta* **34**, 578–579.

Morowitz, H. J., and Tourtellotte, M. E. (1962). The smallest living cells. *Sci. Am.* **206**, 117–126.

Moser, M. J., Holley, W. R., Chatterjee, A., and Mian, I. S. (1997). The proofreading domain of *Escherichia coli* DNA polymerase I and other DNA and/or RNA exonuclease domains. *Nucleic Acids Res.* **25**, 5110–5118.

Mott, R. (2000). Accurate formula for *P*-values of gapped local sequence and profile alignments. *J. Mol. Biol.* **300**, 649–659.

Moult, J. (2005). A decade of CASP: Progress, bottlenecks and prognosis in protein structure prediction. *Curr. Opin. Struct. Biol.* **15**, 285–289.

Mount, D. W. (2004). *Bioinformatics. Sequence and Genome Analysis*, 2nd ed. Cold Spring Harbor Laboratory Press, Cold Spring Harbor, NY.

Munoz-Dorado, J., Inouye, S., and Inouye, M. (1991). A gene encoding a protein serine/threonine kinase is required for normal development of *M. xanthus*, a gram-negative bacterium. *Cell* **67**, 995–1006.

Murphy, W. J., Pevzner, P. A., and O'Brien, S. J. (2004). Mammalian phylogenomics comes of age. *Trends Genet.* **20**, 631–639.

Murzin, A. G. (1998). How far divergent evolution goes in proteins. *Curr. Opin. Struct. Biol.* **8**, 380–387.

Mushegian, A. (1999). The minimal genome concept. *Curr. Opin. Genet. Dev.* **9**, 709–714.

Mushegian, A. (2004). Evolution and function of processosome, the complex that assembles ribosomes in eukaryotes: Clues from comparative sequence analysis. In *Practical Bioinformatics*, Vol., 15, p. 265. Springer, New York.

Mushegian, A. (2005). Protein content of minimal and ancestral ribosome. *RNA* **11**, 1400–1406.

Mushegian, A.R. (2000). Annotations of biochemically uncharacterized open reading frames (ORFs). *Mol. Microbiol.* **35**, 697–698.

Mushegian, A. R. (2002). Refining structural and functional predictions for secretasome components by comparative sequence analysis. *Proteins* **47**, 69–74.

Mushegian, A. R., and Koonin, E. V. (1993). The proposed plant connexin is a protein kinase-like protein. *Plant Cell* **5**, 998–999.

Mushegian, A. R., and Koonin, E. V. (1996a). A minimal gene set for cellular life derived by comparison of complete bacterial genomes. *Proc. Natl. Acad. Sci. USA* **93**, 10268–10273.

Mushegian, A. R., and Koonin, E. V. (1996b). Gene order is not conserved in bacterial evolution. *Trends Genet.* **12**, 289–290.

Mushegian, A. R., Bassett, D. E., Jr., Boguski, M. S., Bork, P., and Koonin, E. V. (1997). Positionally cloned human disease genes: Patterns of evolutionary conservation and functional motifs. *Proc. Natl. Acad. Sci. USA* **94**, 5831–5836.

Myllykallio, H., Lipowski, G., Leduc, D., Filee, J., Forterre, P., and Liebl, U. (2002). An alternative flavin-dependent mechanism for thymidylate synthesis. *Science* **297,** 105–107.

Nakahigashi, K., Kubo, N., Narita, S., *et al.* (2002). HemK, a class of protein methyl transferase with similarity to DNA methyl transferases, methylates polypeptide chain release factors, and hemK knockout induces defects in translational termination. *Proc. Natl. Acad. Sci. USA* **99,** 1473–1478.

Nakao, M., Barrero, R. A., Mukai, Y., Motono, C., Suwa, M., and Nakai, K. (2005). Large-scale analysis of human alternative protein isoforms: Pattern classification and correlation with subcellular localization signals. *Nucleic Acids Res.* **33,** 2355–2363.

Needleman, S.B. and Wunsch, C.D. (1970) A general method applicable to the search for similarities in the amino acid sequence of two proteins. *J Mol Biol.* **48,** 443–453.

Neidhardt, F. C. (ed.-in-chief), Curtiss R., III, Ingraham, J. L., *et al.* (eds.) (1996). *Escherichia coli and Salmonella: Cellular and Molecular Biology*. American Society for Microbiology, Washington, DC.

Nelson, K. E., Clayton, R. A., Gill, S. R., *et al.* (1999). Evidence for lateral gene transfer between archaea and bacteria from genome sequence of *Thermotoga maritima. Nature* **399,** 323–329.

Nevers, P., and Saedler, H. (1977). Transposable genetic elements as agents of gene instability and chromosomal rearrangements. *Nature* **268,** 109–115.

Newman, M. E. J. (2003). The structure and function of complex networks. *SIAM Rev.* **45,** 167–256.

Newman, M. E. J. (2005). *Power Laws, Pareto Distributions and Zipf's Law*. Available at http://arxiv.org/abs/cond-mat/0412004. Accessed April 27, 2006.

Nicholson, J. K. (2006). Reviewers peering from under a pile of "omics" data. *Nature* **440,** 992.

Nielsen, P., and Krogh, A. (2005). Large-scale prokaryotic gene prediction and comparison to genome annotation. *Bioinformatics* **21,** 4322–4329.

Ofengand, J. (2002). Ribosomal RNA pseudouridines and pseudouridine synthases. *FEBS Lett.* **514,** 17–25.

Ofengand, J., Malhotra, A., Remme, J., *et al.* (2001). Pseudouridines and pseudouridine synthases of the ribosome. *Cold Spring Harbor Symp. Quant. Biol.* **66,** 147–159.

Ohno, S. (1970). *Evolution by Gene Duplication*. Springer-Verlag, New York.

Olsen, G. J., and Woese, C. R. (1996). Lessons from an archaeal genome: What are we learning from *Methanococcus jannaschii*? *Trends Genet.* **12,** 377–379.

Olsen, G. J., and Woese, C. R. (1997). Archaeal genomics: An overview. *Cell* **89,** 991–994.

Olsen, G. J., Woese, C. R., and Overbeek, R. (1994). The winds of (evolutionary) change: Breathing new life into microbiology. *J. Bacteriol.* **176,** 1–6.

O'Malley, M. A., and Dupre, J. (2005). Fundamental issues in systems biology. *Bioessays* **27,** 1270–1276.

Oparin, A. I. (1953). *The Origin of Life*. Dover, New York.

Orengo, C. A., and Thornton, J. M. (2005). Protein families and their evolution—A structural perspective. *Annu. Rev. Biochem.* **74,** 867–900.

Orengo, C. A., Jones, D. T., and Thornton, J. M. (1994). Protein superfamilies and domain superfolds. *Nature* **372,** 631–634.

Ota, I. M., and Varshavsky, A. (1993). A yeast protein similar to bacterial two-component regulators. *Science* **262,** 566–569.

Ouatas, T., Salerno, M., Palmieri, D., and Steeg, P. S. (2003). Basic and translational advances in cancer metastasis: Nm23. *J. Bioenerg. Biomembr.* **35,** 73–79.

Ougland, R., Zhang, C. M., Liiv, A., *et al.* (2004). AlkB restores the biological function of mRNA and tRNA inactivated by chemical methylation. *Mol. Cell* **16,** 107–116.

Ouzounis, C. (1999). Orthology: Another terminology muddle. *Trends Genet.* **19,** 445.

Ouzounis, C., and Kyrpides, N. (1996). The emergence of major cellular processes in evolution. *FEBS Lett.* **390,** 119–123.

Ouzounis, C., Casari, G., Valencia, A., and Sander, C. (1996). Novelties from the complete genome of *Mycoplasma genitalium. Mol. Microbiol.* **20,** 895–900.

Ouzounis, C. A., Kunin, V., Darzentas, N., and Goldovsky, L. (2006). A minimal estimate for the gene content of the last universal common ancestor—Exobiology from a terrestrial perspective. *Res. Microbiol.* **157,** 57–68.

Ovcharenko, I., Loots, G. G., Giardine, B. M., *et al.* (2005). Mulan: Multiple-sequence local alignment and visualization for studying function and evolution. *Genome Res.* **15,** 184–194.

Overbeek, R., Fonstein, M., D'Souza, M., Pusch, G. D., and Maltsev, N. (1999). The use of gene clusters to infer functional coupling. *Proc. Natl. Acad. Sci. USA* **96,** 2896–2901.

Park, J., Karplus, K., Bartlett, C., *et al.* (1998). Sequence comparisons using multiple sequences detect three times as many remote homologs as pairwise methods. *J. Mol. Biol.* **284,** 1201–1210.

Park, Y. R., Park, C. H., and Kim, J. H. (2005). GOChase: Correcting errors from gene ontology-based annotations for gene products. *Bioinformatics* **21,** 829–831.

Pasek, S., Bergeron, A., Risler, J. L., Louis, A., Ollivier, E., and Raffinot, M. (2005). Identification of genomic features using microsyntenies of domains: Domain teams. *Genome Res.* **15,** 867–874.

Pearl, F., Todd, A., Sillitoe, I., *et al.* (2005). The CATH Domain Structure Database and related resources Gene3D and DHS provide comprehensive domain family information for genome analysis. *Nucleic Acids Res.* **33,** D247–D251.

Pearson, A., Budin, M., and Brocks, J. J. (2003). Phylogenetic and biochemical evidence for sterol synthesis in the bacterium *Gemmata obscuriglobus. Proc. Natl. Acad. Sci. USA* **100,** 15352–15357.

Pearson, W. R. (1994). MIF proteins are not glutathione transferase homologs. *Protein Sci.* **3,** 525–527.

Pearson, W. R. (2000). Flexible sequence similarity searching with the FASTA3 program package. *Methods Mol. Biol.* **132,** 185–219.

Pearson, W. R., and Lipman, D. J. (1988). Improved tools for biological sequence comparison. *Proc. Natl. Acad. Sci. USA* **85,** 2444–2448.

Pei, J., and Grishin, N. V. (2001). GGDEF domain is homologous to adenylyl cyclase. *Proteins* **42,** 210–216.

Pellegrini, M., Marcotte, E. M., Thompson, M. J., Eisenberg, D., and Yeates, T. O. (1999). Assigning protein functions by comparative genome analysis: Protein phylogenetic profiles. *Proc. Natl. Acad. Sci. USA* **96,** 4285–4288.

Penny, D. (2004). Phylogeny in the comfort zone. *Systematic Biol.* **53,** 669–670.

Penny, D., and Poole, A. (1999). The nature of the last universal common ancestor. *Curr. Opin. Genet. Dev.* **9,** 672–677.

Peterson, K. J., and Eernisse, D. J. (2001). Animal phylogeny and the ancestry of bilaterians: Inferences from morphology and 18S rDNA gene sequences. *Evol. Dev.* **3,** 170–205.

Petsko, G. A. (2001). Homologuephobia. *Genome Biol.* **2,** I002.1–I002.2

Petsko, G. A. (2002). No place like Ome. *Genome Biol.* **3,** Comment1010.

Pevzner, P. A. (2000). *Computational Molecular Biology: An Algorithmic Approach.* MIT Press, Cambridge, MA.

Phage and the Origins of Molecular Biology, 40th Anniversary Edition (2006). Cold Spring Harbor Laboratory Press, Cold Spring Harbor, NY.

Philippe, H., Snell, E. A., Bapteste, E., Lopez, P., Holland, P. W., and Casane, D. (2004). Phylogenomics of eukaryotes: Impact of missing data on large alignments. *Mol. Biol. Evol.* **21,** 1740–1752.

Phillips, S. E., and Stockley, P. G. (1996). Structure and function of *Escherichia coli* met repressor: Similarities and contrasts with trp repressor. *Philos. Trans. R. Soc. London B Biol. Sci.* **351,** 527–535.

Pielou, E. C. (1969). *Introduction to Mathematical Ecology.* Wiley–Interscience, New York.

Pollack, J. D., Myers, M. A., Dandekar, T., and Herrmann, R. (2002). Suspected utility of enzymes with multiple activities in the small genome *Mycoplasma* species: The replacement of the missing "household" nucleoside diphosphate kinase gene and activity by glycolytic kinases. *OMICS* **6,** 247–258.

Ponting, C. P, and Dickens, N. J. (2001). Genome cartography through domain annotation. *Genome Biol.* **2,** Comment2006.

Ponting, C. P., Aravind, L., Schultz, J., Bork, P., and Koonin, E. V. (1999). Eukaryotic signalling domain homologues in archaea and bacteria. Ancient ancestry and horizontal gene transfer. *J. Mol. Biol.* **289,** 729–745.

Ponting, C. P., Hutton, M., Nyborg, A., Baker, M., Jansen, K., and Golde, T. E. (2002). Identification of a novel family of presenilin homologues. *Hum. Mol. Genet.* **11,** 1037–1044.

Popov, K. M, Kedishvili, N. Y., Zhao, Y., Shimomura, Y., Crabb, D. W., and Harris, R. A. (1993). Primary structure of pyruvate dehydrogenase kinase establishes a new family of eukaryotic protein kinases. *J. Biol. Chem.* **268,** 26602–26606.

Powell, B. C., and Hutchison, C. A., 3rd (2006). Similarity-based gene detection: Using COGs to find evolutionarily-conserved ORFs. *BMC Bioinformatics* **7,** 31.

Price, M.N., Huang, K.H., Arkin, A.P., and Alm, E.J. (2005) Operon formation is driven by co-regulation and not by horizontal gene transfer. *Genome Res.* **15,** 809–819.

Ptitsyn, O.B., Finkelstein, A.V. (1980) Similarities of protein topologies: evolutionary divergence, functional convergence or principles of folding? *Q Rev Biophys.* **13,** 339–386.

Pushker, R., Mira, A., and Rodriguea-Valera, F. (2004). Comparative genomics of gene-family size in closely related bacteria. *Genome Biol.* **5,** R27.

Putics, A., Filipowicz, W., Hall, J., Gorbalenya, A. E., and Ziebuhr, J. (2005). ADP-ribose-1′-monophosphatase: A conserved coronavirus enzyme that is dispensable for viral replication in tissue culture. *J. Virol.* **79,** 12721–12731.

Ranea, J. A. G., Grant, A., Thornton, J. M., and Orengo, C. A. (2005). Microeconomic principles explain an optimal genome size in bacteria. *Trends Genet.* **21,** 21–25.

Rawlings, N. D., Morton, F. R., and Barrett, A. J. (2006). MEROPS: The peptidase database. *Nucleic Acids Res.* **34,** D270–D272.

Raymond, J., Zhaxybayeva, O., Gogarten, J.P., et al. (2002) Whole-genome analysis of photosynthetic prokaryotes. *Science.* **298,** 1616–1620.

Reche, P. A. (2000). Lipoylating and biotinylating enzymes contain a homologous catalytic module. *Protein Sci.* **9,** 1922–1929.

Redner, S. (1998). How popular is your paper? An empirical study of the citation distribution. *Eur. Phys. J. B* **4,** 131–134.

Reeck, G. R., de Haën, C., Teller, D. C., *et al.* (1987). Homology in proteins and nucleic acids: A terminology muddle and a way out of it. *Cell* **50,** 667.

Remm, M., Storm, C. E., and Sonnhammer, E. L. (2001). Automated clustering of orthologs and in-paralogs from pairwise species comparisons. *J. Mol. Biol.* **314,** 1041–1052.

Richardson, J. S. (1981). The anatomy and taxonomy of protein structure. *Adv. Protein Chem.* **34,** 167–339.

Richardson, J. S., Getzoff, E. D., and Richardson, D. C. (1978). The beta bulge: A common small unit of nonrepetitive protein structure. *Proc. Natl. Acad. Sci. USA* **75,** 2574–2578.

Ridgen, D. J., and Galperin, M. Y. (2004). The DxDxDG motif for calcium binding: Multiple structural contexts and implications for evolution. *J. Mol. Biol.* **343,** 971–984.

Ridley, M. (1986). *Evolution and Classification: The Reformation of Cladism.* Longman, London.

Rivas, E., and Eddy, S. R. (2001). Noncoding RNA gene detection using comparative sequence analysis. *BMC Bioinformatics* **2,** 8.

Rivera, M. C., and Lake, J. A. (2004). The ring of life provides evidence for a genome fusion origin of eukaryotes. *Nature* **431,** 152–155.

Rivera, M. C., Jain, R., Moore, J. E., and Lake, J. A. (1998). Genomic evidence for two functionally distinct gene classes. *Proc. Natl. Acad. Sci. USA* **95,** 6239–6244.

Robards, A., Lucas, W., Pitts, J., Jongsma, H., and Spray, D. (1990). *Parallels in Cell to Cell Junctions in Plants and Animals,* NATO ASI Series. Springer-Verlag, Berlin.

Robson, B. (2001). Fastfinger: A study into the use of compressed residue pair separation matrices for protein sequence comparison. *IBM Systems J.* **40,** 442–463.

Rodionov, D. A., and Gelfand, M. S. (2005). Identification of a bacterial regulatory system for ribonucleotide reductases by phylogenetic profiling. *Trends Genet.* **21,** 385–389.

Rodriguez-Trelles, F., Tarrio, R., and Ayala, F. J. (2001). Erratic overdispersion of three molecular clocks: GPDH, SOD, and XDH. *Proc. Natl. Acad. Sci. USA* **98,** 11405–11410.

Rogozin, I. B., Makarova, K. S., Murvai, J., *et al.* (2002). Connected gene neighborhoods in prokaryotic genomes. *Nucleic Acids Res.* **30,** 2212–2223.

Rosengren, E., Aman, P., Thelin, S., *et al.* (1997). The macrophage migration inhibitory factor MIF is a phenylpyruvate tautomerase. *FEBS Lett.* **417,** 85–88.

Roth, M. J., Forbes, A. J., Boyne, M. T., 2nd, Kim, Y. B., Robinson, D. E., and Kelleher, N. L. (2005). Precise and parallel characterization of coding polymorphisms, alternative splicing, and modifications in human proteins by mass spectrometry. *Mol. Cell. Proteomics* **4,** 1002–1008.

Rouault, T. A., Stout, C. D., Kaptain, S., Harford, J. B., and Klausner, R. D. (1991). Structural relationship between an iron-regulated RNA-binding protein (IRE-BP) and aconitase: Functional implications. *Cell* **64,** 881–883.

Ruiz-Trillo, I., Paps, J., Loukota, M., *et al.* (2002). A phylogenetic analysis of myosin heavy chain type II sequences corroborates that *Acoela* and *Nemertodermatida* are basal bilaterians. *Proc. Natl. Acad. Sci. USA* **99,** 11246–11251.

Russell, R. B., Copley, R. R., and Barton, G. J. (1996). Protein fold recognition by mapping predicted secondary structures. *J. Mol. Biol.* **259,** 349–365.

Sadreyev, R., and Grishin, N. (2003). COMPASS: A tool for comparison of multiple protein alignments with assessment of statistical significance. *J. Mol. Biol.* **326,** 317–336.

Salamon, P. and Konopka, A.K. (1992) A maximum entropy principle for the distribution of local complexity in naturally occurring nucleotide sequences. *Comput. Chem.* **16,** 117–124

Salzberg, S. L., White, O., Peterson, J., and Eisen, J. A. (2001). Microbial genes in the human genome: Lateral transfer or gene loss? *Science* **292,** 1903–1906.

Sankoff, D. (2000). The early introduction of dynamic programming into computational biology. *Bioinformatics* **16,** 41–47.

Sankoff, D., and Blanchette, M. (1998). Multiple genome rearrangement and breakpoint phylogeny. *J. Comput. Biol.* **5,** 555–570.

Sankoff, D., and Blanchette, M. (1999). Phylogenetic invariants for genome rearrangements. *J. Comput. Biol.* **6,** 431–445.

Sankoff, D., and Krushkal, J. (eds.) (1983). *Time Warps, String Edits, and Macromolecules: The Theory and Practice of Sequence Comparison.* Addison-Wesley, Reading, MA.

Sankoff, D., Leduc, G., Antoine, N., Paquin, B., Lang, B. F., and Cedergren, R. (1992). Gene order comparisons for phylogenetic inference: Evolution of the mitochondrial genome. *Proc. Natl. Acad. Sci. USA* **89,** 6575–6579.

Schaffer, A. A., Wolf, Y. I., Ponting, C. P., Koonin, E. V., Aravind, L., and Altschul, S. F. (1999). IMPALA: Matching a protein sequence against a collection of PSI-BLAST-constructed position-specific scoring matrices. *Bioinformatics* **15,** 1000–1011.

Schaffer, A. A., Aravind, L., Madden, T. L., *et al.* (2001). Improving the accuracy of PSI-BLAST protein database searches with composition-based statistics and other refinements. *Nucleic Acids Res.* **29,** 2994–3005.

Scheel, H., Tomiuk, S., and Hofmann, K. (2003). Elucidation of ataxin-3 and ataxin-7 function by integrative bioinformatics. *Hum. Mol. Genet.* **12,** 2845–2852.

Schiebel, W., Pelissier, T., Riedel, L., *et al.* (1998). Isolation of an RNA-directed RNA polymerase-specific cDNA clone from tomato. *Plant Cell* **10,** 2087–2101.

Schneider-Poetsch, H. A., Braun, B., Marx, S., and Schaumburg, A. (1991). Phytochromes and bacterial sensor proteins are related by structural and functional homologies. Hypothesis on phytochrome-mediated signal transduction. *FEBS Lett.* **281,** 245–249.

Schubert, H. L., Blumenthal, R. M., and Cheng, X. (2003). Many paths to methyl transfer: A chronicle of convergence. *Trends Biochem. Sci.* **28,** 329–335.

Schuldiner, M., Collins, S. R., Thompson, N. J., *et al.* (2005). Exploration of the function and organization of the yeast early secretory pathway through an epistatic miniarray profile. *Cell* **123,** 507–519.

Schuler, G. D., Altschul, S. F., and Lipman, D. J. (1991). A workbench for multiple alignment construction and analysis. *Proteins* **9,** 180–190.

Schulze-Kremer, S. (1997). Adding semantics to genome databases: Towards an ontology for molecular biology. *Proc. Int. Conf. Intell. Syst. Mol. Biol.* **5,** 272–275.

Schwartz, S., Kent, W. J., Smit, A., *et al.* (2003). Human–mouse alignments with BLASTZ. *Genome Res.* **13,** 103–107.

Selkov, E., Maltsev, N., Olsen, G. J., Overbeek, R., and Whitman, W. B. (1997). A reconstruction of the metabolism of *Methanococcus janaschii* from sequence data. *Gene* **197,** GC11–GC26.

Semple, C., Daniel, P., Hordijk, W., Page, R. D., and Steel, M. (2004). Supertree algorithms for ancestral divergence dates and nested taxa. *Bioinformatics* **20,** 2355–2360.

Shen, Y., Buick, R., and Canfield, D. E. (2001). Isotopic evidence for microbial sulphate reduction in the early Archaean era. *Nature* **410,** 77–81.

Shen-Orr, S. S., Milo, R., Mangan, S., and Alon, U. (2002). Network motifs in the transcriptional regulation network of *Escherichia coli. Nat. Genet.* **31,** 64–68.

Simon, H. (1955). On a class of skew distribution functions. *Biometrika* **42,** 425–440.

Simons, K. T., Bonneau, R., Ruczinski, I., and Baker, D. (1999). Ab initio protein structure prediction of CASP III targets using ROSETTA. *Proteins* **37,** (Suppl. 3), 171–176.

Sirevag, R., Buchanan, B. B., Berry, J. A., and Troughton, J. H. (1977). Mechanisms of CO_2 fixation in bacterial photosynthesis studied by the carbon isotope fractionation technique. *Arch. Microbiol.* **112,** 35–38.

Sjolander, K. (2004). Phylogenomic inference of protein molecular function: Advances and challenges. *Bioinformatics* **20,** 170–179.

Smit, A., and Mushegian, A. (2000). Biosynthesis of isoprenoids via mevalonate in Archaea: The lost pathway. *Genome Res.* **10,** 1468–1484.

Smith, B., Ceusters, W., Klagges, B., *et al.* (2005). Relations in biomedical ontologies. *Genome Biol.* **6,** R46.

Smith, H. O., Hutchison, C. A., 3rd, Pfannkoch, C., and Venter, J. C. (2003). Generating a synthetic genome by whole genome assembly: phiX174 bacteriophage from synthetic oligonucleotides. *Proc. Natl. Acad. Sci. USA* **100,** 15440–15445.

Smith, T. F., and Waterman, M. S. (1981). Indentification of common molecular subsequences. *J. Mol. Biol.* **147,** 195–197.

Snel, B., Bork, P., and Huynen, M. A. (2002). The identification of functional modules from the genomic association of genes. *Proc. Natl. Acad. Sci. USA* **99,** 5890–5895.

Snel, B., Huynen, M. A., and Dutilh, B. E. (2005). Genome trees and the nature of genome evolution. *Annu. Rev. Microbiol.* **59,** 191–209.

Sober, E. (1991). *Reconstructing the Past: Parsimony, Evolution, and Inference.* MIT Press, Cambridge, MA.

Sober, E. (2004). The contest between likelihood and parsimony. *Systematic Biol.* **53,** 644–653.

Sober, E., and Steel, M. (2002). Testing the hypothesis of common ancestry. *J. Theor. Biol.* **218,** 395–408.

Soding, J. (2005). Protein homology detection by HMM-HMM comparison. *Bioinformatics* **21,** 951–960.

Sonnhammer, E. L., and Koonin, E. V. (2002). Orthology, paralogy, and proposed classification for paralog subtypes. *Trends Genet.* **18,** 619–620.

Speed, R. R., and Winkler, H. H. (1991). Acquisition of thymidylate by the obligate intracytoplasmic bacterium *Rickettsia prowazekii. J. Bacteriol.* **173,** 1704–1710.

Stebbins, C. E., Russo, A. A., Schneider, C., Rosen, N., Hartl, F. U., and Pavletich, N. P. (1997). Crystal structure of an Hsp90–geldanamycin complex: Targeting of a protein chaperone by an antitumor agent. *Cell* **89,** 239–250.

Steitz, T. A., and Moore, P. B. (2003). RNA, the first macromolecular catalyst: The ribosome is a ribozyme. *Trends Biochem. Sci.* **28,** 411–418.

Stewart, C. B., Schilling, J. W., and Wilson, A. C. (1987). Adaptive evolution in the stomach lysozymes of foregut fermenters. *Nature* **330,** 401–404.

Storey, J. D., and Siegmund, D. (2001). Approximate *p*-values for local sequence alignments: Numerical studies. *J. Comput. Biol.* **8,** 549–556.

Storm, C. E., and Sonnhammer, E. L. (2002). Automated ortholog inference from phylogenetic trees and calculation of orthology reliability. *Bioinformatics* **18,** 92–99.

Stormo, G. D. (2000). DNA binding sites: Representation and discovery. *Bioinformatics* **16,** 16–23.

Stumpf, M. P., Wiuf, C., and May, R. M. (2005). Subnets of scale-free networks are not scale-free: Sampling properties of networks. *Proc. Natl. Acad. Sci. USA* **102,** 4221–4224.

Sugimoto, H., Taniguchi, M., Nakagawa, A., Tanaka, I., Suzuki, M., and Nishihira, J. (1999). Crystal structure of human D-dopachrome tautomerase, a homologue of macrophage migration inhibitory factor, at 1.54 A resolution. *Biochemistry* **38,** 3268–3279.

Swofford, D. L., and Madison, W. P. (1992). Parsimony, character-state reconstructions, and evolutionary inferences. In *Systematics, Historical Ecology, and North American Freshwater Fishes* (R. L. Mayden, ed.). Stanford University Press, Stanford, CA.

Swofford, D. L., Olsen, G. J., Waddell, P. J., and Hillis, D. M. (1996). Phylogenetic inference. In *Molecular Systematics* (D. M. Hillis, C. Moritz, and B. K. Mable, eds.), 2nd ed. Sinauer, Sunderland, MA.

Swofford, D. L., Waddell, P. J., Huelsenbeck, J. P., Foster, P. G., Lewis, P. O., and Rogers, J. S. (2001). Bias in phylogenetic estimation and its relevance to the choice between parsimony and likelihood methods. *Systematic Biol.* **50,** 525–539.

Tamames, J. (2001). Evolution of gene order conservation in prokaryotes. *Genome Biol.* **2,** Research0020.

Tanaka, R., Yi, T. M., and Doyle, J. (2005). Some protein interaction data do not exhibit power law statistics. *FEBS Lett.* **579,** 5140–5144.

Tanaka, T., Tateno, Y., and Gojobori, T. (2005). Evolution of vitamin B_6 (pyridoxine) metabolism by gain and loss of genes. *Mol. Biol. Evol.* **22,** 243–250.

Tanaka, T., Ikeo, K., and Gojobori, T. (2006). Evolution of metabolic networks by gain and loss of enzymatic reaction in eukaryotes. *Gene* **365C,** 88–94.

Tannenbaum, E., and Shakhnovich, E. I. (2004). Error and repair catastrophes: A two-dimensional phase diagram in the quasispecies model. *Phys. Rev. E. Stat. Nonlin. Soft Matter Phys.* **69,** 011902.

Tanner, N. K., and Linder, P. (2001). DExD/H box RNA helicases: From generic motors to specific dissociation function. *Mol. Cell* **8,** 251–262.

Tatusov, R. L., Mushegian, A. R., Bork, P., *et al.* (1996). Metabolism and evolution of *Haemophilus influenzae* deduced from a whole-genome comparison with *Escherichia coli. Curr. Biol.* **6,** 279–291.

Tatusov, R. L., Koonin, E. V., and Lipman, D. J. (1997). A genomic perspective on protein families. *Science* **278,** 631–637.

Tatusov, R. L., Natale, D. A., Garkavtsev, I. V., *et al.* (2001). The COG database: New developments in phylogenetic classification of proteins from complete genomes. *Nucleic Acids Res.* **29,** 22–28.

Taylor, W. R. (2002). A "periodic table" for protein structures. *Nature* **416,** 657–660.

Telford, M. J. (2004). The multimeric beta-thymosin found in nematodes and arthropods is not a synapomorphy of the *Ecdysozoa. Evol. Dev.* **6,** 90–94.

Terakita, A. (2005). The opsins. *Genome Biol.* **6,** 213.

Thomson, J. M., Gaucher, E. A., Burgan, M. F., *et al.* (2005). Resurrecting ancestral alcohol dehydrogenases from yeast. *Nat. Genet.* **37,** 630–635.

Tipton, K., and Boyce, S. (2000). History of the enzyme nomenclature system. *Bioinformatics* **16,** 34–40.

Tong, A. H., Evangelista, M., Parsons, A. B., *et al.* (2001). Systematic genetic analysis with ordered arrays of yeast deletion mutants. *Science* **294,** 2364–2368.

Travers, J., and Milgram, S. (1969). An experimental study of the small world problem. *Sociometry* **32,** 425–443.

Trifonov, E. N. (2000). Consensus temporal order of amino acids and evolution of the triplet code. *Gene* **261,** 139–151.

Trifonov, E. N. (2004). The triplet code from first principles. *J. Biomol. Struct. Dyn.* **22,** 1–11.

Valencia, A. (2005). Automatic annotation of protein function. *Curr. Opin. Struct. Biol.* **15,** 267–274.

Varshavsky, A. (2004). "Spalog" and "sequelog": Neutral terms for spatial and sequence similarity. *Curr. Biol.* **14,** R181–R182.

Verhees, C. H., Kengen, S. W. M., Tuininga, J. E., *et al.* (2003). The unique features of glycolytic pathways in Archaea. *Biochem. J.* **375,** 231–246.

Vesterstrom, J., and Taylor, W. R. (2006). Flexible secondary structure based protein structure comparison applied to the detection of circular permutation. *J. Comput. Biol.* **13,** 43–63.

Vetsigian, K., Woese, C., and Goldenfeld, N. (2006). Collective evolution and the genetic code. *Proc. Natl. Acad. Sci. USA* **103,** 10696–10701.

Vinayagam, A., Shi, J., Pugalenthi, G., Meenakshi, B., Blundell, T. L., and Sowdhamini, R. (2003). DDBASE2.0: Updated domain database with improved identification of structural domains. *Bioinformatics* **19,** 1760–1764.

Vintsyuk, T. K. (1968). Speech discrimination by dynamic programming. *Cybernetics* **4,** 52–57. [Russian: *Kibernetika* **4,** 81–88]

Wagner, A. (2005). *Robustness and Evolvability in Living Systems.* Princeton University Press, Princeton, NJ.

Wagner, A., and Fell, D. A. (2001). The small world inside large metabolic networks. *Proc. Biol. Sci.* **268,** 1803–1810.

Walker, J. E., Saraste, M., Runswick, M. J., and Gay, N. J. (1982). Distantly related sequences in the alpha- and beta-subunits of ATP synthase, myosin, kinases and other ATP-requiring enzymes and a common nucleotide-binding fold. *EMBO J.* **1,** 945–951.

Wan, H., and Wootton, J. C. (2000). A global compositional complexity measure for biological sequences: AT-rich and GC-rich genomes encode less complex proteins. *Comput. Chem.* **24,** 71–94.

Wang, L., Xie, J., and Schultz, P. G. (2006). Expanding the genetic code. *Annu. Rev. Biophys. Biomol. Struct.* **35,** 225–249.

Watts, D. J., and Strogatz, S. H. (1998). Collective dynamics of "small-world" networks. *Nature* **393,** 440–442.

Wei, C., Gershowitz, A., and Moss, B. (1975). N6, O_2'-dimethyladenosine—A novel methylated ribonucleoside next to the 5′ terminal of animal cell and virus mRNAs. *Nature* **257,** 251–253.

Wei, C. M., and Moss, B. (1974). Methylation of newly synthesized viral messenger RNA by an enzyme in vaccinia virus. *Proc. Natl. Acad. Sci. USA* **71,** 3014–3018.

Wheelis, M. L., Kandler, O., and Woese, C. R. (1992). On the nature of global classification. *Proc. Natl. Acad. Sci. USA* **89,** 2930–2934.

Winkler, W. C., Nahvi, A., Sudarsan, N., Barrick, J. E., and Breaker, R. R. (2003). An mRNA structure that controls gene expression by binding *S*-adenosylmethionine. *Nat. Struct. Biol.* **10,** 701–707.

Winter, W. P., Walsh, K. A., and Neurath, H. (1968). Homology as applied to proteins. *Science* **162,** 1433.

Woese, C. R. (1998a). The universal ancestor. *Proc. Natl. Acad. Sci. USA* **95,** 6854–6859.

Woese, C. R. (1998b). Default taxonomy: Ernst Mayr's view of the microbial world. *Proc. Natl. Acad. Sci. USA* **95,** 11043–11046.

Woese, C. R. (2000). Interpreting the universal phylogenetic tree. *Proc. Natl. Acad. Sci. USA* **97,** 8392–8396.

Woese, C. R. (2002). On the evolution of cells. *Proc. Natl. Acad. Sci. USA* **99,** 8742–8747.

Woese, C. R., Magrum, L. J., and Fox, G. E. (1978). Archaebacteria. *J. Mol. Evol.* **11,** 245–251.

Woese, C. R., Maniloff, J., and Zablen, L. B. (1980). Phylogenetic analysis of the mycoplasmas. *Proc. Natl. Acad. Sci. USA* **77,** 494–498.

Woese, R. (2004). The archaeal concept and the world it lives in: A retrospective. *Photosynth. Res.* **80,** 361–372.

Wolf, Y. I., Aravind, L., Grishin, N. V., and Koonin, E. V. (1999). Evolution of aminoacyl-tRNA synthetases—Analysis of unique domain architectures and phylogenetic trees reveals a complex history of horizontal gene transfer events. *Genome Res.* **9,** 689–710.

Wolf, Y. I., Grishin, N. V., and Koonin, E. V. (2000a). Estimating the number of protein folds and families from complete genome data. *J. Mol. Biol.* **299,** 897–905.

Wolf, Y. I., Kondrashov, A. S., and Koonin, E. V. (2000b). Interkingdom gene fusions. *Genome Biol.* **1,** Research0013.

Wolf, Y. I., Rogozin, I. B., Grishin, N. V., Tatusov, R. L., and Koonin, E. V. (2001a). Genome trees constructed using five different approaches suggest new major bacterial clades. *BMC Evol. Biol.* **1,** 8.

Wolf, Y. I., Rogozin, I. B., Kondrashov, A. S., and Koonin, E. V. (2001b). Genome alignment, evolution of prokaryotic genome organization, and prediction of gene function using genomic context. *Genome Res.* 11, 356-372.

Wolfe, K. H., Morden, C. W., and Palmer, J. D. (1991). Ins and outs of plastid genome evolution. *Curr. Opin. Genet. Dev.* **1,** 523–529.

Wootton, J. C. (1994). Non-globular domains in protein sequences: Automated segmentation using complexity measures. *Comput. Chem.* **18,** 269–285.

Wootton, J.C., and Federhen, S. (1996) Analysis of compositionally biased regions in sequence databases. *Methods Enzymol.* **266,** 554–571.

Wu, N., Deiters, A., Cropp, T. A., King, D., and Schultz, P. G. (2004). A genetically encoded photocaged amino acid. *J. Am. Chem. Soc.* **126,** 14306–14307.

Xia, Y., and Levitt, M. (2004). Funnel-like organization in sequence space determines the distributions of protein stability and folding rate preferred by evolution. *Proteins* **55,** 107–114.

Xie, T., and Ding, D. (2000). Investigating 42 candidate orthologous protein groups by molecular evolutionary analysis on genome scale. *Gene* **261,** 305–310.

Yamashita, R., Suzuki, Y., Wakaguri, H., Tsuritani, K., Nakai, K., and Sugano, S. (2006). DBTSS: Database of human transcription start sites, progress report 2006. *Nucleic Acids Res.* **34,** D86–D89.

Yan, Y., and Moult, J. (2005). Protein family clustering for structural genomics. *J. Mol. Biol.* **353,** 744–759.

Yanai, I., Wolf, Y. I., and Koonin, E. V. (2002). Evolution of gene fusions: Horizontal transfer versus independent events. *Genome Biol.* **3,** Research0024.

Yang, Z. (1997). PAML: A program package for phylogenetic analysis by maximum likelihood. *Comput. Appl. Biosci.* **13,** 555–556.

Yona, G., and Levitt, M. (2002). Within the twilight zone: A sensitive profile–profile comparison tool based on information theory. *J. Mol. Biol.* **315,** 1257–1275.

Yu, Y. K., Wootton, J. C., and Altschul, S. F. (2003). The compositional adjustment of amino acid substitution matrices. *Proc. Natl. Acad. Sci. USA* **100,** 15688–15693.

Yule, G. U. (1925). A mathematical theory of evolution based on the conclusions of Dr. J. C. Willis, F.R.S. *Philos. Trans. R. Soc. London B* **213,** 21–87.

Zachariah, M. A., Crooks, G. E., Holbrook, S. R., and Brenner, S. E. (2005). A generalized affine gap model significantly improves protein sequence alignment accuracy. *Proteins* **58,** 329–338.

Zamore, P. D., and Haley, B. (2005). Ribo-gnome: The big world of small RNAs. *Science* **309,** 1519–1524.

Zaretsky, K. (1965). Reconstruction of a tree from the distances between its pendant vertices. *Uspekhi Math Nauk (Russian Mathematical Survey)* **20,** 90–92.

Zdobnov, E. M., von Mering, C., Letunic, I., and Bork, P. (2005). Consistency of genome-based methods in measuring Metazoan evolution. *FEBS Lett.* **579,** 3355–3361.

Zeldovich, K. B., Berezovsky, I. N., and Shakhnovich, E. I. (2006). Physical origins of protein superfamilies. *J. Mol. Biol.* **357,** 1335–1343.

Zhai, Y., Heijne, W. H., Smith, D. W., and Saier, M. H., Jr. (2001). Homologues of archaeal rhodopsins in plants, animals and fungi: Structural and functional predications for a putative fungal chaperone protein. *Biochim. Biophys. Acta* **1511,** 206–223.

Zhang, C., and DeLisi, C. (1998). Estimating the number of protein folds. *J. Mol. Biol.* **284,** 1301–1305.

Zhang, C. T. (1997). Relations of the numbers of protein sequences, families and folds. *Protein Eng.* **10,** 757–761.

Zharkikh, A. (1994). Estimation of evolutionary distances between nucleotide sequences. *J. Mol. Evol.* **39,** 315–329.

Zhou, X., Kao, M. C., and Wong, W. H. (2002). Transitive functional annotation by shortest-path analysis of gene expression data. *Proc. Natl. Acad. Sci. USA* **99,** 12783–12788.

Zipkas, D., and Riley, M. (1975). Proposal concerning mechanism of evolution of the genome of *Escherichia coli. Proc. Natl. Acad. Sci. USA* **72,** 1354–1358.

Zmasek, C. M., and Eddy, S. R. (2001). A simple algorithm to infer gene duplication and speciation events on a phylogenetic tree. *Bioinformatics* **17,** 821–828.

Zmasek, C. M., and Eddy, S. R. (2002). RIO: Analyzing proteomes by automated phylogenomics using resampled inference of orthologs. *BMC Bioinformatics* **3,** 14.

Zozulya, S., Echeverri, F., and Nguyen, T. (2001). The human olfactory receptor repertoire. *Genome Biol.* **2,** Research0018.

Zuckerkandl, E., and Pauling, L. (1965). Molecules as documents of evolutionary history. *J. Theor. Biol.* **8,** 357–366.

Index

CPSIA information can be obtained
at www.ICGtesting.com
Printed in the USA
LVOW02*1409290716
497892LV00014BA/139/P

9 780120 887941